Lecture Notes in Computer Science 14382

The series Lecture Notes in Computer Science (LNCS), including its subseries Lecture Notes in Artificial Intelligence (LNAI) and Lecture Notes in Bioinformatics (LNBI), has established itself as a medium for the publication of new developments in computer science and information technology research, teaching, and education.

LNCS enjoys close cooperation with the computer science R & D community, the series counts many renowned academics among its volume editors and paper authors, and collaborates with prestigious societies. Its mission is to serve this international community by providing an invaluable service, mainly focused on the publication of conference and workshop proceedings and postproceedings. LNCS commenced publication in 1973.

Fernando Ortiz-Rodriguez ·
Boris Villazón-Terrazas · Sanju Tiwari ·
Carlos Bobed
Editors

Knowledge Graphs and Semantic Web

5th Iberoamerican Conference and 4th Indo-American Conference, KGSWC 2023 Zaragoza, Spain, November 13–15, 2023 Proceedings

Editors
Fernando Ortiz-Rodriguez
Autonomous University of Tamaulipas
Ciudad Victoria, Mexico

Sanju Tiwari
Autonomous University of Tamaulipas
Ciudad Victoria, Mexico

Boris Villazón-Terrazas
University of La Rioja
Madrid, Spain

Carlos Bobed
University of Zaragoza
Zaragoza, Spain

ISSN 0302-9743 ISSN 1611-3349 (electronic)
Lecture Notes in Computer Science
ISBN 978-3-031-47744-7 ISBN 978-3-031-47745-4 (eBook)
https://doi.org/10.1007/978-3-031-47745-4

This Springer imprint is published by the registered company Springer Nature Switzerland AG
The registered company address is: Gewerbestrasse 11, 6330 Cham, Switzerland

Paper in this product is recyclable.

Preface

This volume contains the main proceedings of the Fifth Iberoamerican and the Fourth Indo-American Knowledge Graphs and Semantic Web Conference (KGSWC 2023), held jointly during November 13–15, 2023, at the University of Zaragoza in Zaragoza, Spain. KGSWC is established as a yearly venue for discussing the latest scientific results and technology innovations related to Knowledge Graphs and the Semantic Web. At KGSWC, international scientists, industry specialists, and practitioners meet to discuss knowledge representation, natural language processing/text mining, and machine/deep learning research. The conference's goals are (a) to provide a forum for the AI community, bringing together researchers and practitioners in industry to share ideas about innovative projects, and (b) to increase the adoption of AI technologies in these domains.

KGSWC 2023 followed on from successful past events in 2019, 2020, 2021, and 2022. It was also a venue for broadening the focus of the Semantic Web community to span other relevant research areas in which semantics and web technology play an important role and for experimenting with innovative practices and topics that deliver extra value to the community.

The main scientific program of the conference comprised 20 papers: 18 full research papers and two short research papers selected out of 50 reviewed submissions, which corresponds to an acceptance rate of 40%. The program was completed with four workshops sessions, and a Winter School where researchers could present their latest results and advances and learn from experts. The program also included five high-profile experts as invited keynotes (Claudia d'Amato, Università degli Studi di Bari, Italy; Valentina Presutti, University of Bologna, Italy; Deborah McGuinness, Rensselaer Polytechnic Institute, USA; Pascal Hitzler, Kansas State University, USA; and Steffen Staab, Universität Stuttgart, Germany), with novel Semantic Web topics. A few industry sessions were also organized, with contributors including Ultipa, IET, Catalink and the Building Digital Twin Association.

The General and Program Committee chairs would like to thank the many people involved in making KGSWC 2023 a success. First, our thanks go to the four co-chairs of the main event and the 60-plus reviewers for ensuring a rigorous 150 blind review process, which led to an excellent scientific program, with an average of three reviews per article.

Further, we note the kind support of all people from University of Zaragoza, Zaragoza. We are thankful for the kind support of all people from Springer. We also thank Fatima Zahra, Juan Pablo Martínez, Rubén Barrera and Gerardo Haces, who administered the website and helped to make strong publicity. We finally thank our sponsors and our community for their vital support of this edition of KGSWC.

The editors would like to close the preface with warm thanks to our supporting keynotes, the program committee for rigorous commitment in carrying out reviews, and last but not least, our enthusiastic authors who made this event truly international.

November 2023

Fernando Ortiz-Rodriguez
Boris Villazón-Terrazas
Sanju Tiwari
Carlos Bobed

Organization

Chairs

Fernando Ortiz-Rodríguez	Universidad Autónoma de Tamaulipas, Mexico
Boris Villazón-Terrazas	EY wavespace/UNIR, Spain
Sanju Tiwari	Universidad Autónoma de Tamaulipas, Mexico

Local Chairs

Carlos Bobed	University of Zaragoza, Spain
Fernando Bobillo	University of Zaragoza, Spain
Eduardo Mena Nieto	University of Zaragoza, Spain

Workshop

Sven Groppe	University of Lübeck, Germany
Shishir Shandilya	VIT Bhopal University, India
Shikha Mehta	JIIT Noida, India

Winter School

Sanju Tiwari	Universidad Autónoma de Tamaulipas, Mexico
Fernando Ortiz-Rodríguez	Universidad Autónoma de Tamaulipas, Mexico
Shishir Shandilya	VIT Bhopal University, India
Fatima Zahra	University of Khenchela, Algeria

Publicity

Fatima Zahra	University of Khenchela, Algeria
Yusniel Hidalgo Delgado	Universidad de las Ciencias Informáticas, Cuba

Program Committee Chairs

Sanju Tiwari	Universidad Autónoma de Tamaulipas, Mexico
Fernando Ortiz-Rodríguez	Universidad Autónoma de Tamaulipas, Mexico
Boris Villazón-Terrazas	EY wavespave/UNIR, Spain

Program Committee

Ahlem Rhayem	Universidad Politécnica de Madrid, Spain
Alba Fernandez-Izquierdo	BASF Digital Solutions, Spain
Alberto Fernández	University Rey Juan Carlos, Spain
Alex Mircoli	Università Politécnica delle Marche, Italy
Alejandro Rodríguez	Universidad Politécnica de Madrid, Spain
Amit Sheth	University of South Carolina, USA
Antonella Carbonaro	University of Bologna, Italy
Adolfo Anton-Bravo	MPVD.es, Spain
Amed Abel Leiva Mederos	Universidad Central de las Villas, Cuba
Ana B. Ríos-Alvarado	Universidad Autónoma de Tamaulipas, Mexico
Bishnu Sarker	Meharry Medical College, France
Beyza Yaman	ADAPT Centre - Trinity College Dublin, Ireland
Boris Villazón-Terrazas	EY wavespace/UNIR, Spain
Boyan Brodaric	Geological Survey of Canada, Canada
Carlos Bobed	everis/NTT Data - University of Zaragoza, Spain
Cogan Shimizu	Kansas State University, USA
C. Maria Keet	University of Cape Town, South Africa
Diego Collarana	Enterprise Information System (EIS), Germany
Diego Rubén Rodríguez Regadera	Universidad Camilo José Cela, Spain
Dimitris Kontokostas	Diffbot, Greece
Edgar Tello Leal	Universidad Autonóma de Tamaulipas, Mexico
Edgard Marx	Leipzig University of Applied Sciences (HTWK), Germany
Eduardo Mena	University of Zaragoza, Spain
Erick Antezana	Norwegian University of Science and Tech, Norway
Eric Pardede	La Trobe University, Australia
Fatima N. Al-Aswadi	Universiti Sains Malaysia, Malaysia; Hodeidah University, Yemen
Fatima Zahra Amara	University of Khenchela, Algeria
Federica Rollo	University of Modena and Reggio Emilia, Italy
Fernando Bobillo	University of Zaragoza, Spain
Fernando Ortiz-Rodríguez	Universidad Autónoma de Tamaulipas, Mexico

Francesco Osborne	Open University, UK
Gerardo Haces-Atondo	Universidad Autónoma de Tamaulipas, Mexico
Gustavo De Assis Costa	Federal Institute of Education, Goiás, Brazil
Hong Yung Yip	University of South Carolina, USA
Ismael Navas-Delgado	University of Malaga, Spain
Janneth Alexandra Chicaiza Espinosa	Universidad Técnica Particular de Loja, Ecuador
Jennifer D'Souza	TIB Leibniz Information Centre for Science and Technology University Library, Germany
José L. Martínez-Rodríguez	Universidad Autónoma de Tamaulipas, Mexico
José [illegible] Medina-[illegible]	Universidad Autónoma de Tamaulipas, Mexico
José [illegible]	Universidad de [illegible], Spain
Lu Zhou	Kansas State University, USA
Luis Ramos	TIB Hannover, Germany
Manas Gaur	University of Maryland, Baltimore, USA
Manuel [illegible]	Universidad de las Ciencias Informáticas, Cuba
Mayank Kejriwal	Information Sciences Institute, USC, USA
[illegible]	TIB Hannover, Germany
Md Kamruzzaman Sarker	University of Hartford, USA
Miguel Ángel Sicilia	University of Alcalá, Spain
Nandana Mihindukulasooriya	IBM Research, USA
[illegible]	Rensselaer Polytechnic Institute, USA
[illegible]	Elsevier B.V., The Netherlands
Pascal Hitzler	Kansas State University, USA
[illegible]	University of [illegible]
[illegible]	Universidad [illegible], Spain
[illegible]	University of Dar es Salaam, Kenya
[illegible]	Cognizant, India
[illegible]	KIT Karlsruhe, Germany
[illegible]	[illegible], France
[illegible]	Universidad Autónoma de Tamaulipas, Mexico
[illegible]	[illegible]
[illegible]	[illegible]
[illegible]	University of [illegible], Germany
[illegible]	[illegible], Japan
[illegible]	University of [illegible], China
[illegible]	University of Leipzig, Germany
[illegible]	Universidad de las Ciencias Informáticas, Cuba

Contents

An Ontology for Tuberculosis Surveillance System

Azanzi Jiomekong[1(✉)], Hippolyte Tapamo[1], and Gaoussou Camara[2]

[1] Department of Computer Science, University of Yaounde 1, Yaounde, Cameroon
{fidel.jiomekong,hippolyte.tapamo}@facsciences-uy1.cm
[2] University Alioune Diop de Bambey, EIR-IMTICE, Bambey, Senegal
gaoussou.camara@uadb.edu.sn

Abstract. Existing epidemiological surveillance systems use relational databases to store data and the SQL language to get information and automatically build statistics tables and graphics. However, a lack of logical and machine-readable relations among relational databases prevent computer-assisted automated reasoning and useful information may be lost. To overcome this difficulty, we propose the use of an ontology based-approach. Given that existing ontologies for epidemiological surveillance of TB does not exist, in this article, we present how we developed with the help of an epidemiologist an ontology for TB Surveillance System (O4TBSS). Currently, this ontology contains 807 classes, 117 ObjectProperties, 19 DataProperties.

Keywords: Ontologies · Knowledge Graph · Epidemiological surveillance · Tuberculosis · O4TBSS · EPICAM

1 Introduction

Tuberculosis (TB) is a global scourge, responsible for millions of deaths yearly. Most recently, the global concerns about the emergence of multidrug-resistant and extensively drug resistant TB (MDR and XDR-TB) caused by the bacterium's resistance to the usual drugs complicated the management of TB. Caused by an inadequate treatment, MDR and XDR-TB treatment is costly, less effective and resistant strains can propagate to other individuals [6]. Effective management of this disease requires developing systems such as epidemiological surveillance systems which have to provide all needed information to stakeholders [5]. Data collected by epidemiological surveillance systems are usually stored in relational databases. Although these databases have proven their effectiveness in data representation and management, the absence of a knowledge model describing the semantic of data limits the discovery of new knowledge through semantic reasoning and inference mechanisms [15]. In order to overcome such difficulties, researchers have been working on knowledge representation through ontologies [7,17].

F. Ortiz-Rodriguez et al. (Eds.): KGSWC 2023, LNCS 14382, pp. 1–15, 2023.
https://doi.org/10.1007/978-3-031-47745-4_1

Studer et al. [24] defined an ontology as "a formal, explicit specification of a shared conceptualization." In the medical domain, digitized information management has greatly improved medical practice and medical ontologies have become the standard means of recording and accessing conceptualized medical knowledge [7,13,27]. Many researchers have addressed the modeling of TB information [1,12,18,19,21]. However, the ontologies proposed are too large [21] and the details provided in a disease-specific ontology such as ours may get lost in larger ontologies. On the other hand, the ontologies specific to TB [1,12,18,19] do not cover all the aspects of tuberculosis surveillance.

In a previous work [4,9], we developed a platform which is used for epidemiological surveillance of tuberculosis in Cameroon. It uses the PostgreSQL database management system to store data, get information and build statistical tables and graphics. During the use of this platform, we realized that a lack of logical and machine-readable relations among PostgreSQL tables prevents computer-assisted automated reasoning and useful information may be lost. This remark is supported by existing literature [15]. In this paper, we present the development of an ontology for Tuberculosis Surveillance that we named O4TBSS which can be used for the semantic annotation of tuberculosis data stored in relational databases.

The rest of the article is organized as follows: the Sect. 2 presents the methodology we used to develop the ontology, Sect. 3 reports the ontology development, Sect. 4 presents a use case and Sect. 5 presents the conclusion.

2 Ontology Development Methodology

During the development of the Ontology for Tuberculosis Surveillance System, we have followed a methodology made up of a set of principles, designed activities and phases based on scrum [22] and NeOn [25] methodologies.

Scrum [22] describes how the team members should be organized in order to develop a system in three phases: The pre-development phase (planning and design), the development phase (system development or enhancement) and the post-development phase (system integration and documentation). The scrum members composed of the scrum team (developers and users) and the scrum master develop the system in many sprints (iterative development). At the beginning of the development, the product backlog (list of tasks to proceed) is created and divided into many sprints backlogs (list of tasks at each sprint). With scrum, the meetings called scrum meetings take place at the beginning and the end of the project and each sprint.

NeOn methodology [25] for ontology building is composed of a set of scenarios that the knowledge engineer can combine in different ways, and any combination should include Scenario 1. These scenarios present how an ontology can be developed from the specification to the implementation by reusing and re-engineering non-ontological resources, reusing ontological resources (after a re-engineering if necessary), reusing ontology design patterns. The ontology obtained can be adapted to one or various languages and cultural communities to obtain a multilingual ontology.

Based on the combination of scrum and NeOn, the methodology we used to develop O4TBSS is composed of the pre-development step, the development step and the post-development step.

2.1 The Pre-development Step

The Pre-development step involves the specification, the analysis and the design of the application in which the ontology will be integrated. The specification permits us to obtain the application specification document (ASD). It contains the users' needs and all the features of the software to develop. The analysis activity uses the ASD to understand the system in order to delineate, and identify its features. During the software design, the software architecture, the different modules and the relations among these modules are defined. If the ontology is necessary, it will be specified in the software architecture and its role will be clearly defined.

At the end of the pre-development step, the first version of the application specification (containing competencies questions), analysis and design is produced. The product backlog of the ontology to be built is also produced and a scrum meeting will allow us to define the list of tasks to be executed to build the ontology.

2.2 The Development and Post-development Steps

During the development, the tasks contained in the product backlog are organized in many sprint backlogs and executed. Then, the ontology is developed through repeated cycles (iteratively) and in modules (incrementally), allowing the scrum team to take advantage of what was learned during the development of earlier versions. This step is composed of two main phases: the development of the first version of the ontology and the development of the next versions.
First Version. The first phase consists of the development of the first version of the ontology. It is composed of three activities and proceeds as follows:

1. **The identification of knowledge sources:** an inventory of existing knowledge sources (human experts, domain resources, existing ontologies) is made. If existing ontologies match the needs, they are adopted. Else, the identified resources will be use to build the ontology;
2. **Knowledge acquisition:** during knowledge acquisition, knowledge is acquired from domain experts, existing resources (ontologies/domain resources) or both;
3. **Knowledge representation:** during this activity, the knowledge obtained is serialized in a machine readable form.

After the development of the first version of the ontology, the evaluation is performed. The feedback of this evaluation is presented at the scrum meeting. This feedback will allow us to define the next steps of ontology development.

The Next Versions. The second phase is an iterative and incremental phase in which each increment consists of exploiting the evaluation feedback in order to complete specifications, analysis, design and to develop the new versions of the ontology. Each increment involves the sprint planning meeting which will result in a set of features that the ontology must meet; knowledge identification; knowledge acquisition; and knowledge representation. At the end of each sprint, a sprint review meeting allow us to evaluate the ontology given the specifications, analysis, design and competencies questions. Note that at each review, a reasoner is used to check the ontology consistency.

Post-development Step. The Post-development step involves the population of the ontology with instances to obtain a Knowledge Graph (KG) [8]. Thereafter, the integration of this knowledge graph in related software should be done. For example, a semantic search engine [10] may be developed to navigate in the Knowledge Graph.

3 Ontology Building

This section presents the development of an Ontology for Tuberculosis Surveillance System (O4TBSS). Thus, the pre-development step and development steps are presented in the following subsections.

3.1 Pre-development

During the pre-development, the specifications, analysis and design of the application which will integrate the ontology will allow us to determine the need and role of an ontology.

Software Specifications and Analysis. In a previous work [4,9], we present a platform for tuberculosis surveillance. This platform allowed the National Tuberculosis Control Program (NTCP) in Cameroon to obtain data for tuberculosis management. These data are stored in the PostgreSQL database management system and interfaces (example of Fig. 1) allows users to search information using multiple searching criteria. During the use of this system on the field, we realized that a lack of logical and machine-readable relations among PostgreSQL tables prevents computer-assisted automated reasoning and useful information may be lost. Then, a new module of the platform which enables users to access all needed information is required. The main functionalities of this module are:

- Provide a way (e.g., semantic search engine) to stakeholder to get all needed information;
- Discovering new knowledge from existing ones. For example, to get the correct answers to the queries like "does patient x be at risk to become TB-MDR", the system must have access to patients' knowledge (e.g., patient characteristics and treatment behavior) and be able to reason based on this knowledge to infer patients characteristics.

The new module must allow doctors, epidemiologists and decision makers to get access to all the relevant knowledge. The use case these actors will execute is given by the Fig. 2.

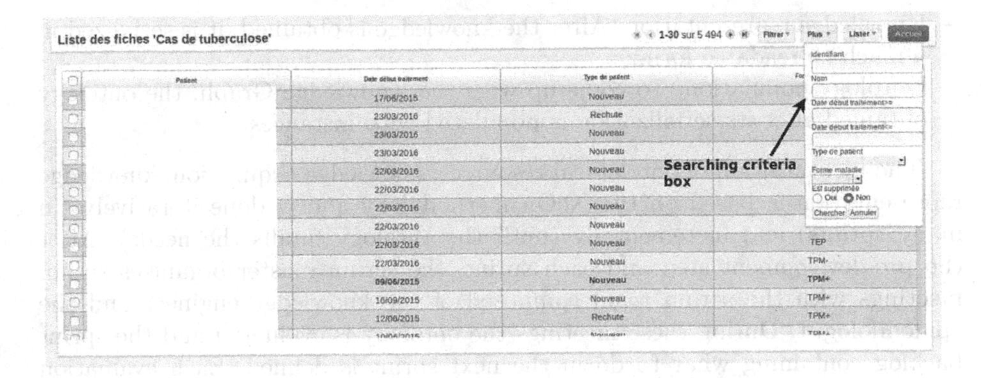

Fig. 1. Searching for patients using multiple criteria

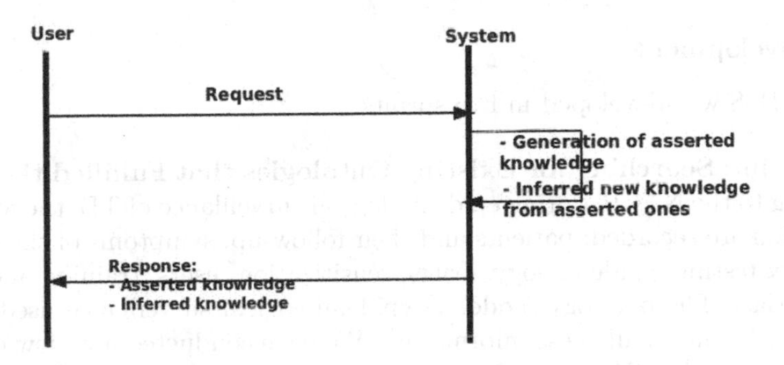

Fig. 2. The general use case executed by all users

System Design. To allow users to have access to all knowledge, the data must be stored using a data structure supporting inferences. As many researchers have proved that ontologies is the best choice for knowledge modeling [15,23], we have chosen to use an ontology.

Product Backlog Definition for the Development of the Ontology. The product backlog comprises the list of tasks to be executed in order to develop the ontology. These are:

- Identification and evaluation of existing ontologies. This task consists of finding existing ontologies that can be used in the system;
- Identification of domain resources. During the identification of domain resources, existing knowledge sources will be identified. In case existing ontologies are not sufficient, these resources will be useful resources for ontology enrichment;
- Knowledge acquisition from ontological knowledge sources. This task consists of using existing methodologies to acquire knowledge from existing ontologies and domain resources;

- Knowledge representation. After the knowledge is obtained, it is serialized in a machine readable form;
- Ontology population. To come up with the Knowledge Graph, the ontology obtained after its serialization is populated with instances.

The identification of ontological resources, knowledge acquisition, knowledge representation is based on the NeOn methodology and is done iteratively (in many sprints) and incrementally (until the ontology fulfills the needs). After the pre-development step and each sprint, the scrum master organizes scrum meetings with the scrum team composed of the knowledge engineer and the epidemiologist. During these meetings the ontology is evaluated and the sprint backlog containing what to do in the next sprint is defined. Each evaluation allowed us to determine the ontology consistency using the Pellet reasoner, and to what extent the ontology developed fulfills the requirements.

3.2 Development

The O4TBSS was developed in five sprints.

First Sprint: Searching for Existing Ontologies that Fulfilled the Need. According to the NTCP, during epidemiological surveillance of TB, the following information are recorded: patients and their follow-up, symptoms of the disease, laboratory testing, epidemiology, drugs, sensitization, users, training and training materials. The ontology modeling epidemiological surveillance used by the NTCP must contain all these information. We have conducted a review of existing ontologies using Bioportal [27] and Google's Search Engine. Keywords such as "tuberculosis", "tuberculosis surveillance", "ontology for tuberculosis surveillance" and "tuberculosis ontology" were used to carry out searches. A total of 38 ontologies were found on BioPortal repository using the keyword "tuberculosis" and 48 ontologies were found using the keyword "tuberculosis surveillance." These ontologies were examined and all excluded because they did not focus on epidemiological surveillance of tuberculosis. A total of 12 scientific papers were identified from Google Search Engine using the keywords "ontology for tuberculosis surveillance" and "tuberculosis ontology". Nine of these papers were excluded because they did not focus on tuberculosis ontology and four papers were retained. The first one entitled "A Tuberculosis Ontology for Host Systems Biology" [12] focuses on clinical terminology. It aims at providing a standard vocabulary to TB investigators of systems biology approaches and omics technologies. The ontology presented has been made available in a csv format; "RepTB: a gene ontology based drug repurposing approach for tuberculosis" [19] focuses on drug repurposing. "An ontology for factors affecting tuberculosis treatment adherence behavior in sub-Saharan Africa" [18] focuses on the factors that influence TB treatment behavior in sub-Saharan Africa; and "An Ontology based Decision support for Tuberculosis Management and Control in India" [1] which presents the use of an ontology for TB management in India. Although these papers are about ontologies of TB, only one ontology is available for download in a csv format and this ontology covers just the clinical aspects of epidemiological surveillance.

At the end of the first sprint, we have noted that no existing ontology covers the domain that we want to represent. This justified the development of a new ontology.

Second Sprint: Knowledge Extraction from Knowledge Sources. In a previous work [2,3], we extracted knowledge from some knowledge sources of tuberculosis management. This knowledge was used to construct an ontology for tuberculosis surveillance. This ontology is composed of 329 terms with 97 classes, 117 DataProperties and 115 ObjectProperties. Given that this ontology models the epidemiological surveillance system of tuberculosis, it is yet to be evaluated to see if it is complete. That is why, we evaluated this ontology given two criteria:

1. The completeness of the modeled domains, which measures if all the domains covered by epidemiological surveillance are well covered by the ontology;
2. The completeness of the ontology for each domain involved in the epidemiological surveillance, which measures if each domain of interest is appropriately covered in this ontology.

The keywords were identified from this ontology and used to carry out searches of existing ontologies on Bioportal repository and Google Search Engine. We found 275 ontologies. For each term, we noted the list of ontologies obtained. For the ontologies found in the BioPortal repository, the BioPortal ontology visualization tool (see Fig. 3) was used to visualize the terms that are presented in the ontology. If an ontology contains the relevant terms, it is selected. In many cases, two ontologies have the same terms when searching using certain keywords e.g., "patient", "doctor", "nurse", "tuberculosis", etc. Then, the most complete were selected. The ontologies not present in the BioPortal repository were examined using Protege. The ontology in the csv file was examined using LibreOffice Calc. The Table 1 presents the ontologies selected for our purpose.

Comparing the ontologies selected with the ontology constructed with TB domain resources, we found that only the ontology constructed using TB resources takes into account all the aspects of epidemiological surveillance. However, by considering the completeness of each domain covered by epidemiological surveillance, we remark that the ontologies selected are more complete. In the next sprint, we will show how knowledge has been extracted from the ontologies presented in table 1 and combined with the ontologies constructed using domain resources to build O4TBSS.

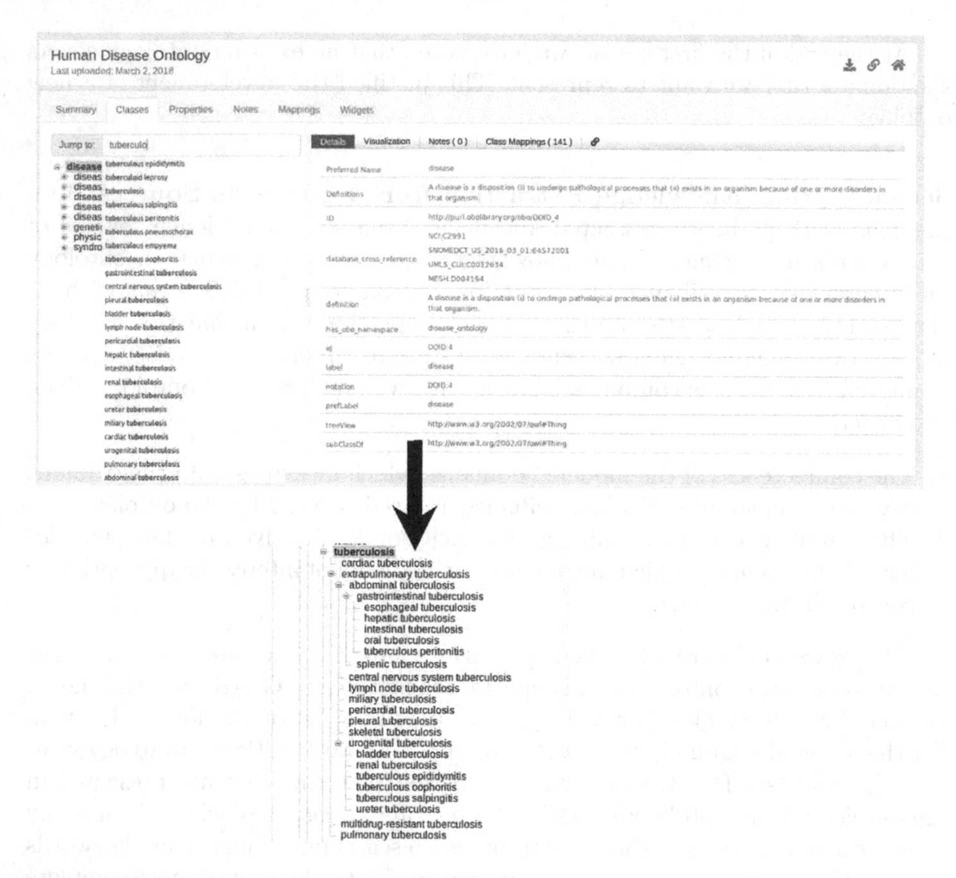

Fig. 3. Example of browsing Human Disease Ontology (DOID) using Bioportal visualization tool

Third Sprint: Ontology Construction The ontologies selected during the second sprint were used to construct O4TBSS. To this end, ontological knowledge was extracted using either ontofox [28] or Protege. For the ontologies not presented in ontofox such as "Adherence and Integrated Care ontology", Protege software was used for their examination, identification of irrelevant terms and the deletion of the latter. Knowledge obtained was imported in Protege and examined term by term with the help of an epidemiologist to evaluate each term and identify redundancies. Redundant terms identified were removed. Additional terms were extracted from the ontology constructed using domain resources to enrich the ontology obtained. The Pellet reasoner in Protege allowed us to verify the consistency of the ontology obtained. The Table 2 and the Fig. 4 respectively present the metric and a part of the ontology obtained.

Table 1. List of ontologies selected for our purpose. The keyword column presents the keywords that were used to find the ontology. The ontology column presents the ontology selected given the keyword. The covered domain column presents the domain of epidemiological surveillance covered by the ontology and the description column presents a brief description of the ontology

Keywords	Ontology	Covered domains	Ontology description
Epidemiological surveillance	Epidemiology Ontology (EPO)	Epidemiology	This is an ontology describing the epidemiological, demographics and infection transmission process [20]
Tuberculosis symptoms	Symptom Ontology (Symp)	Tuberculosis sign and symptoms	Symp aims to understand the relationship between signs and symptoms and capture the terms relative to the signs and symptoms of a disease [16]
Tuberculosis	Human Disease Ontology (DOID)	Patients and their follow-up, Epidemiology	Human Disease Ontology is an ontology that represents a comprehensive hierarchically controlled vocabulary for human disease representation [21]
Tuberculosis ontology	A Tuberculosis Ontology for Host Systems Biology	Patients, symptoms, laboratory testing	Tuberculosis Ontology for Host Systems Biology focuses on clinical terminology of tuberculosis diagnosis and treatment. It is available in a csv format [12]
Patient	Adherence and Integrated Care	Patient and their follow-up	This ontology is an ontology that defines the medication adherence of patient [26]
Patient	Presence Ontology (PREO)	Patient and their follow-up	This ontology defines relationships that model the encounters taking place every day among providers, patients, and family members or friends in environments such as hospitals and clinics [14]
Patient	Mental Health Management Ontology (MHMO)	Patient and their follow-up	The Mental Health Management Ontology is an ontology for mental health-care management [29]

Fourth Sprint: Ontology Enrichment. After building the ontology, we decided to populate it with some data gathered from the TB database. But we remarked that some data contained in the database can be considered as concepts/property. With SQL queries, we extracted these knowledge composed of 70 classes and enriched our ontology. The actual version of the ontology is composed of 943 terms, with 807 classes, 117 ObjectProperties and 19 DataProperties.

Fifth Sprint: Ontology Population. The main purpose of this work was to construct O4TBSS. However, we decided to populate it with instances to obtain a Knowledge Graph and make some inference on it. Thus, during the population, a flat view of the TB database was created by making a simple SQL query. This query allowed us to gain access to information and the information obtained was populated in the ontology. To keep the relation between the tables in the database, the tuples identification in the database were used as the identification of these instances in the ontology. For example, the TB case with ID *"TBCASE_14f7ee"* is linked to its appointment with ID *"RDV_14f5e7a"* in the database. Then, in the ontology, their identifications will also be *"TBCASE_14f7ee"* and *"RDV_14f5e7a"*. The complete ontology and the source code written for its population is avail able on github. It should be noted that given the confidentiality of medical data, we did not use the surveillance data of NTCP.

Table 2. O4TBSS terms and terms imported from 7 other ontologies sources and enriched with TB database terms

#	Ontologies	Classes	DataProperties	ObjectProperties	Total
	O4TBSS	807	117	19	943
1	Epidemiology Ontology	95	0	0	95
2	Symptom Ontology	12	0	0	12
3	Adherence and Integrated Care	246	12	2	260
4	Presence Ontology	205	25	5	235
5	Human Disease Ontology	22	14	0	36
6	Mental Health Management Ontology	143	64	0	217
7	A Tuberculosis Ontology For Host Systems Biology	125	0	0	125
8	Ontology constructed with TB domain resources	17	8	6	31

Fig. 4. A screenshot of O4TBSS obtained after the third sprint

4 Use Case: Automatic Detection of TB-MDR Susceptible Patients by Reasoning on O4TBSS

Our main purpose in this paper was to present how we built O4TBSS. After this presentation, we are going to show how reasoning mechanisms can be used to infer new knowledge more easily than SQL Joins queries and replies to the competencies questions such as: "which are the characteristics of patients at risk of TB-MDR?". This section presents use cases in which the reasoning mechanism allowed us to derive new knowledge from existing knowledge. Given that the NTCP did not give the right to use the data to populate the ontology, because of their sensitivity, only fake data are used to demonstrate the capabilities of the ontology once populated. From the 5000 patients' information stored in the tuberculosis database, the first 50 were selected and used to generate fake data and populate the ontology. The Description Logic (DL) queries were used to define simple axioms and save in the ontology so that these axioms can be reused further for inference. The advantage of using this feature of protégé is that it allows us to ensure that the axiom is inferring the right knowledge. The Pellet reasoner implemented in the Protege software were used for use cases execution.

The TB-MDR is generally caused by an inadequate treatment of tuberculosis, which can give rise to an epidemic of TB difficult to cure [18]. According to the National TB Control Program, the patients who did not come to their appointments to get the medications are those who will later develop the resistance to drugs and come back with TB-MDR. Health workers revealed that often, patients will follow the first 4 months of treatment, once they start feeling better, they don't come back for the last two months of their treatment. Consequently they come later with TB-MDR. According to epidemiologists, these patients and their characteristics must be identified in time and the right actions must be made.

The current version of the TB platform does not consider the TB-MDR patients. However, the information on the follow-up of appointments of the patients are stored in the database. To get access to this information, a SQL query must be made with the current version. Given that the database is flat, to get access to other information with the link to the patients, a joined query with many tables must be done. The example of Fig. 5 shows how this task can be difficult to make and prone to errors. The current use case (Fig. 6) shows how an axiom allows us to get all the patients at risk of becoming TB-MDR. The inference mechanism allows us to infer all patients characteristics and the links with other entities (e.g., the link with the TB exams) will allow us to have more information on the patient.

This use case shows that the ontology can be used to classify patients according to their behavior. It can also be used to automatically identify other categories of patients, by using the inference mechanism.

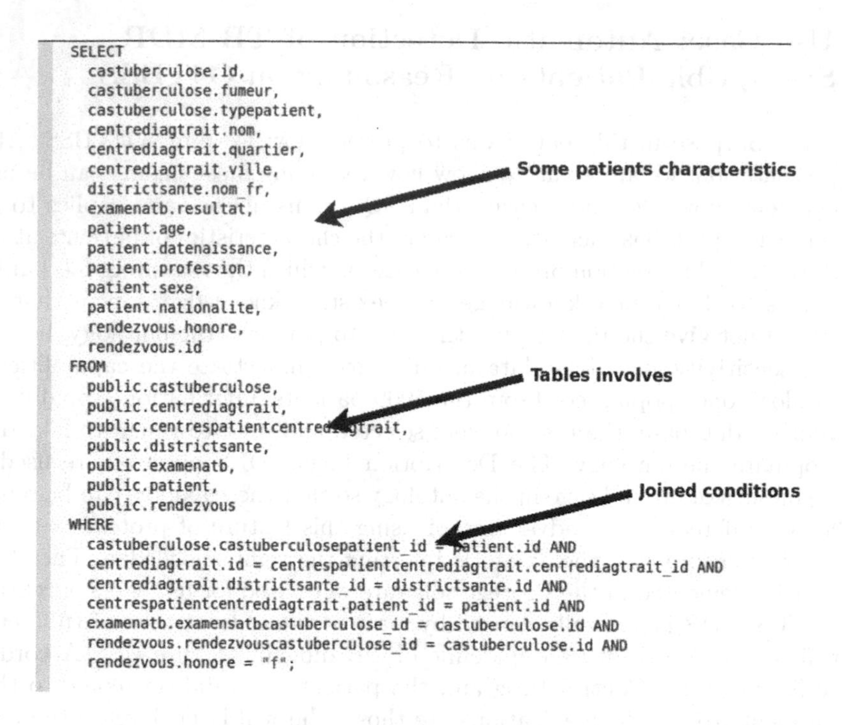

```
SELECT
  castuberculose.id,
  castuberculose.fumeur,
  castuberculose.typepatient,
  centrediagtrait.nom,
  centrediagtrait.quartier,
  centrediagtrait.ville,
  districtsante.nom_fr,
  examenatb.resultat,
  patient.age,
  patient.datenaissance,
  patient.profession,
  patient.sexe,
  patient.nationalite,
  rendezvous.honore,
  rendezvous.id
FROM
  public.castuberculose,
  public.centrediagtrait,
  public.centrespatientcentrediagtrait,
  public.districtsante,
  public.examenatb,
  public.patient,
  public.rendezvous
WHERE
  castuberculose.castuberculosepatient_id = patient.id AND
  centrediagtrait.id = centrespatientcentrediagtrait.centrediagtrait_id AND
  centrediagtrait.districtsante_id = districtsante.id AND
  centrespatientcentrediagtrait.patient_id = patient.id AND
  examenatb.examensatbcastuberculose_id = castuberculose.id AND
  rendezvous.rendezvouscastuberculose_id = castuberculose.id AND
  rendezvous.honore = "f";
```


Fig. 5. An example of a SQL query used to find patients at risk of TB-MDR and their characteristics.

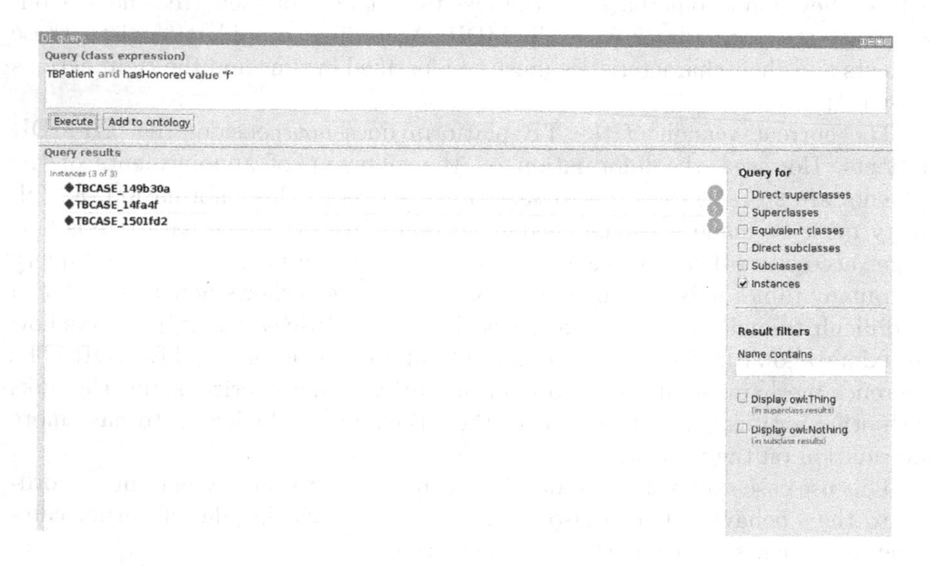

Fig. 6. Inferring the patients at risk of TB-MDR using Protege and DL query. A simple click on a patient permits access to the patient's characteristics. This type of knowledge can be saved in the ontology by clicking on the "add to ontology" button.

5 Conclusion

In this article we reported the development of an ontology for tuberculosis surveillance. This ontology can be used for the annotation of clinical and epidemiological data of tuberculosis. Our motivation was to provide a knowledge model for epidemiological data which permits stakeholders involved in epidemiological surveillance of TB to easily have access to all needed information using the reasoning mechanism. Even if we demonstrated only the reasoning feature of the ontology, it should be noted that the ontology developed can also be used for data exchange and integration. It can easily be integrated with other ontologies. This ontology may serve as an example to model epidemiological surveillance of other infectious diseases.

Data collected during epidemiological surveillance may become very large and the users may face the problem of information overload which is the well known problem in the medical domain [11]. On the other hand, there are many users (epidemiologists, decision makers, doctors, etc.) of epidemiological surveillance systems and they have different preferences on data. In the future work, we plan to populate the ontology to obtain a KG and to build a semantic search engine on top of this KG in order to filter information reaching users according to their profile.

References

1. Abhishek, K., Singh, M.P.: An ontology based decision support for tuberculosis management and control in India. Int. J. Eng. Tech. **8**, 2860–2877 (2016). https://doi.org/10.21817/ijet/2016/v8i6/160806247
2. Azanzi, F.J., Camara, G.: Knowledge extraction from source code based on hidden markov model: application to EPICAM. In: 14th IEEE/ACS International Conference on Computer Systems and Applications, AICCSA 2017, Hammamet, Tunisia, October 30 - November 3, 2017, pp. 1478–1485 (2017). https://doi.org/10.1109/AICCSA.2017.99
3. Azanzi, F.J., Camara, G., Tchuente, M.: Extracting ontological knowledge from java source code using hidden markov models. Open Comput. Sci. **9**(1), 181–199 (2019). https://doi.org/10.1515/comp-2019-0013
4. Azanzi, J., Tapamo, H., Camara, G.: Combining scrum and model driven architecture for the development of an epidemiological surveillance software. Revue Africaine de Recherche en Informatique et Mathématiques Appliquées Volume 39–2023 (Jul 2023). https://doi.org/10.46298/arima.9873
5. Choi, B.C.K.: The past, present, and future of public health surveillance. Scientifica **2012**, 875253 (2012)
6. Glaziou, P., Floyd, K., Raviglione, M.: Global epidemiology of tuberculosis. In: Seminars in Respiratory and Critical Care Medicine, vol. 39 (2018)
7. Hauer, T., et al.: An architecture for semantic navigation and reasoning with patient data - experiences of the health-e-child project. In: Sheth, A., et al. (eds.) ISWC 2008. LNCS, vol. 5318, pp. 737–750. Springer, Heidelberg (2008). https://doi.org/10.1007/978-3-540-88564-1_47
8. Ivanov, Y.: What is an enterprise knowledge graph and why do i want one? https://www.ontotext.com/knowledgehub/fundamentals/what-is-a-knowledge-graph/

9. Jiomekong, A., Camara, G.: Model-driven architecture based software development for epidemiological surveillance systems. Studi. Health Tech. Inf. **264**, 531–535 (2019). https://doi.org/10.3233/shti190279
10. Kassim, J.M., Rahmany, M.: Introduction to semantic search engine. In: 2009 International Conference on Electrical Engineering and Informatics, vol. 02, pp. 380–386 (2009). https://doi.org/10.1109/ICEEI.2009.5254709
11. Kolusu, H.R.: Information overload and its effect on healthcare. Oregon Health & Science University, Master of biomedical informatics (2015)
12. Levine, D., et al.: A tuberculosis ontology for host systems biology. Tuberculosis **95**(5), 570–574 (2015). https://doi.org/10.1016/j.tube.2015.05.012
13. Lin, Y., Xiang, Z., He, Y.: Brucellosis ontology (IDOBRU) as an extension of the infectious disease ontology. J. Biomed. Semantics **2**, 9 (2011). https://doi.org/10.1186/2041-1480-2-9
14. Maitra, A., Kamdar, M.: Presence ontology (2019). http://bioportal.bioontology.org/ontologies/PREO
15. Martinez-Cruz, C., Blanco, I.J., Vila, M.A.: Ontologies versus relational databases: are they so different? a comparison. Artif. Intell. Rev. **38**(4), 271–290 (2012). https://doi.org/10.1007/s10462-011-9251-9
16. MediaWiki: symptom ontology (2009). https://symptomontologywiki.igs.umaryland.edu/mediawiki/index.php
17. Munir, K., Anjum, M.S.: The use of ontologies for effective knowledge modelling and information retrieval. Appl. Comput. Inf. **14**(2), 116–126 (2018). https://doi.org/10.1016/j.aci.2017.07.003
18. Ogundele, O.A., Moodley, D., Seebregts, C.J., Pillay, A.W.: An ontology for tuberculosis treatment adherence behaviour. In: Proceedings of the 2015 Annual Research Conference on South African Institute of Computer Scientists and Information Technologists, pp. 30:1–30:10. SAICSIT 2015, ACM, New York, NY, USA (2015). https://doi.org/10.1145/2815782.2815803
19. Passi, A., Rajput, N., Wild, D., Bhardwaj, A.: RepTB: a gene ontology based drug repurposing approach for tuberculosis. J. Cheminf. 10 (2018). https://doi.org/10.1186/s13321-018-0276-9
20. Pesquita, C., Ferreira, J., Couto, F., Silva, M.J.: The epidemiology ontology: an ontology for the semantic annotation of epidemiological resources. J. Biomed. Semant. **5**, 4 (2014)
21. Schriml, L.M., et al.: Human Disease Ontology 2018 update: classification, content and workflow expansion. Nucleic Acids Res. **47**(D1), D955–D962 (2018). https://doi.org/10.1093/nar/gky1032
22. Schwaber, K.: Agile Project Management With Scrum. Microsoft Press, Redmond, WA, USA (2004)
23. Shen, F., Lee, Y.: Knowledge discovery from biomedical ontologies in cross domains. PLOS ONE **11**(8), 1–34 (2016). https://doi.org/10.1371/journal.pone.0160005
24. Studer, R., Benjamins, V.R., Fensel, D.: Knowledge engineering: principles and methods. Data Knowl. Eng. **25**(1–2), 161–197 (1998). https://doi.org/10.1016/S0169-023X(97)00056-6
25. Suárez-Figueroa, M.C., Gómez-Pérez, A., Fernández-López, M.: The NeOn methodology for ontology engineering. In: Suárez-Figueroa, M.C., Gómez-Pérez, A., Motta, E., Gangemi, A. (eds.) Ontology Engineering in a Networked World, pp. 9–34. Springer, Heidelberg (2012). https://doi.org/10.1007/978-3-642-24794-1_2

26. Villaran, E.R., Parra, C.L.: Adherence and integrated care (2019). https://bioportal.bioontology.org/ontologies/ADHER_INTCARE_EN
27. Whetzel, P.L., et al.: Bioportal: enhanced functionality via new web services from the national center for biomedical ontology to access and use ontologies in software applications. Nucleic Acids Res. **39**(Web-Server-Issue), 541–545 (2011). https://doi.org/10.1093/nar/gkr469
28. Xiang, Z., Courtot, M., Brinkman, R.R., Ruttenberg, A., He, Y.: Ontofox: web-based support for ontology reuse. BMC. Res. Notes **3**(1), 175 (2010). https://doi.org/10.1186/1756-0500-3-175
29. Yamada, D.B., et al.: Proposal of an ontology for mental health management in brazil. Procedia Comput. Sci. **138**, 137–142 (2018). https://doi.org/10.1016/j.procs.2018.10.020

Should I Stay or Should I Go

A New Reasoner for Description Logic

Aaron Eberhart[1,2(✉)], Joseph Zalewski[1], and Pascal Hitzler[1]

[1] DaSe Lab, Kansas State University, Manhattan, KS 66506, USA
aaron.eberhart@gmail.com
[2] metaphacts GmbH, 69190 Walldorf, Germany

Abstract. We present the Emi reasoner, based on a new interpretation of the tableau algorithm for reasoning with Description Logics with unique performance characteristics and specialized advantages. Emi turns the tableau inside out, solving the satisfiability problem by adding elements to expressions rather than adding expressions to element node labels. This strategy is inspired by decidable reasoning algorithms for Horn Logics and $\mathcal{EL}^{++}$ that run on a loop rather than recursive graph-based strategies used in a tableau reasoner. Because Emi solves the same problem there will be a simple correspondence with tableaux, yet it will feel very different during execution, since the problem is inverted. This inversion makes possible many unique and straightforward optimizations, such as paralellization of many parts of the reasoning task, concurrent ABox expansion, and localized blocking techniques. Each of these optimizations contains a design trade-off that allows Emi to perform extremely well in certain cases, such as instance retrieval, and not as well in others. Our initial evaluations show that even a naive and largely un-optimized implementation of Emi is performant with popular reasoners running on the JVM such as Hermit, Pellet, and jFact.

1 Introduction

Knowledge graph schema are complex artifacts that can be very useful, but are often difficult and expensive to produce and maintain. This is especially true when encoding them in OWL (the Web Ontology Language) as ontologies for data management or reasoning. The high expressivity of OWL is a boon, in that it makes it possible to describe complex relationships between classes, roles,[1] and individuals in an ontology. At the same time, however, this high expressivity is often an obstacle to its correct usage that can limit adoption, and can hamper any practical reasoning applications by adding complexity to the reasoning process.

To manage the complexity of OWL some prefer to accept hardness as it is and develop tools to manage or simplify the tricky parts. However, it occasionally is the case that problems appear difficult when they are actually quite intuitive

[1] We refer to properties as **roles**, unless a distinction is relevant, as this is the standard description logic term. These include both object properties and data properties.

F. Ortiz-Rodriguez et al. (Eds.): KGSWC 2023, LNCS 14382, pp. 16–31, 2023.
https://doi.org/10.1007/978-3-031-47745-4_2

when expressed differently. When this happens it can be helpful to start again from the beginning and re-imagine what is possible by trying something entirely different. This will of course not alter the fundamental proven computational complexity of any reasoning or modeling problem. But it can lead to algorithms and design patterns that are more easily understandable, and potentially uncover unique use cases and optimizations.

In this spirit we have used a new API for OWL called (f OWL) [2] for our experiment that can represent and better facilitate ontology data for the reasoning tasks we want it to perform. A custom reasoning algorithm called Emilia[2] (Emi) was implemented specifically to leverage the unique advantages of the new API, and with time the two will be able to seamlessly work within a single framework. The Emi reasoner will be able solve all of the same problems that current reasoners are able to solve, but its execution follows an iterative rule-like path rather than recursive tableau.

This type of experiment may seem equivalent and redundant to logicians and mathematicians, and looking at it only in a purely formal sense this can seem to be the case, however Emi works entirely differently from current systems and presents many opportunities for new research. The reasoner presented here is a prototype which demonstrates that the technique is valid and no less efficient than other comparable Java Virtual Machine (JVM) reasoners – the potential for new unique research directions and optimization techniques using this method will be the subject of future works. Initial testing is underway of Emi, and it is already performing competitively with other state-of-the-art systems in use today, and is particularly efficient with instance retrieval, despite being a naive prototype system written in a high level language with only the most basic and straightforward of optimizations.

2 $\mathcal{ALCH}$

The Emi algorithm currently supports $\mathcal{ALCH}$ reasoning, and the syntax and semantics of that logic are given below. Future extensions will likely expand expressivity for additional description logics.

2.1 Syntax

The signature Σ for the Description Logic $\mathcal{ALCH}$ is defined as $\Sigma = \langle N_I, N_C, N_R \rangle$ where:

- N_I is a set of individual element names.
- N_C is a set of class names that includes $\top$ and $\bot$.
- N_R is a set of role names.
- N_I, N_C, N_R are pairwise disjoint

Expressions in $\mathcal{ALCH}$ use the following grammar:

$\mathbf{R} ::= N_R$

$\mathbf{C} ::= N_C \mid \neg\mathbf{C} \mid \mathbf{C} \sqcap \mathbf{C} \mid \mathbf{C} \sqcup \mathbf{C} \mid \exists\mathbf{R}.\mathbf{C} \mid \forall\mathbf{R}.\mathbf{C}$

[2] **E**ager **M**aterializing and **I**terating **L**ogical **I**nference **A**lgorithm.

2.2 Semantics

An interpretation $\mathcal{I} = (\Delta^{\mathcal{I}}, \cdot^{\mathcal{I}})$ maps N_I, N_C, N_R to elements, sets, and relations in $\Delta^{\mathcal{I}}$ with function $\cdot^{\mathcal{I}}$. An axiom A in $\mathcal{ALCH}$ is satisfiable if there is an interpretation where $\cdot^{\mathcal{I}}$ maps all elements, sets, and relations in A to $\Delta^{\mathcal{I}}$. An ontology O is a set of axioms formed from $\mathcal{ALCH}$ expressions, and is satisfiable if there is an interpretation that satisfies all axioms it contains, this interpretation being a model for O. $\cdot^{\mathcal{I}}$ is defined in Table 1 below.

Table 1. $\mathcal{ALCH}$ Semantics

Description	Expression	Semantics
Individual	x	$x^{\mathcal{I}} \in \Delta^{\mathcal{I}}$
Top	$\top$	$\Delta^{\mathcal{I}}$
Bottom	$\bot$	$\emptyset$
Class	B	$B^{\mathcal{I}} \subseteq \Delta^{\mathcal{I}}$
Role	R	$R^{\mathcal{I}} \subseteq \Delta^{\mathcal{I}} \times \Delta^{\mathcal{I}}$
Negation	$\neg B$	$\Delta^{I} \setminus B^{\mathcal{I}}$
Conjunction	$B \sqcap C$	$B^{\mathcal{I}} \cap C^{\mathcal{I}}$
Disjunction	$B \sqcup C$	$B^{\mathcal{I}} \cup C^{\mathcal{I}}$
Existential Restriction	$\exists R.B$	$\{ x \mid \text{there is } y \in \Delta^{\mathcal{I}} \text{ such that } (x, y) \in R^{\mathcal{I}} \text{ and } y \in B^{\mathcal{I}} \}$
Universal Restriction	$\forall R.B$	$\{ x \mid \text{for all } y \in \Delta^{\mathcal{I}} \text{ where } (x, y) \in R^{\mathcal{I}}, \text{ we have } y \in B^{\mathcal{I}} \}$
Class Assertion	$B(a)$	$a^{\mathcal{I}} \in B^{\mathcal{I}}$
Role Assertion	$R(a, b)$	$(a, b) \in R^{\mathcal{I}}$
Negated Role Assertion	$\neg R(a, b)$	$(a, b) \notin R^{\mathcal{I}}$
Class Subsumption	$B \sqsubseteq C$	$B^{\mathcal{I}} \subseteq C^{\mathcal{I}}$
Class Equivalence	$B \equiv C$	$B^{\mathcal{I}} = C^{\mathcal{I}}$
Role Subsumption	$R \sqsubseteq S$	$R^{\mathcal{I}} \subseteq S^{\mathcal{I}}$

3 (f OWL)

We use (f OWL)[2] because it includes many novel optimizations that streamline ontology and reasoner development. True to the functional paradigm (f OWL) uses functions and shuns classes. Indeed all OWL ontology objects can be created directly with (f OWL) functions. And since these functions are not bound up in an arbitrary class hierarchy, they can operate on independent data structures that are internally typed to represent OWL semantics. A (f OWL) ontology itself is simply another data structure made of collections of smaller similar structures. This means that in most cases an expression, an axiom, even an ontology can be traversed recursively as-is without writing more functions. The use of immutable data structures by (f OWL) in standard Clojure style also permits straightforward implementation of concurrent processes that operate on ontology data.

3.1 (f OWL) and Emi

Emi makes use of many features of (f OWL), some of which are uniquely advantageous in that they have no comparable alternative in other reasoners or APIs. One major advantage with using (f OWL) is the ontology-as-data-structure approach, which means that once (f OWL) has loaded the ontology into memory, Emi can efficiently traverse and manipulate expressions in the ontology without worrying about concurrency, since a (f OWL) ontology is immutable. Once the ontology is processed and elements in the signature are assigned mutable memory for use during reasoning, it is straightforward to define a partially parallelizable reasoning process with Clojure built-in functions that can be precisely lazy or eager when needed. The optimal configuration for when to choose each strategy is subject to many design trade-offs and does not have an objective best answer, though we believe we have developed a decent strategy. Emi processes axioms eagerly, but sets of axioms lazily and in parallel when possible due to the general observation that many non-synthetic ontologies contain a large number of axioms, most of which are rather small. This can affect performance on synthetic datasets where the occurrence of large or complex axioms is potentially higher and a different strategy may work better.

4 Emi Reasoner

As mentioned in the introduction, the new reasoner we are developing works iteratively without a graph, and is more like reasoning in $\mathcal{EL}^{++}$ or datalog where the process runs on a loop that eventually terminates rather than an expansive graph traversal. To reason iteratively we effectively need to turn the tableau inside out so we can work directly with axioms instead of a graph of elements. This means we will need to look at each axiom individually and add elements to or remove elements from expressions according to the tableau expansion rules as if they were backwards. This process slowly builds up a model for the ontology by partially realizing expressions until they become satisfiable and no longer need to be modified. Since every element is inspected for every axiom and the inverted tableau rules are followed in expressions we know this will be consistent. We assume that all expressions in our algorithm will be converted into Negation Normal Form (NNF); it is important to emphasize that absolutely no other logical preprocessing is used in the algorithm. In this section we assume that readers are familiar with the basics of tableau reasoning, if not they can consult [1] for further information.

4.1 Partial Interpretations

This algorithm works by directly attempting to realize an ontology and eventually build a consistent interpretation using the axioms as they are written in NNF, by adding and removing elements until all axioms are satisfiable or the ontology is shown to be unsatisfiable. We refer to the changing states of a program as partial interpretations.

Before continuing to the definition, it is important to stop and emphasize the subtle difference between partial interpretations and standard interpretations. A partial interpretation *may* be equivalent to some interpretation, and indeed when Emi terminates and determines we have a satisfiable ontology this final partial interpretation is a model. However partial interpretations are not necessarily models and represent the states of the program as different assignments are attempted in order to find a model. This difference may be confusing at first for logicians who are used to only seeing standard interpretations, but the distinction is important not just for showing a simple proof but also for describing the actual implementation of the algorithm as well. Our terminology deliberately corresponds much more closely to how an actual implemented algorithm, rather than a theoretical proof of one, would function by avoiding whenever possible notions of infinity or concepts that would need to be represented by global variables that are pervasive in proofs and inconvenient or impossible to implement. We do this intentionally so that it can be understood by programmers as well as logicians; we hope not just to show correctness but that it is clear to anyone how they could actually write a reasoner such as this.

Definition 1. *A* partial interpretation *representing the current state of the algorithm when a function is called is* $\mathcal{I}^* = (\mathcal{R}^{\mathcal{I}^*}, \cdot^{\mathcal{I}^*})$ *where* $\cdot^{\mathcal{I}^*}$ *is a function that maps elements, sets, and relations in an ontology* O *to a realization* $\mathcal{R}^{\mathcal{I}^*}$ *of* O.

Definition 2. *A* realization *of ontology* O *is a set of assertions that states for every class name* A *and element* x *in* O *that* $x \in A^{\mathcal{I}^*}$ *or* $x \notin A^{\mathcal{I}^*}$ *and for every role name* R *and element pair* (x, y) *in* O *that* $(x, y) \in R^{\mathcal{I}^*}$ *or* $(x, y) \notin R^{\mathcal{I}^*}$.

Each partial interpretation in this algorithm corresponds to some realization with a function $\cdot^{\mathcal{I}^*}$. Note that there is no restriction against inconsistent realizations and partial interpretations, only that they must be complete. A partial interpretation $\mathcal{I}^*$ with realization $\mathcal{R}^{\mathcal{I}^*}$ of ontology O is said to be a model of O iff there is an interpretation $\mathcal{I}$ of $O \cup \mathcal{R}^{\mathcal{I}^*}$.

As mentioned previously, while $\mathcal{I}$ often denotes a single consistent interpretation, $\mathcal{I}^*$ denotes the *current interpretation* when a function is evaluated in the algorithm, and as such may change over time and contain information that is later proved to be incorrect. Frequently we will say things like "Add x to $E^{\mathcal{I}^*}$", and this indicates that the partial interpretation $\mathcal{I}^*$ will henceforth be modified in the program state to represent the described change to $\cdot^{\mathcal{I}^*}$ that produces the desired realization.

4.2 Inverted Tableau Expansion Rules

Inverting the tableau rules is straightforward when we turn the standard tableau terminology, such as occurs in [1,6], on its head and assume that "labels" represent the action of the $\cdot^{\mathcal{I}^*}$ function on class and role names in an ontology to produce a realization, and that for any class or role name E that $E^{\mathcal{I}^*} = \mathcal{L}(E)$. To connect this idea with labelling terminology in a graph we use a labelled object, which can represent both notions.

Definition 3. *A* labelled object *E is an object that is associated with some set $\mathcal{L}(E)$, or* label.

First we note that nodes and labels do not necessarily need to be connected in a graph to represent the semantics of $\mathcal{ALCH}$, and can be reinterpreted as freely associating labelled objects. A labelled object can be a node in a tableau graph or simply an isolated named object with no inherent graph connections. This means that a label for a complex expression can be defined recursively to match the semantics of a partial interpretation $\mathcal{I}^*$, e.g. when an expression $C \sqcup D$ is satisfiable after applying the inverted tableau rules we have $\mathcal{L}(C \sqcup D) = (C \sqcup D)^{\mathcal{I}^*} = C^{\mathcal{I}^*} \cup D^{\mathcal{I}^*}$. We can represent axioms equivalently as labelled objects without actually needing to make a graph as long as all names in the signature are not duplicated in expressions but actually reference the same labelled objects.

For this algorithm we consider the inverted $\mathcal{ALC}$ tableau rules sufficient for deciding $\mathcal{ALCH}$ satisfiability, since the addition of the $\mathcal{H}$ fragment does not permit complex role expressions and they do not require expansion to check. The RBox axioms that are not included here will be described in Table 4 with the TBox axioms. In Table 2 we show the tableau expansion rules for $\mathcal{ALC}$ [1] and assume the standard notion of blocking for them, which can be found with the original proofs, then show the inversions in Table 3 where blocking is defined in Subsect. 4.4.

Table 2. $\mathcal{ALC}$ Tableau Expansion Rules

		the following expansion rules apply to element x when x is not blocked
$\sqcap$-rule	if 1.	$(C \sqcap D) \in \mathcal{L}(x)$
	2.	$\{C, D\} \not\in \mathcal{L}(x)$
	then	$\mathcal{L}(x) \rightarrow \mathcal{L}(x) \cup \{C, D\}$
$\sqcup$-rule	if 1.	$(C \sqcup D) \in \mathcal{L}(x)$
	2.	$\{C, D\} \cap \mathcal{L}(x) = \emptyset$
	then	either $\mathcal{L}(x) \rightarrow \mathcal{L}(x) \cup \{C\}$
		or else $\mathcal{L}(x) \rightarrow \mathcal{L}(x) \cup \{D\}$
$\exists$-rule	if 1.	$\exists R.C \in \mathcal{L}(x)$
	2.	there is no y s.t. $\mathcal{L}((x, y)) = R$ and $C \in \mathcal{L}(x)$
	then	create a new node y and edge (x, y) with $\mathcal{L}(y) = \{C\}$ and $\mathcal{L}((x, y)) = R$
$\forall$-rule	if 1.	$\forall R.C \in \mathcal{L}(x)$
	2.	there is some y s.t. $\mathcal{L}((x, y)) = R$ and $C \not\in \mathcal{L}(x)$
	then	$\mathcal{L}(y) \rightarrow \mathcal{L}(y) \cup \{C\}$

The labelling terminology is used in this subsection to show tableau correspondence, though in general we avoid it since it is more obvious what is happening to non-logicians when we show our algorithm as an implementable process that produces interpretations, rather than a convoluted mathematical abstraction. Specifically, we believe it is more natural to think of a predicate

Table 3. Inverted $\mathcal{ALC}$ Tableau Expansion Rules

Rule		
$\sqcap$-rule	if 1.	$x \in \mathcal{L}(C \sqcap D)$
	2.	$x \notin \mathcal{L}(C) \cap \mathcal{L}(D)$
	then	$\mathcal{L}(C) \rightarrow \mathcal{L}(C) \cup \{x\}$
		$\mathcal{L}(D) \rightarrow \mathcal{L}(D) \cup \{x\}$
$\sqcup$-rule	if 1.	$x \in \mathcal{L}(C \sqcup D)$
	2.	$x \notin \mathcal{L}(C) \cup \mathcal{L}(D)$
	then	either $\mathcal{L}(C) \rightarrow \mathcal{L}(C) \cup \{x\}$
		or else $\mathcal{L}(D) \rightarrow \mathcal{L}(D) \cup \{x\}$
$\exists$-rule	if 1.	$x \in \mathcal{L}(\exists R.C)$
	2.	there is no y s.t. $(x, y) \in \mathcal{L}(R)$ and $y \in \mathcal{L}(C)$
	3.	there is no z s.t. x is blocked by z
	then	create a new element y with $y \in \mathcal{L}(C)$ and $(x, y) \in \mathcal{L}(R)$
$\forall$-rule	if 1.	$x \in \mathcal{L}(\forall R.C)$
	2.	there is some y s.t. $(x, y) \in \mathcal{L}(R)$ and $y \notin \mathcal{L}(C)$
	then	$\mathcal{L}(C) \rightarrow \mathcal{L}(C) \cup \{x\}$

that represents a set as a labelled object where the label represents the set associated with the predicate, than it is to invert this and represent all the elements as labels for connected sets of predicate names.

4.3 ABox Expansion

One of the more useful optimizations that this algorithm permits is the expansion of ABox axioms into many additional, unstated, expressions which must hold in every possible model. This expansion begins while the algorithm evaluates the ABox and detects any immediate clashes. Later while evaluating the TBox the expansion detects any clashes that cannot possibly be fixed to allow the algorithm to terminate quickly, and it can be done concurrently whenever the algorithm adds or removes elements from expressions. In Definition 4 we maintain the assumption that all expressions are NNF and use $\neg$ to simplify the appearance of equivalent expressions.

Definition 4. *For partial interpretation $\mathcal{I}^*$, expression E, and element/pair x:*

1. *A* clash *means $x \in E^{\mathcal{I}^*}$ and $x \in \neg E^{\mathcal{I}^*}$*
2. *x* must be in *E iff there is no $\mathcal{I}^*$ where modifying $\cdot^{\mathcal{I}^*}$ to obtain $x \in \neg E^{\mathcal{I}^*}$ does not cause a clash*
3. *x* must not be in *E iff there is no $\mathcal{I}^*$ where modifying $\cdot^{\mathcal{I}^*}$ to obtain $x \in E^{\mathcal{I}^*}$ does not cause a clash*
4. *An* unresolvable clash *means x must be in $E^{\mathcal{I}^*}$ and x must not be in $E^{\mathcal{I}^*}$*

The set of all elements that must be in an expression $E^{\mathcal{I}^*}$, denoted $E_m^{\mathcal{I}^*}$, is a subset of the elements of $E^{\mathcal{I}^*}$, and the set of all elements that must not be in $E^{\mathcal{I}^*}$, written $E^{\mathcal{I}^*}_{\overline{m}}$, is a subset of the elements of $\neg E^{\mathcal{I}^*}$. Like the expressions themselves, the exact members of $E^{\mathcal{I}^*}_m$ and $E^{\mathcal{I}^*}_{\overline{m}}$ are unknown at the beginning of the algorithm, so $E^{\mathcal{I}^*}_m$ and $E^{\mathcal{I}^*}_{\overline{m}}$ indicate the additional knowledge of constraints on satisfiability that grow as the algorithm runs and are thus referred to with the $\mathcal{I}^*$ as well. When we talk about element(s) being "added" to these sets (nothing is ever removed), it is to indicate the change in $\cdot^{\mathcal{I}^*}$ that must hold in all future $\mathcal{I}^*$. Sets of elements that must (not) be in a complex expression can be computed in a straightforward way from the components of the expression they are in, for example $(B \sqcap C)^{\mathcal{I}^*}_m \equiv B^{\mathcal{I}^*}_m \sqcap C^{\mathcal{I}^*}_m$ and $(B \sqcap C)^{\mathcal{I}^*}_{\overline{m}} \equiv B^{\mathcal{I}^*}_{\overline{m}} \sqcup C^{\mathcal{I}^*}_{\overline{m}}$ (note the use of DeMorgan for $\overline{m}$).

Expansion occurs during TBox evaluation by, whenever possible, propagating the elements that must (not) be in expressions from antecedent to consequent, e.g. if we have axioms $\{A(x), A \sqsubseteq B\}$ then $x \in A_m$, and when we check $A \sqsubseteq B$ it is often possible to conclude $x \in B_m$. This is not always possible for expressions containing disjunction and negation except in certain very specific cases. Regardless the effect is powerful. Maintaining these sets allows us to explore only potentially correct solutions by preemptively avoiding actions that will be inconsistent in every model and also detect unresolvable clashes so the evaluation can terminate more quickly.

4.4 Local Blocking

The notion of blocking is also required for termination due to the new elements[3] created by the existential function. Our notion of blocking corresponds to the usual definition in that it references the same cyclic patterns in roles, however Emi cannot use blocking in reference to the global role hierarchy which is not computed explicitly, so cycles are detected locally by tracing the dependency paths that emerge as new elements are created to satisfy expressions.

Definition 5. *A new element x was* created to satisfy *expression $\exists R.B$ in an algorithm A if there are partial interpretations $\mathcal{I}^*, \mathcal{I}^{*\prime}$ for A where for some y we have $(y, x) \in R^{\mathcal{I}^*}, x \in B^{\mathcal{I}^*}, y \in \exists R.B^{\mathcal{I}^*}$ and $y \notin \exists R.B^{\mathcal{I}^{*\prime}}$ if $x \notin \Delta^{\mathcal{I}^{*\prime}}$.*

Definition 6. *For expression $\exists R.B^{\mathcal{I}^*}$, an element x is* blocked *by element z if $(z, x) \in R^{\mathcal{I}^*}$ and x was created to satisfy $\exists R.B$, or if for $(z, y_0) \in R^{\mathcal{I}^*}$ where y_0 was created to satisfy $\exists R.B$ we have $n \geq 0$ pairs of elements such that $\bigcup_{k=1}^{n}(y_{k-1}, y_k) \cup \{(y_n, x)\} \subseteq \Delta^{\mathcal{I}^*} \times \Delta^{\mathcal{I}^*}$.*

Fortunately this restriction is rather intuitive to implement: we simply need the function for an existential to not create new elements when checking new elements that exist as a result of a prior application of the same function on the same existential, since this will induce a cycle. It is effectively as if elements have a 'history' represented by the chain of pairs that they inherit from the

[3] When we say 'new element' this is equivalent to a fresh element symbol. It is written this way because we are avoiding terminology that refers to an infinite set of names and instead refer to what a program really does here, i.e. create something new.

element that created them and which also contains their own origin. An element will therefore not generate new elements in a function with the existential that created it, and not if the function for the existential created an element that it depends on for its existence, or that is in its 'history'.

4.5 Termination Condition

The final component necessary for the algorithm to work is a termination condition. Unlike a tableau, this algorithm will run on a loop and will actually attempt to directly realize the ontology. How then do we know that the algorithm has correctly realized all of the axioms? This is actually rather simple. If the algorithm begins an iteration of checking all axioms with initial state $\mathcal{I}_a^*$ and ends this iteration with state $\mathcal{I}_b^*$ without encountering any unresolvable clashes, then we know it can terminate successfully if $\mathcal{I}_a^* = \mathcal{I}_b^*$. Since we know that each $\mathcal{I}^*$ represents a possible interpretation, it is clear that $\mathcal{I}^*$ is a model when no clash occurs after every axiom is checked and all elements remain the same in all expressions.

Definition 7. *A partial interpretation $\mathcal{I}^*$ is said to* equal *partial interpretation $\mathcal{I}^{*\prime}$, or $\mathcal{I}^* = \mathcal{I}^{*\prime}$, iff $\forall E \in N_C \cup N_R$ we have $E^{\mathcal{I}^*} = E^{\mathcal{I}^{*\prime}}$.*

Equality can be verified by checking if the names in the signature have not changed any of their memberships. If an element has been added to the signature, or has moved into or out of any class or role, the algorithm must check the axioms one more time to see if these changes cause side effects. Otherwise the algorithm will have checked every axiom and found them to be satisfiable without making any changes and has produced a model.

4.6 Algorithm Definition

An outline of Emi is shown in Algorithm 1 that uses functions from Table 4.

Backtracking. Explicit backtracking in this algorithm is handled by 'reporting' clashes and unresolvable clashes.

Definition 8. *A* report *for Algorithm 1 immediately terminates execution of whatever process is occurring and returns to the function that handles this report.*

A report is meant to act like a programming exception. For example, when an unresolvable clash is directly reported while evaluating the ABox the behavior is clear, the ontology is unsatisfiable so all reasoning stops and the algorithm immediately exits by returning False.

Standard clashes occur when the inverted tableau rules require an action that is known to be inconsistent, but which could potentially be resolved by making changes elsewhere. In this case the immediate action is to stop trying the obviously inconsistent task and instead do something different. This type of clash is normally handled in Emi by the code that solves the inverted tableau rules in the same way that a standard tableau might. However, there are cases where the *sat* function must intervene, for instance when the antecedent of an axiom can neither lack nor contain an element and both assignments have been attempted to exhaustion. In this case we have indirectly found an unresolvable clash. If a clash is reported but is able to be resolved, the algorithm merely proceeds along as normal, and any problems that a removal causes will themselves be reported and dealt with in the same way. Emi internally ensures that backtracking does not return to an identical previous program state so that termination is not a concern. ABox expansion greatly impacts this functionality by automatically removing many impossible solutions from consideration, allowing these reports to happen faster.

Parallelization. Because we are not working with one single expanded concept representing the entire set of axioms like a standard tableau, it is possible in this algorithm to parallelize many operations on separate axioms. In the simplest case, it is unproblematic for the algorithm to process more than one axiom at the same time so long as they do not share any class or role names. Clashes that occur as a result of two independent axioms will in either case still be detected elsewhere and are dealt with regardless of the ordering. Additionally, this idea can be extended in the implementation to allow for axioms that share names to be evaluated concurrently as well, so long as shared names are not modified in ways that can cause a clash and begin backtracking. Emi makes use of the simplest case already and the implementation of more complex parallelization is in development.

Table 4. *sat* function behavior for $\mathcal{ALCH}$ satisfiability of ontology O, class name A, class expressions B, C, and roles R, S

Axiom	Action
$A(a)$	If $a \in A^{\mathcal{I}^*}_{\overline{m}}$ report an unresolvable clash, otherwise add a to $A^{\mathcal{I}^*}$ and $A^{\mathcal{I}^*}_{m}$
$\neg A(a)$	If $a \in A^{\mathcal{I}^*}_{m}$ report an unresolvable clash, otherwise add a to $A^{\mathcal{I}^*}_{\overline{m}}$
$B(a)$	Add a new class A and the axiom $A \sqsubseteq B$ to O and replace $B(a)$ with $A(a)$
$R(a, b)$	If $(a, b) \in R^{\mathcal{I}^*}_{\overline{m}}$ report an unresolvable clash, otherwise add (a, b) to $R^{\mathcal{I}^*}$ and $R^{\mathcal{I}^*}_{m}$
$\neg R(a, b)$	If $(a, b) \in R^{\mathcal{I}^*}_{m}$ report an unresolvable clash, otherwise add (a, b) to $R^{\mathcal{I}^*}_{\overline{m}}$
$B \sqsubseteq C$	For all $x \in B^{\mathcal{I}^*}$ add x to C using the inverted tableau expansion rules If a clash is reported, instead bactrack to remove x from B If both report a clash, report an unresolvable clash
$B \equiv C$	Do $sat(B \sqsubseteq C)$ and $sat(C \sqsubseteq B)$ and report any unresolvable clashes
$R \sqsubseteq S$	For all $(x, y) \in R^{\mathcal{I}^*}$ add (x, y) to S. If a clash is reported, instead backtrack to remove (x, y) from R. If both report a clash, report an unresolvable clash

Algorithm 1: $\mathcal{ALCH}$ satisfiability for ontology O

```
Result: A realization that is a model of O when True, otherwise False
        when an unresolvable clash occurs
// Initialize an empty set to represent a realization for O
I* ← {};
// Assume all E/E_m/E_m̄ are empty and add them to I*
for E ∈ N_C ∪ N_R do
    E ← {};
    E_m ← {};
    E_m̄ ← {};
    I* ← I* ∪ {E, E_m, E_m̄};
// Obtain NNF of all axioms
F ← NNF of ABox of O;
A ← NNF of TBox of O;
// Process ABox
for F ∈ F do
    sat(F);
    if an unresolvable clash is reported then
        return False
// Process TBox
repeat
    I*_old ← I*;
    for A ∈ A do
        sat(A);
        if an unresolvable clash is reported then
            return False
until I* = I*_old;
return True
```

4.7 Correctness

The correctness of this algorithm follows from the direct correspondence between the tableau rules and their inversions along with a few minor details. First we will examine the inverted tableau rules. Each rule contains two distinct states in both the standard and inverted rules, the antecedent 'if' clause where certain conditions must hold, and the consequent 'then' clause where changes are made to the state in order to find a model. Thus it will be sufficient to show that each 'if' and 'then' clause from the inverted tableau rules is re-writable into the corresponding clause in the standard tableau rules.

$\sqcap$-rule 'if' This state in the inverted tableau rules corresponds to the standard tableau rules, since in each case we know x should be in $C \sqcap D$, but it is either not in C or D or both.

'then' The change made in response to the 'if' clause is also equivalent, since the inversion adding x to $C^{\mathcal{I}^*}$ and $D^{\mathcal{I}^*}$ represents the notion that we now have $x \in C^{\mathcal{I}^*}$ and $x \in D^{\mathcal{I}^*}$ which is identical to the standard tableau rules adding $\{C, D\}$ to $\mathcal{L}(x)$.

$\sqcup$-rule 'if' This state in the inverted tableau rules corresponds to the standard tableau rules, since in each case we know x should be in $C \sqcup D$, but it is neither in C nor in D.

'then' The change made in response to the 'if' clause is also equivalent, since the inversion adding x to $C^{\mathcal{I}^*}$ or $D^{\mathcal{I}^*}$ represents the notion that we now have $x \in C^{\mathcal{I}^*}$ or $x \in D^{\mathcal{I}^*}$ which is identical to the standard tableau rules either adding $\{C\}$ to $\mathcal{L}(x)$ or adding $\{D\}$ to $\mathcal{L}(x)$.

$\exists$-rule 'if' This state in the inverted tableau rules corresponds to the standard tableau rules, since in each case we know x should be in $\exists R.C$, but the current state means that there is either no $(x, y) \in R$ such that $y \in C$ or there is no $y \in C$ such that $(x, y) \in R$. The blocking condition (3) only applies when x is blocked because it is a new individual that exists as a (possibly indirect) result of an element previously created to satisfy this expression. When x is blocked in this way it is clear that we have completed the first iteration of a cycle and do not need to continue.

'then' The change made in response to the 'if' clause is also equivalent, since the inversion creating some y such that $y \in C^{\mathcal{I}^*}$ and $(x, y) \in R^{\mathcal{I}^*}$ represents the notion that $x \in \exists R.C$, which is identical to the standard tableau rules creating a new node and edge for y with $\mathcal{L}(y) = \{C\}$ and $\mathcal{L}((x, y)) = R$.

$\forall$-rule 'if' This state in the inverted tableau rules corresponds to the standard tableau rules, since in each case we know x should be in $\forall R.C$, but the current state means that there is some y such that $(x, y) \in R$ and $y \notin C$.

'then' The change made in response to the 'if' clause is also equivalent, since the inversion adding every y to $C^{\mathcal{I}^*}$ wherever $(x, y) \in R^{\mathcal{I}^*}$ represents the notion that $x \in \forall R.C$, which is identical to the standard tableau rules adding $\{C\}$ to any y so that $\mathcal{L}(y) = \{C\}$.

The only remaining thing to discuss is that Algorithm 1 and the *sat* function are sound and complete. For this it is simple to outline without a lengthy proof, since as we mentioned previously, the sat functions will test all elements in all expressions with the tableau expansion rules, exactly as you would have in a tableau. It is unnecessary and in fact unhelpful in this algorithm to combine all expressions into a connected whole since each axiom is itself evaluated consistently.

The subsumption axioms are solved in such a way that they implicitly compute a class and role hierarchy, so any inference dependent on a hierarchy will still be entailed. Next, the stop condition of Algorithm 1 when the interpretation has not changed is identical with a completed tableau that no longer requires expansion or backtracking since both are models. And of course all cases where unresolvable clashes can occur are sound because they indicate instances where an element simultaneously must be in and must not be in an expression so the algorithm should terminate. As for completeness, it is again clear that whenever an unresolvable clash occurs it is identical with a case when a tableau discovers a clash but cannot backtrack to correct it.

5 Evaluation

The implementation of the Emi reasoner is evaluated against other popular JVM reasoners such as Hermit[4], Pellet[5], and jFact[6] [4,7,8] using Leiningen[7], which can run Clojure as well as Java programs. 500 ontologies were randomly sampled, half from ODPs and Biomedical ontologies in [3], and half from the Ontology Reasoner Evaluation 2015 competition dataset. These files were altered by removing any axiom that is not expressible in $\mathcal{ALCH}$ for testing. Among the 500 ontologies, 56 of them either timed out on all reasoners or caused our evaluation program to crash in some unexpected way due to lack of heap. It is not entirely clear from our logs which reasoner may be breaking in this way so we omit them, though we are confident that Emi has no heap issues or memory leaks (it is quite difficult to even *intentionally* create memory leaks with Clojure) and it can load all files by itself without error.

All testing was done on a computer running Ubuntu 20.04.1 64-bit with an Intel Core i7-9700K CPU@3.60 GHz x 8, 47.1 GiB DDR4, and a GeForce GTX 1060 6 GB/PCIe/SSE2.

5.1 Results

Testing and development is ongoing for the Emi reasoner and these results are a first example of the potential for this type of algorithm. A preliminary example of the reasoning time can be found in Fig. 1. In this example, 3 tests are performed and timed on each reasoner: initialization from a file in memory shown in Fig. 2, satisfiability and consistency checking shown in Fig. 3, and retrieval of the elements in all non-empty class and role names after reasoning has completed shown in Fig. 4.[8] Each test was given a timeout of 5 min, and when a test timed out more than 5 tests in a row for a reasoner it was prevented from testing again on the same file to save time in the evaluation. Across all tests, when we compare the percent difference between the reasoner time and the average time for all reasoners we see that Emi is overall 9.8% faster than average, Hermit is 4.2% slower than average, Pellet is 42.3% faster than average, and jFact is 55.9% slower than average. No reasoner was ever more than 100% better than average or 300% worse than average. Emi, Hermit, and Pellet each had 2 files when they

[4] Hermit Version 1.4.5.519.
[5] Openllet Version 2.6.5.
[6] jFact Version 5.0.3.
[7] https://leiningen.org/.
[8] Raw data and charts are available for inspection https://tinyurl.com/kgswc2023.

failed due to an exception, while jFact failed on 70 files, though this is largely due to an issue it has with anonymous individuals in the ORE dataset.

As you can see in Figs. 2 and 3, the Emi reasoner has a relatively slow initialize time and time for satisfiability on small ontologies. This is partially due to the fact that Emi solves all ABox axioms without complex expressions while it loads the ontology. This allows it to preemptively reject any naively unsatisfiable ABox, and the cost of this is usually only a few milliseconds. Emi seems to scale better than other systems and its performance improves in comparison as the number of axioms increases. Also, Emi is definitively faster in every single test when asked to compute the elements of all non-empty classes and roles, except the two tests where it had a timeout when computing satisfiability. Emi finishes reasoning with this information already computed, it is in effect computing both things at once, and only needs a variable amount of time to answer in our tests because it has to sort out the anonymous elements from every expression for comparison with the other reasoners where these elements are hidden by the OWL API [5]. Otherwise it could answer this in constant time.

Looking more closely at the data, there are three large clusters at 163, 325, and approximately 5500 axioms due to the synthetic ontologies in the ORE dataset. As you can see, all reasoners appear to behave similarly across multiple files of the same size in the clusters. Except for these clusters there appears to be a mostly stochastic distribution in the performance of each reasoner across different files with jFact doing great on small files but not scaling well, while Hermit, Pellet, and Emi start off a bit slower and seem to scale much better on very large files.

An interesting pattern we have noticed in the evaluation is that Emi usually outperforms other reasoners when it checks large ontologies with empty or nearly-empty ABoxes. This case makes sense when you notice that in $\mathcal{ALCH}$, as long as all axioms do not contain either negation or Top in the antecedent, there will always be a model where every predicate is empty. This is the default state when Emi starts, so it simply checks everything and the initial realization turns out to be fine after the first iteration.

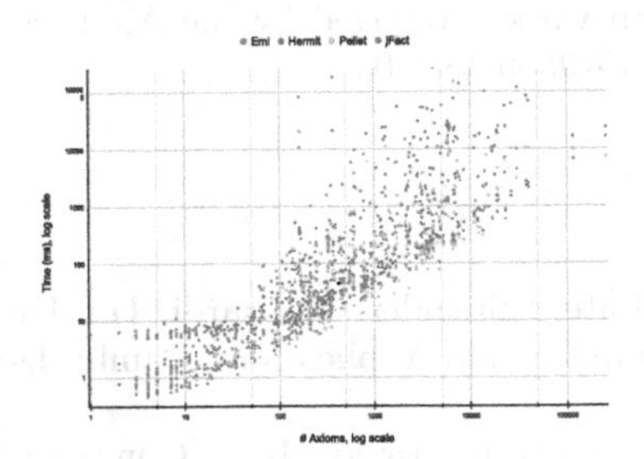

Fig. 1. All Reasoning Tests

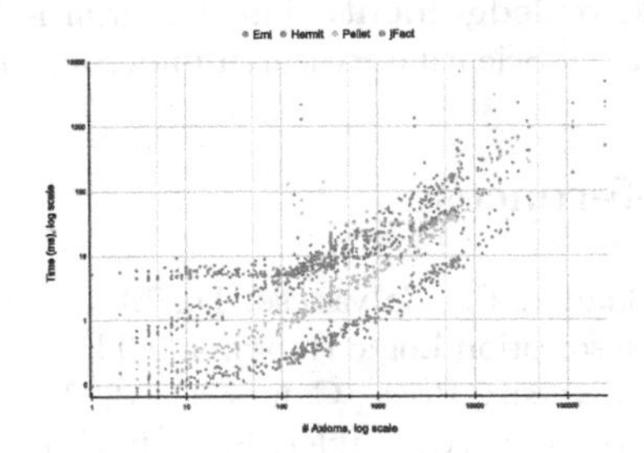

Fig. 2. Initialize

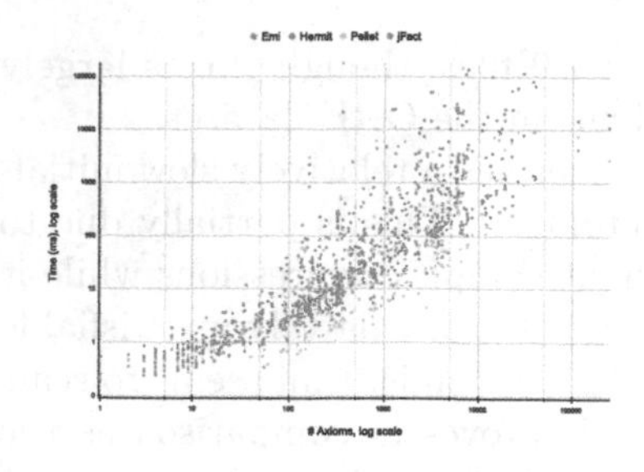

Fig. 3. Satisfiability

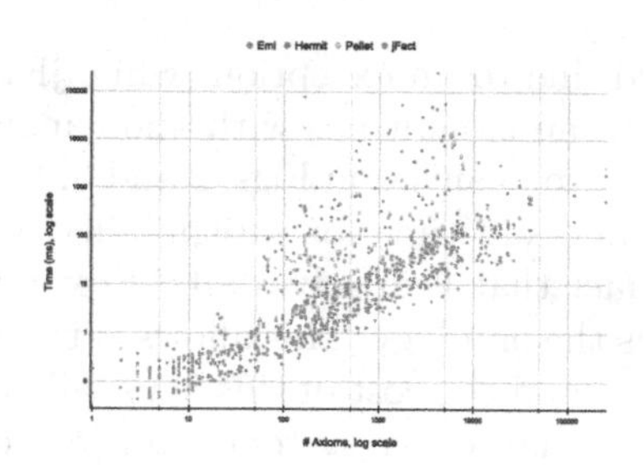

Fig. 4. Non-Empty Classes and Roles

6 Future Work

In the future there is quite a bit of work left to finalize and sufficiently verify Emi. These improvements will be ongoing as part of any new extensions. Parallelization in particular will likely prove difficult to formally pin down, though empirically its use is not as yet seeming to be problematic when comparing against the behavior of other reasoners. Some obvious extensions that are planned for the near future are inverse roles, nominals, and cardinality expressions which can work quite directly with some of the already existing code. Role chains are another interesting and useful addition we are considering, though these will be difficult to implement efficiently so we will be cautious.

An interesting observation we have made while considering extensions is that some of the common difficulty with nominals in reasoning may turn out to be trivial in many cases for the Emi algorithm because nominals are not necessarily connected to anything as long as we maintain sufficient information about (in)equality. Once this is implemented and our hypothesis is checked it would also be straightforward to extend nominals to nominal schemas, since Emi does not normalize or pre-process away the original axioms. Initial testing suggest there is an intuitive way to bind nominal variables within axioms to solve this directly as Emi reasons.

Acknowledgement. This material is based upon work supported by the Air Force Office of Scientific Research under award number FA9550-18-1-0386.

References

1. Baader, F., Calvanese, D., McGuinness, D., Patel-Schneider, P., Nardi, D.: The Description Logic Handbook: Theory, Implementation and Applications. Cambridge University Press, Cambridge (2003)
2. Eberhart, A., Hitzler, P.: A functional API for OWL. In: Taylor, K.L., Gonçalves, R.S., Lécué, F., Yan, J. (eds.) Proceedings of the ISWC 2020 Demos and Industry Tracks: From Novel Ideas to Industrial Practice co-located with 19th International Semantic Web Conference (ISWC 2020), Globally online, November 1-6, 2020 (UTC). CEUR Workshop Proceedings, vol. 2721, pp. 315–320. CEUR-WS.org (2020). http://ceur-ws.org/Vol-2721/paper580.pdf

3. Eberhart, A., Shimizu, C., Chowdhury, S., Sarker, M.K., Hitzler, P.: Expressibility of OWL axioms with patterns. In: Verborgh, R., et al. (eds.) ESWC 2021. LNCS, vol. 12731, pp. 230–245. Springer, Cham (2021). https://doi.org/10.1007/978-3-030-77385-4_14
4. Glimm, B., Horrocks, I., Motik, B., Stoilos, G., Wang, Z.: hermit: an owl 2 reasoner. J. Autom. Reason. **53**(3), 245–269 (2014). https://doi.org/10.1007/s10817-014-9305-1, https://doi.org/10.1007/s10817-014-9305-1
5. Horridge, M., Bechhofer, S.: The OWL API: a java API for working with OWL 2 ontologies. In: Proceedings of the 6th International Conference on OWL: Experiences and Directions, vol. 529, pp. 49–58. OWLED 2009, CEUR-WS.org, Aachen, DEU (2009)
6. Horrocks, I., Kutz, O., Sattler, U.: The even more irresistible sroiq. In: Proceedings of the Tenth International Conference on Principles of Knowledge Representation and Reasoning, pp. 57–67. KR'06, AAAI Press (2006)
7. Sirin, E., Parsia, B., Grau, B.C., Kalyanpur, A., Katz, Y.: Pellet: a practical owl-dl reasoner. Web Semant. **5**(2), 51–53 (2007). https://doi.org/10.1016/j.websem.2007.03.004
8. Tsarkov, D., Horrocks, I.: FaCT++ description logic reasoner: system description. In: Furbach, U., Shankar, N. (eds.) IJCAR 2006. LNCS (LNAI), vol. 4130, pp. 292–297. Springer, Heidelberg (2006). https://doi.org/10.1007/11814771_26

An Intelligent Article Knowledge Graph Formation Framework Using BM25 Probabilistic Retrieval Model

Jasir Mohammad Zaeem, Vibhor Garg, Kirti Aggarwal(✉), and Anuja Arora

Jaypee Institute of Information Technology, Noida, India
aggarwalkirti25@gmail.com

Abstract. Knowledge graphs are considered as the best practice to illustrate semantic relationships among the collection of documents. The research article presents an intelligent article knowledge graph formation framework that utilizes the BM25 probabilistic retrieval model for constructing a knowledge graph to express conditional dependency structure between articles. The framework generates conditional probability features by employing the Okapi BM25 Score and rank documents according to the search query. Also, the CRF model is trained to perform Named Entity Recognition to empower semantic relationships. This indicates that a BM25 score feature matrix is created of all named entity in the collection which comes out as high dimensional. This problem is rectified using singular value decomposition dimension reduction technique and pairwise cosine similarity is calculated on reduced dimensions. Based on a similarity value, the documents are connected and construct the knowledge graph structure. The results represent the outcome of the search engine for different types of queries on IMDB database and constructed knowledge graphs based on similarity between articles in BBC and IMDB Dataset.

Keywords: Knowledge graph · Okapi BM25 · Search engine · Ranking

1 Introduction and Background Study

Knowledge graphs are considered as an effective and the most significant representation to showcase the information in a structured way. It is basically a graph database that stores knowledge in such a way that can be conveniently analyzed and queried [1]. The fundamentals of graphs comprise of nodes, edges, and properties were,

- Entity (people/places/objects/concepts) are represented by nodes in a knowledge graph [2, 3]. For example, a person node might be used to represent a person;
- Edges represent relationships (friendship, professional contact, event attendees, etc.)
- Properties/ features are additional information associated with nodes and edges, such as attributes or metadata [4]. For example, a person node might have properties such as name, age, or occupation.

F. Ortiz-Rodriguez et al. (Eds.): KGSWC 2023, LNCS 14382, pp. 32–43, 2023.
https://doi.org/10.1007/978-3-031-47745-4_3

This research introduces an innovative approach to knowledge graph construction, leveraging the BM25 probabilistic retrieval model in a novel manner. The framework focuses on revealing conditional dependency structures between articles, powered by the unique generation of conditional probability features using the Okapi BM25 Score. This combination of BM25 and conditional probability features represents a fresh perspective on knowledge graph development, promising to enhance our understanding of semantic relationships within document collections. The framework holds potential applications in diverse domains, such as information retrieval, knowledge management, and recommendation systems [5].

Nowadays, Knowledge graph is used to enhance search engine outcomes by providing more extensive and precise results [6]. By structuring information, a knowledge graph facilitates search engines in comprehending the associations between diverse entities and concepts [7]. For instance, when a user conducts a search on a specific topic, a knowledge graph can be utilized by a search engine to pinpoint the pertinent entities and relationships linked with that topic [8]. This can enhance the accuracy and relevance of the search engine results, including related entities and concepts that might be of interest to the user.

Knowledge graphs can significantly enhance the ranking of search outcomes as well. Search engines can benefit from knowledge graphs by comprehending the interrelations between various entities and concepts [9]. This, in turn, facilitates the presentation of more precise and relevant search outcomes to users. For instance, a search engine can use a knowledge graph to recognize the most relevant entities and concepts associated with a search query and organize search outcomes in order of priority [10]. Moreover, a knowledge graph can empower search engines to offer personalized suggestions and recommendations by considering the user's search history and preferences. By scrutinizing the user's interactions with the graph, the search engine can identify patterns and relationships that can be used to provide personalized suggestions [11]. Considering the background studies conducted, several limitations have been identified within the realm of existing work.

- A limitation lies in the depth of semantic relationships within traditional knowledge graphs, particularly in cases where Named Entity Recognition (NER) is not integrated. The framework takes strides to address this concern by incorporating NER, thereby enhancing the richness of semantic relationships and offering a more detailed understanding.
- Furthermore, many extant knowledge graphs remain static and may not adapt adeptly to diverse query types. In contrast, the proposed framework presents a solution wherein knowledge graphs can be dynamically generated to align with specific search queries. This adaptability holds promise for enhancing the relevance and precision of query results.

We directed this research to propose a generic framework to build knowledge graph of the most relevant documents for any specific search query. The research contributions of the research work are as follows

- An Effective knowledge graph framework is proposed to support the relevant search query.

- A probabilistic best matching model Okapi BM25 score and Named entity is applied to apply conditional dependency structure relationship to form knowledge graph representing semantic relationship of search query;
- Singular Value decomposition technique is applied to reduce high dimensional Okapi BM25 score document and generate an efficacious feature vector.
- A knowledge graph generation technique using pairwise cosine similarity measure is proposed to present hidden relationships between diverse documents utilizing BM25 content matching score.

The remainder of the paper comprises of an intelligent knowledge graph formation framework that is discussed in Sect. 2. Section 3 discusses the probabilistic models for search outcome ranking. Section 4 details feature vector generation and dimensionality reduction methodologies in support of illustrating the significant information. Graph construction is discussed in Sect. 5. Section 6 depicts dataset, intermediate results, and knowledge graph research outcome of search engine query document utilizing discussed framework. At the end conclusion and future scopes are discussed in Sect. 7.

2 Knowledge Graph Formation Framework

The knowledge graph formation framework is a methodical approach that provides a sequence of processes and techniques for developing a knowledge graph. The main objective is to construct a well-organized knowledge graph from different data sources by recognizing pertinent entities, their properties, and the connections among them [12]. The framework usually encompasses multiple phases, such as collecting data, preparing data, identifying named entities, extracting relationships, linking entities, and constructing a knowledge graph. Each phase involves different algorithms and methods that are utilized to convert unstructured data into a structured knowledge graph as shown in Fig. 1.

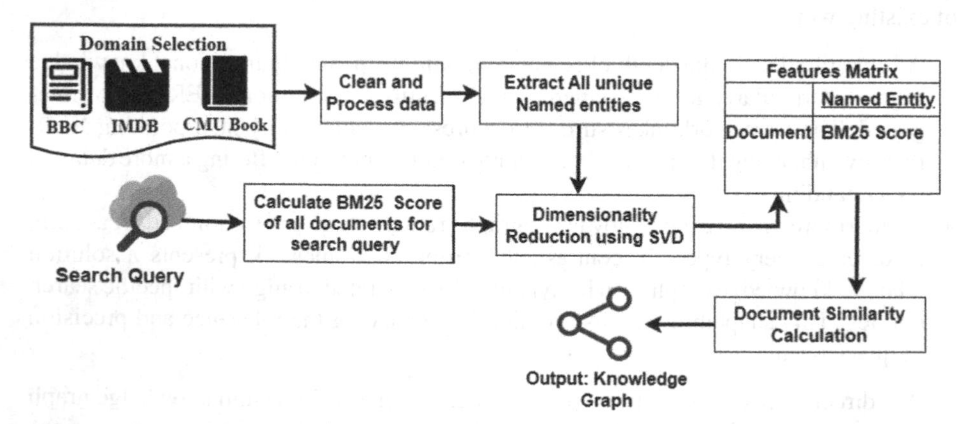

Fig. 1. Knowledge Graph Formation Framework using Probabilistic Ranking Function BM25

The proposed knowledge graph framework is shown in Fig. 1 and its intermediate steps are detailed herewith.

- ***Identify the domain and scope of the knowledge graph:*** The process begins to determine the specific domain or subject for which knowledge graph needs to construct (Abu-Salih, 2021). Therefore, the process starts with domain selection and user search query input.
- ***Collect and preprocess data:*** Gather data from various sources and preprocess it to ensure that it is clean, consistent, and structured in a way that is compatible with the knowledge graph [13]. Here, process cleans the domain data and process it to standard set of document id, document text, and additional metadata tuples.
- ***Extract entities and relationships:*** Use natural language processing (NLP) techniques to identify named entities and relationships from the preprocessed data. Named entity recognition is a technique to organize data at abstract level which helps in defining the types of entities and relationships [14, 15].
- ***Calculate Documents ranking and dimensionality reduction:*** The probabilistic ranking function BM25 is used to rank the documents based on their relative proximity. Further, to generate knowledge graph, feature vectors are formed which will store named entity of terms along with BM25 score. Although, the BM25 generated feature vector consists of a high dimension feature vector of each document in collection. This issue has been rectified using a singular value decomposition feature reduction approach.
- ***Document similarity and knowledge graph generation:*** Generate a graph database to create the actual knowledge graph of documents, including nodes for entities and edges for relationships based on documents similarity score calculated using BM25 score as feature vector [4]. Once the graph is created, it must be populated with data from the preprocessed data sources [13, 16].

3 Probabilistic Model for Search Outcome Ranking

Probabilistic models used in search outcome ranking utilize probability theory to assess the relevance of search results. These models evaluate the probability of a document being relevant to a user's query and assign a score accordingly [17]. To calculate relevance scores, probabilistic models often incorporate statistical measures like term frequency, inverse document frequency, and document length. Other factors, such as query terms, document metadata, and user behavior, may also be considered [18].

TF-IDF. Term Frequency-Inverse Document Frequency (TF-IDF) is a technique used in natural language processing and information retrieval to determine the importance of a word in a document or collection of documents [19]. The term frequency (TF) refers to how often a term appears in a document, while the inverse document frequency (IDF) measures the significance of a term in a document collection. To calculate the TF-IDF $W(t, d)$ score of a term, we multiply the term frequency $TF(t, d)$ of the word t in the document d by the inverse document frequency of that term. This way, we can identify the most relevant words in the document by looking at the frequency of the word in the document and the rarity of the word across the corpus.

$$W(t, d) = TF(t, d) \times \log\left(\frac{N}{df_t}\right) \tag{1}$$

where N is the total number of documents and df_x is number of documents containing x.

BM25. The Best Match 25 (BM25) is a ranking function used in information retrieval that extends the principles of TF-IDF. While both methods compute the relevance of a document to a query, BM25 is considered to be a more advanced model as it considers other factors like the document length and the average length of documents in the collection. The BM25 score of a document d in Collection D of documents can be calculated using the following equation:

$$core(t, d) = \left(\frac{TF(t, d) \times IDF(t, D) \times (k1 + 1)}{TF(t, d) + k1 \times \left(1 - b + \frac{b \times l(d)}{avgdl}\right)} \right) \tag{2}$$

where t is the term, $TF(t, d)$ is the term frequency of the term in the document d, $IDF(t, D)$ is the inverse document frequency of the term in the collection D, $k1$ and b are parameters that control the relative importance of term frequency and document length, $l(d)$ is the length of the document d, and $avgdl$ is the average length of the documents in the collection D. To build the search engine, a search query is retrieved from the user. After receiving the query, BM25 score of the query is calculated using Eq. 2, against all the documents in the dataset. Then the search results are ranked based on these scores to display the most relevant documents to the user.

4 Feature Vector Generation and Dimensionality Reduction

In this section, the methodology to construct the graph of linked document is presented. For this purpose, first each dataset is cleaned and processed to get the standard set of id, name, text and additional metadata tuples. Then, Named Entity Recognition (NER) is used to extract a total set of unique named entities from a dataset.

Named Entity Recognition (NER). NER refers to a branch of natural language processing that emphasizes on detecting and categorizing entities that have been named in text. In this research paper we have used sequence labeling, which involves assigning labels (such as "person" or "location") to each word in a sentence. A Conditional Random Field (CRF) model is trained to predict the most likely sequence of labels for a given sequence of words. A labeled training set is employed to train a CRF, where each word in the sequence is associated with its corresponding named entity type. By examining the patterns and relationships between words and their labels in the training set, the CRF acquires knowledge to anticipate the labels of new, unseen sequences. After training, the CRF is utilized to label entities in fresh text sequences. This involves feeding the sequence into the CRF, which utilizes the learned patterns and relationships to assign a label to each word. The resulting sequence of labels is used to extract named entities from the text. Figure 2 shows the named entities after applying NER on a document.

After applying the NER, a BM25 relevance score for each of the named entity in the document is calculated and a feature matrix is constructed to describe the document. Encoding of documents set and named entity is performed to show data in numeric

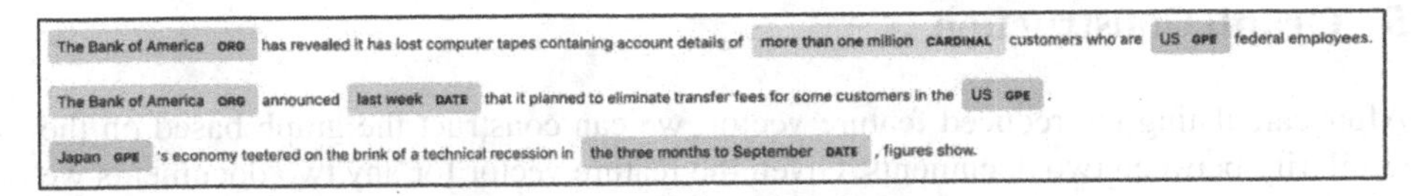

Fig. 2. Named Entities after Applying NER on a Document

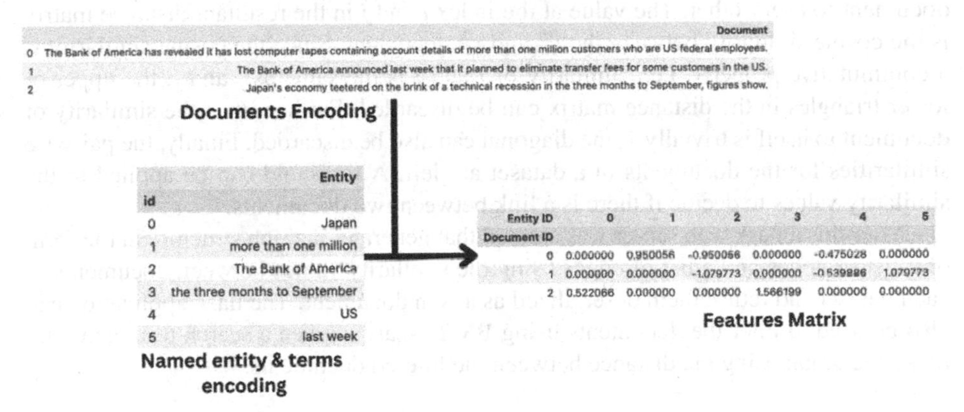

Fig. 3. Process of Feature Vector Generation

form. Finally, BM25 is computed for each entity across the dataset which gives a feature matrix. This complete process is shown in Fig. 3.

In the case of a dataset with a couple thousand documents the feature vector can be a few dozen thousand numbers long. So, for further processing Singular Value Decomposition (SVD) was used to reduce this dimension to a smaller number. The SVD of a matrix A is defined as:

$$A = U\Sigma V^T \tag{3}$$

where U and V are orthogonal matrices, and Σ is a diagonal matrix containing the singular values of A. V^T is transpose of matrix V. To conduct dimensionality reduction with SVD, we can calculate the truncated SVD of a matrix A by retaining solely the first k most significant singular values and their corresponding columns in U and V. Consequently, we obtain a decreased matrix A', which contains only k columns, in contrast to the initial n columns:

$$A\prime = U_k \Sigma_k (V_k)^T \tag{4}$$

where U_k and V_k are the first k columns of U and V, respectively, and Σ_k is the diagonal matrix containing the first k singular values. The matrix A' obtained through truncated SVD represents a reduced-dimensional version of the original matrix A, where only the columns corresponding to the k most significant dimensions or features are retained.

After experimentation, 500 dimensions were decided for this research work because it accounts for 80+ % variance while reducing the dimensions by almost 39,000 in case of the IMDb dataset (defined in Sect. 6) and 24,000 in case of the BBC dataset (defined in Sect. 6).

5 Graph Construction

After calculating the reduced feature vector, we can construct the graph based on the similarity between two documents. Given the feature vector for any two documents we can then gauge the similarity between them using the cosine similarity, i.e., the cosine of the angle between the vectors. So, the pairwise cosine similarity is calculated of each document to every other. The value at the index i and j in the resultant distance matrix is the cosine similarity between the ith and j^{th} document. Since the cosine similarity is a commutative property, i.e., similarity of i and j is the same as j and i, the upper or lower triangles in the distance matrix can be discarded. Further since the similarity of document to itself is trivially 1, the diagonal can also be discarded. Finally, the pairwise similarities for the documents in a dataset are left. A threshold can be applied to the similarity values to decide if there is a link between two documents.

After this a flask web server was created that generates a graph structure in the form of a list of nodes and a list of edges using the similarity values between documents in each dataset and return them deserialized as a json document. The flask application can also be used to filter the documents using BM25 search, given a search term and only make the graph using the distance between the filtered documents.

6 Dataset and Research Outcome

Table 1. Dataset Details

Dataset Name	Description	# of records
IMDB Top 1500[1]	Movies Information consist of movie name, release date, imdb id, genres, rating, description, plot synopsis	1500 Movies
BBC[2]	News articles from BBC between 2004 and 2005 across five categories, business, entertainment, politics, sport, technology	2225 News Articles
CMU Book Summary Dataset[3]	Dataset of 16,559 books and their plot summaries extracted from Wikipedia. It contains summaries and metadata such as author and genres	2000 Books selected from the dataset

[1] https://datasets.imdbws.com/.
[2] http://mlg.ucd.ie/datasets/bbc.html.
[3] https://www.cs.cmu.edu/~dbamman/booksummaries.html.

Table 1 details of the dataset used in this research work. The generated system ingests the documents in form of text documents. So, in the case of the IMDb movies dataset, a document is the concatenation of the various data points available for the movie, including description, summary, and cast. Once the document corpus is indexed, search queries can be performed on it. BBC and CMU Book Summary Dataset has news articles and book summaries as text documents, respectively.

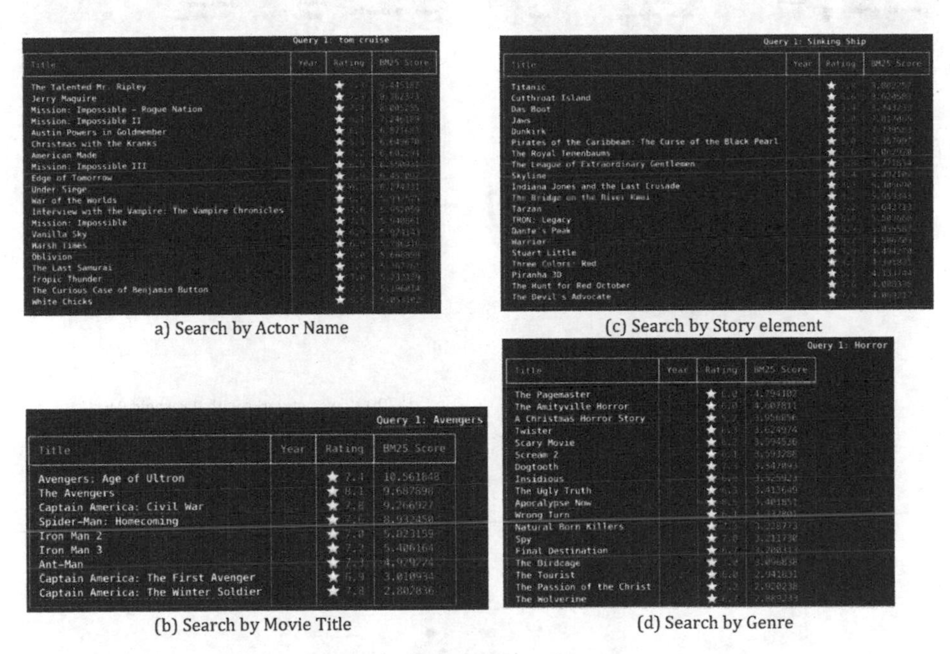

Fig. 4. BM25 Ranking Outcome

6.1 BM25 Articles Ranking Outcome

Figure 4 displays the results of the implemented Okapi BM25 algorithm, and the results are ranked using the BM25 score. Figure 4(a–d) shows the results of executing different queries on the IMDb dataset. Queries can be made involving movie titles, cast members, genres and plot points. Figure 4(a) shows the result of a query performed on actor name as "tom cruise". Figure 4(b) shows the result of a query performed on story element as "Sinking Ship". Figure 4(c) shows the result of a query performed on movie title as "Avengers". Figure 4(d) shows the result of a query performed in the genre "Horror".

6.2 Graph Formation Outcome

A web application was created in JavaScript using the React UI Library to visualize the data in the graph format. This visualization application was used to display the documents of a corpus in the form of nodes. Hovering over the nodes shows more information about the documents and a link to the original source. Document linking is user configurable through a threshold selector. Cosine Similarity was chosen as the linkage criteria between documents i.e., a higher threshold will result in fewer linkages. In the extreme case where the threshold is 1, the maximum selection, there will be no links. The threshold can be lowered to discover new connection by changing the threshold value.

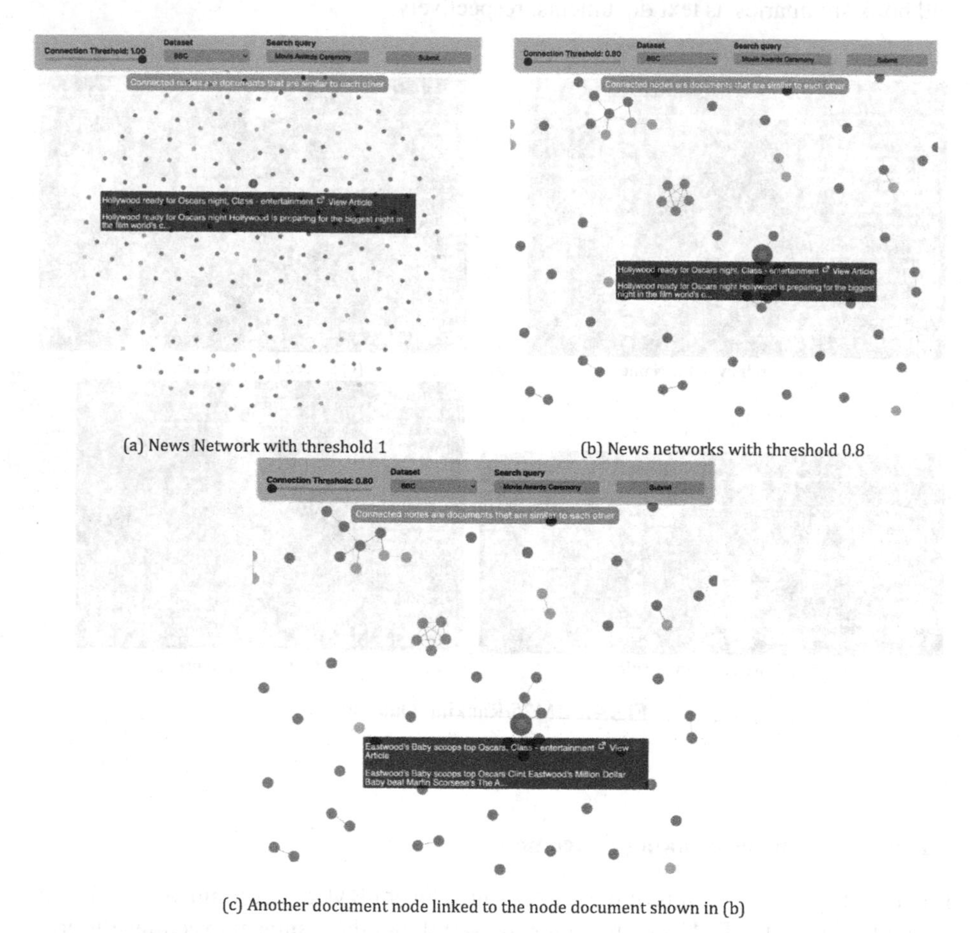

(a) News Network with threshold 1

(b) News networks with threshold 0.8

(c) Another document node linked to the node document shown in (b)

Fig. 5. BBC Graph Formation outcome for Query "Movie Awards Ceremony"

Figure 5 shows the constructed knowledge graphs on BBC dataset for search query 'Movie Awards Ceremony' using the different threshold values. Highlighted node in Fig. 5(a–b) represents a news article about the Oscar awards. Figure 5(a) shows the graph when threshold is 1 and it can be seen from the figure that there is no linkage between any document. Figure 5(b) shows the graph when threshold is 0.8 and it results in formation of new connections with similar articles. Figure 5(c) is like Fig. 5(b), but the highlighted node is different, and it can be seen in the description, that this highlighted node is also a news article about Oscar awards, and it is linked with previous highlighted node. Hence, both the nodes have high similarity.

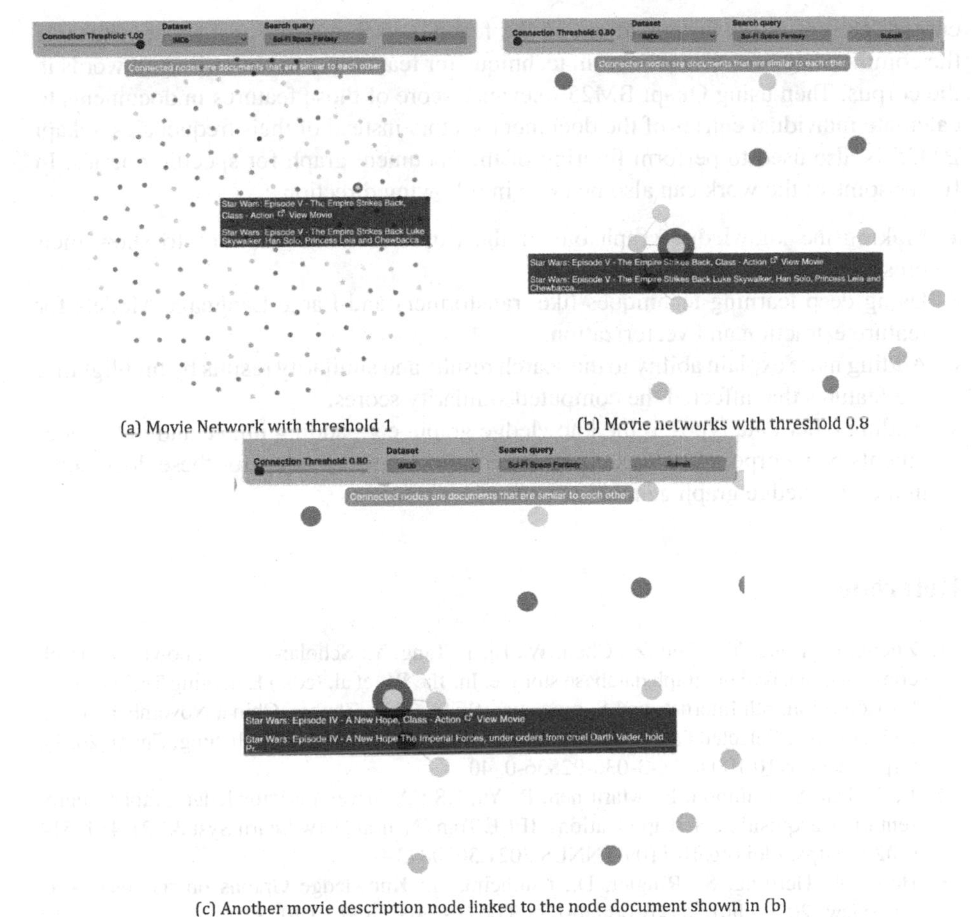

(a) Movie Network with threshold 1

(b) Movie networks with threshold 0.8

(c) Another movie description node linked to the node document shown in (b)

Fig. 6. IMDb Dataset search outcome for the Query "Sci-Fi Space Fantasy"

Figure 6 demonstrated the query "Sci-Fi Space Action" performed on the IMDb movies dataset. The nodes represent various movies that involve the themes Sci-Fi, Space or Action. The highlighted node in Fig. 6(a–b), is a movie in the Star Wars franchise, "Star Wars: Episode V". Figure a and Figure b show the graph when threshold is 1 and 0.8 respectively. Decreasing the threshold creates new connection between related movies. Highlighted node in Fig. 6(b) is connected to highlighted node in Fig. 6(c) "Star Wars: Episode IV" as they bear similarities being part of the same franchise and continuity.

7 Conclusion and Future Scope

In this research work a system is introduced that automatically generates a knowledge graph linking unstructured text documents based on their similarity. Document similarity is computed by the cosine similarity of the vector representations of documents. A novel method of vectorizing text documents is implemented to achieve this. Using named entities present in the corpus to create the features in the vector as a refinement over the common bag of words or n-gram technique for feature generation using all words in the corpus. Then using Okapi BM25 relevance score of those features in documents to calculate individual entries of the document vectors instead of their frequencies. Okapi BM25 is also used to perform filtering of the document graph for specific queries. In future some of the work can also be done in following directions:

- Making the knowledge graph out of the extracted named entities to show their presence and relation in documents.
- Using deep learning techniques like transformers and Large Language Models for feature extraction and vectorization.
- Adding more explain ability to the search results and similarity results by highlighting the features that affected the computed similarity scores.
- Adding other data types to the knowledge graph, e.g., adding image and video documents by incorporating the outputs from vision to text models for these documents in the knowledge graph generation process.

References

1. Zheng, D., Long, Y., Zhou, Z., Chen, W., Li, J., Tang, Y.: Scholar-course knowledge graph construction based on graph database storage. In: Jia, W., et al. (eds.) Emerging Technologies for Education: 6th International Symposium, SETE 2021, Zhuhai, China, November 11–12, 2021, Revised Selected Papers, pp. 448–459. Springer International Publishing, Cham (2021). https://doi.org/10.1007/978-3-030-92836-0_40
2. Ji, S., Pan, S., Cambria, E., Marttinen, P., Yu, P.S.: A survey on knowledge graphs: representation, acquisition, and applications. IEEE Trans Neural Netw Learn Syst **33**(2), 494–514 (2022). https://doi.org/10.1109/TNNLS.2021.3070843
3. Heist, N., Hertling, S., Ringler, D., Paulheim, H.: Knowledge Graphs on the Web – an Overview (2020). http://arxiv.org/abs/2003.00719. Accessed 29 2023
4. Hogan, A., et al.: Knowledge graphs. ACM Comput. Surv. **54**(4) (2021). https://doi.org/10.1145/3447772
5. Vakaj, E., Tiwari, S., Mihindukulasooriya, N., Ortiz-Rodríguez, F., Mcgranaghan, R.: NLP4KGC: natural language processing for knowledge graph construction. In: ACM Web Conference 2023 - Companion of the World Wide Web Conference, WWW 2023, p. 1111 (2023). https://doi.org/10.1145/3543873.3589746
6. Khorashadizadeh, H., Tiwari, S., Groppe, S.: A Survey on Covid-19 Knowledge Graphs and Their Data Sources, pp. 142–152 (2023). https://doi.org/10.1007/978-3-031-35078-8_13
7. Pan, J.Z., Vetere, G., Gomez-Perez, J.M., Wu, H.: Exploiting linked data and knowledge graphs in large organisations (2017). https://doi.org/10.1007/978-3-319-45654-6
8. Sellami, S., Zarour, N.E.: Keyword-based faceted search interface for knowledge graph construction and exploration. Int. J. Web Inform. Syst. **18**(5–6), 453–486 (2022). https://doi.org/10.1108/IJWIS-02-2022-0037
9. Paulheim, H.: Knowledge graph refinement: a survey of approaches and evaluation methods. Semant Web **8**(3), 489–508 (2017). https://doi.org/10.3233/SW-160218
10. Smith, C.L., Rieh, S.Y.: Knowledge-context in search systems: toward information-literate actions. In: CHIIR 2019 - Proceedings of the 2019 Conference on Human Information Interaction and Retrieval, pp. 55–62. Association for Computing Machinery, Inc (2019). https://doi.org/10.1145/3295750.3298940

11. Wu, C., Wu, F., Huang, Y., Xie, X.: personalized news recommendation: methods and challenges. ACM Trans. Inf. Syst. **41**(1) (2023). https://doi.org/10.1145/3530257
12. Weikum, G.: Knowledge graphs 2021: a data odyssey. In: Proceedings of the VLDB Endowment, PVLDB, pp. 3233–3238 (2021). https://doi.org/10.14778/3476311.3476393
13. Cimiano, P., Paulheim, H.: Knowledge graph refinement: a survey of approaches and evaluation methods. Semant Web, p. 1 (2016). http://www.geonames.org/. Accessed 5 May 2023
14. Bizer, C.: The emerging web of linked data. IEEE Intell. Syst. **24**(5), 87–92 (2009). https://doi.org/10.1109/MIS.2009.102
15. Lehmann, J., et al.: DBpedia - a large-scale, multilingual knowledge base extracted from Wikipedia. Semant Web **6**(2), 167–195 (2015). https://doi.org/10.3233/SW-140134
16. Ebisu, T., Ichise, R.: Generalized translation-based embedding of knowledge graph. IEEE Trans. Knowl. Data Eng. **32**(5), 941–951 (2020). https://doi.org/10.1109/TKDE.2019.2893920
17. Dang, V., Croft, W.B.: Diversity by proportionality: An election-based approach to search result diversification. In: SIGIR 2012 - Proceedings of the International ACM SIGIR Conference on Research and Development in Information Retrieval, pp. 65–74 (2012). https://doi.org/10.1145/2348283.2348296
18. Bifet, A., Castillo, C., Chirita, P.A., Weber, I.: An analysis of factors used in search engine ranking. In: Proceedings of the 1st International Workshop on Adversarial Information Retrieval on the Web, AIRWeb 2005 - Held in Conjunction with the 14th International World Wide Web Conference (2005), pp. 48–57. https://citeseerx.ist.psu.edu/document?repid=rep1&type=pdf&doi=2a50f8048bcafa83c5780170104c2160422ed705. Accessed 29 Apr 2023
19. Ramos, J.: Using TF-IDF to determine word relevance in document queries. In: Proceedings of the First Instructional Conference on Machine Learning, vol. 242, no. 1, pp. 29–48 (2003). https://citeseerx.ist.psu.edu/document?repid=rep1&type=pdf&doi=b3bf6373ff41a115197cb5b30e57830c16130c2c. Accessed 1 May 2023

Defeasible Reasoning with Knowledge Graphs

Dave Raggett(✉)

W3C/ERCIM, Sophia Antipolis, France
dsr@w3.org

Abstract. Human knowledge is subject to uncertainties, imprecision, incompleteness and inconsistencies. Moreover, the meaning of many everyday terms is dependent on the context. That poses a huge challenge for the Semantic Web. This paper introduces work on an intuitive notation and model for defeasible reasoning with imperfect knowledge, and relates it to previous work on argumentation theory. PKN is to N3 as defeasible reasoning is to deductive logic. Further work is needed on an intuitive syntax for describing reasoning strategies and tactics in declarative terms, drawing upon the AIF ontology for inspiration. The paper closes with observations on symbolic approaches in the era of large language models.

Keywords: defeasible reasoning · argumentation theory · knowledge graphs

1 Defeasible Reasoning

1.1 Introduction

The accepted wisdom for knowledge graphs presumes deductive logic as the basis for machine reasoning. In practice, application logic is usually embedded in conventional programming, exploiting scripting APIs and graph query languages, which make it costly to develop and update as application needs evolve.

Declarative approaches to reasoning hold out the promise of increased agility for applications to cope with frequent change. Notation 3 (N3) is a declarative assertion and logic language [1] that extends the RDF data model with formulae, variables, logical implication, functional predicates and a lightweight notation. N3 is based upon traditional logic, which provides mathematical proof for deductive entailments for knowledge that is certain, precise and consistent.

Unfortunately, knowledge is rarely perfect, but is nonetheless amenable to reasoning using guidelines for effective arguments. This paper introduces the Plausible Knowledge Notation (PKN) as an alternative to N3 that is based upon defeasible reasoning as a means to extend knowledge graphs to cover imperfect everyday knowledge that is typically uncertain, imprecise, incomplete and inconsistent.

"PKN is to N3 as defeasible reasoning is to deductive logic"

The work described in this paper was supported by the European Union's Horizon RIA research and innovation programme under grant agreement No. 101092908 (SMARTEDGE)

F. Ortiz-Rodriguez et al. (Eds.): KGSWC 2023, LNCS 14382, pp. 44–51, 2023.
https://doi.org/10.1007/978-3-031-47745-4_4

Defeasible reasoning creates a presumption in favour of the conclusions, which may need to be withdrawn in the light of new information. Reasoning develops arguments in support of, or counter to, some supposition, building upon the facts in the knowledge graph or the conclusions of previous arguments.

As an example, consider the statement: *if it is raining then it is cloudy*. This is generally true, but you can also infer that it is somewhat likely to be raining if it is cloudy. This is plausible based upon your rough knowledge of weather patterns. In place of logical proof, we have multiple lines of argument for and against the premise in question just like in courtrooms and everyday reasoning (Fig. 1).

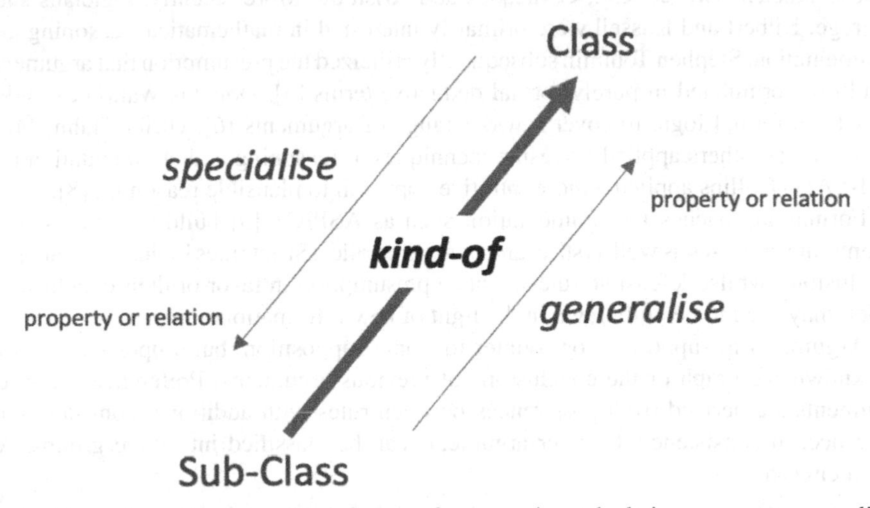

Fig. 1. Illustration of how plausible inferences for properties and relations can act as generalizations or specializations of existing knowledge.

The above figure shows how properties and relations involving a class may be likely to apply to a sub-class as a specialization of the parent class. Likewise, properties and relations holding for a sub-class may be likely to apply to the parent class as a generalization. The likelihood of such inferences is influenced by the available metadata. Inferences can also be based on implication rules, and analogies between concepts with matching structural relationships. PKN [2] further supports imprecise concepts:

- fuzzy terms, e.g., cold, warm and hot, which form a scalar range with overlapping meanings.
- fuzzy modifiers, e.g., very old, where such terms are relative to the context they apply to.
- fuzzy quantifiers, e.g., few and many, for queries akin to SPARQL.

PKN represents an evolution from graph databases to cognitive databases, that can more flexibly support reasoning over everyday knowledge. For a web-based demonstrator, see [3].

1.2 Relation to Previous Work

The Stanford Encyclopedia of Philosophy entry on argument and argumentation [4] lists five types of arguments: deduction, induction, abduction, analogy and fallacies. Argumentation can be adversarial where one person tries to beat down another, or cooperative where people collaborate on seeking a better joint understanding by exploring arguments for and against a given supposition. The latter may further choose to focus on developing a consensus view, with the risk that argumentation may result in group polarization when people's views become further entrenched.

Studies of argumentation have been made by a long line of philosophers dating back to Ancient Greece, e.g., Carneades and Aristotle. More recently, logicians such as Frege, Hilbert and Russell were primarily interested in mathematical reasoning and argumentation. Stephen Toulmin subsequently criticized the presumption that arguments should be formulated in purely formal deductive terms [5]. Douglas Walton extended tools from formal logic to cover a wider range of arguments [6]. Ulrike Hahn, Mike Oaksford and others applied Bayesian techniques to reasoning and argumentation [7], whilst Alan Collins applied a more intuitive approach to plausible reasoning [8].

Formal approaches to argumentation such as ASPIC+ [9] build arguments from axioms and premises as well as strict and defeasible rules. Strict rules logically entail their conclusions, whilst defeasible rules create a presumption in favor of their conclusions, which may need to be withdrawn in the light of new information.

Arguments in support of, or counter to, some supposition, build upon the facts in the knowledge graph or the conclusions of previous arguments. Preferences between arguments are derived from preferences between rules with additional considerations in respect to consistency. Counter arguments can be classified into three groups. An argument can:

- ***undermine*** another argument when the conclusions of the former contradict premises of the latter.
- ***undercut*** another argument by casting doubt on the link between the premises and conclusions of the latter argument.
- ***rebut*** another argument when their respective conclusions can be shown to be contradictory.

AIF [10] is an ontology intended to serve as the basis for an interlingua between different argumentation formats. It covers information (such as propositions and sentences) and schemes (general patterns of reasoning). The latter can be used to model lines of reasoning as argument graphs that reference information as justification. The ontology provides constraints on valid argument graphs, for example:

> ***Scheme for Argument from Expert Opinion:***
>
> *premises*: E asserts that A is true (false), E is an expert in domain D containing A; *conclusion*: A is true (false); *presumptions*: E is a credible expert, A is based on *evidence*; exceptions: E is not reliable, A is not consistent with what other experts assert.

Conflict schemes model how one argument conflicts with another, e.g., if an expert is deemed unreliable, then we cannot rely on that expert's opinions. Preference schemes define preferences between one argument and another, e.g., that expert opinions are preferred over popular opinions. The AIF Core ontology is available in a number of standard ontology formats (RDF/XML, OWL/XML, Manchester OWL Syntax).

PKN defines a simple notation and model for imperfect knowledge. Arguments for and against a supposition are constructed as chains of plausible inferences that are used to generate explanations. PKN draws upon Alan Collins core theory of plausible reasoning [COLLINS] in respect to statement metadata corresponding to intuitions and gut feelings based upon prior experience. This is in contrast to Bayesian techniques that rely on the availability of rich statistics, which are unavailable in many everyday situations.

Recent work on large language models (LLMs), such as GPT-4, have shown impressive capabilities in respect to reasoning and explanations. However, there is a risk of hallucinations, where the system presents convincing yet imaginary results. Symbolic approaches like PKN are expected to play an important and continuing role in supporting semantic interoperability between systems and knowledge graphs. LLMs are trained on very large datasets, and in principle, could be exploited to generate symbolic models in a way that complements traditional approaches to knowledge engineering.

2 Plausible Knowledge Notation (PKN)

The Plausible Knowledge Notation is an intuitive lightweight syntax designed to support defeasible reasoning. PKN documents use data types restricted to numbers (as in JSON) and names with optional prefixes.

2.1 PKN Statements

PKN supports several kinds of statements: properties, relations, implications and analogies. These optionally include a scope and a set of parameters as metadata. The scope is one or more names that indicate the context in which the statement applies, e.g., ducks are similar to geese in that they are birds with relatively long necks when compared to other bird species. Each parameter consists of a name and a value. Parameters represent prior knowledge as an informal qualitative gut feeling based upon prior experience. Predefined parameters include:

certainty - the confidence in the associated statement being true.

strength - the confidence in the consequents being true for an implication statement, i.e., the likelihood of the consequents holding if the antecedents hold.

inverse - the confidence in the antecedents being true when using an implication statement in reverse, i.e., the likelihood of the antecedents holding if the consequents hold.

typicality - the likelihood that a given instance of a class is typical for that class, e.g., that a Robin is a typical song bird.

similarity – the extent to which one thing is similar to another, e.g., the extent that they have some of the same properties.

dominance - the relative importance of an instance of a class as compared to other instances. For a country, for instance, this could relate to the size of its population or the size of its economy.

multiplicity - the number of items in a given range, e.g., how many different kinds of flowers grow in England, remembering that parameters are qualitative not quantitative.

This paper is too short to provide detailed information, so here are a few examples of PKN statements, starting with properties:

```
flowers of Netherlands includes daffodils, tulips (certainty high)
```

Here "flowers" is the descriptor, "Netherlands" is the argument, "includes" is the operator, and "daffodils, tulips" is the referent. In other words, daffodils and tulips are amongst the flowers found in the Netherlands. The metadata indicates that this statement has a high certainty. Next here are two examples of relation statements:

```
Belgium similar-to Netherlands for latitude
Paul close:friend-of John
```

Next here is an implication statement with a locally scoped variable:

```
weather of ?place includes rainy implies weather of ?place includes
cloudy (strength high, inverse low)
```

This example has a single antecedent and a single consequent. Note the use of "?place" as a variable, and metadata for the confidence in using the statement for forward and backward inferences. Next is a couple of examples of analogy statements:

```
leaf:tree::petal:flower
dog:puppy::cat:?
```

Next, here are some examples of queries, which are akin to SPARQL:

```
which ?x where ?x is-a person and age of ?x is very:old
count ?x where age of ?x greater-than 20 from ?x is-a person
few ?x where color of ?x includes yellow from ?x kind-of rose
```

The first query lists the people in the PKN graph who are considered to be very old. The second query counts the number of people older than 20. The third query checks whether there are few yellow roses in the PKN graph.

PKN allows statements to embed sub-graphs for statements about statements, e.g.

```
Mary believes {{John says {John loves Joan}} is-a lie}
```

which models "Mary thinks John is lying when he says he loves Joan.

2.2 Fuzzy Knowledge

Plausible reasoning subsumes fuzzy logic as expounded by Lotfi Zadeh in his 1965 paper on fuzzy logic, see [11]. Fuzzy logic includes four parts: fuzzification, fuzzy rules, fuzzy inference and defuzzification.

Fuzzification maps a numerical value, e.g., a temperature reading, into a fuzzy set, where a given temperature could be modelled as 0% cold, 20% warm and 80% hot. This involves transfer functions for each term, and may use a linear ramp or some kind of smooth function for the upper and lower part of the term's range.

Fuzzy rules relate terms from different ranges, e.g., if it is hot, set the fan speed to fast, if it is warm, set the fan speed to slow. The rules can be applied to determine the desired fan speed as a fuzzy set, e.g., 0% stop, 20% slow and 80% fast. Defuzzification maps this back to a numeric value.

Fuzzy logic works with fuzzy sets in a way that mimics Boolean logic in respect to the values associated with the terms in the fuzzy sets. Logical AND is mapped to selecting the minimum value, logical OR is mapped to selecting the maximum value, and logical NOT to one minus the value, assuming values are between zero and one.

Plausible reasoning expands on fuzzy logic to support a much broader range of inferences, including context dependent concepts, and the means to express fuzzy modifiers and fuzzy quantifiers.

Here is an example of a scalar range along with the definition of the constituent terms:

```
range of age is infant, child, adult for person
age of infant is 0, 4 for person
age of child is 5, 17 for person
age of adult is 18, age-at-death for person
```

The range property lists the terms used for different categories. The age property for the terms then specifies the numerical range. Additional properties can be used to define the transfer function.

PKN allows terms to be combined with one or more fuzzy modifiers, e.g., "very:old" where very acts like an adjective when applied to a noun. The meaning of modifiers can be expressed using PKN statements for relations and implications, together with scopes for context sensitivity. In respect to old, "very" could either be defined by reference to a term such as "geriatric" as part of a range for "age", or with respect to the numerical value, e.g., greater than 75 years old.

Fuzzy quantifiers have an imprecise meaning, e.g., include few, many and most. Their meaning can be defined in terms of the length of the list of query variable bindings that satisfy the conditions. *Few* signifies a small number, *many* signifies a large number, and *most* signifies that the number of bindings for the *where* clause is a large proportion of the number of bindings for the *from* clause.

2.3 PKN and RDF

The Resource Description Framework (RDF), see [12], defines a data model for labelled directed graphs, along with exchange formats such as Turtle, query expressions with SPARQL, and schemas with RDF-S, OWL and SHACL. RDF identifiers are either globally scoped (IRIs) or locally scoped (blank nodes). RDF literals include numbers, booleans, dates and strings. String literals can be tagged with a language code or a data type IRI.

The semantics of RDF graphs is based upon Description Logics, see [13] and [14]. RDF assumes that everything that is not known to be true should treated as unknown. This can be contrasted with closed contexts where the absence of some statement implies that it is not true.

Description Logics are based upon deductive proof, whereas, PKN is based upon defeasible reasoning which involves presumptions in favor of plausible inferences, and estimating the degree to which the conclusions hold true. As such, when PKN graphs are translated into RDF, defeasible semantics are implicit and dependent on how the resulting graphs are interpreted. Existing tools such as SPARQL don't support defeasible reasoning.

Consider PKN property statements. The descriptor, argument, operator and referent, along with any statement metadata can be mapped to a set of RDF triples where the subject of the triples is a generated blank node corresponding to the property statement. Comma separated lists for referents and scopes can be mapped to RDF collections.

PKN relations statements can be handled in a similar manner. It might be tempting to translate the relation's subject, relationship and object into a single RDF triple, but this won't work when the PKN relation is constrained to a scope, or is associated with statement metadata. Recent work on RDF 1.2 [15] should help.

PKN implication statements are more complex to handle as they involve a sequence of antecedents and a sequence of consequents, as well as locally scoped variables. One possible approach is to first generate a blank node for the statement, and use it as the subject for RDF collections for the variables, antecedents and consequents.

PKN analogy statements are simpler, although there is a need to be able to distinguish variables from named concepts, e.g. as in "dog:puppy::cat:?".

3 Plausible Reasoning and Argumentation

Following the work of Allan Collins, PKN uses qualitative metadata in place of detailed reasoning statistics, which are challenging to obtain. Heuristic algorithms are used to estimate the combined effects of different parameters on the estimated certainty of conclusions. Reasoning generally starts from the supposition in question and seeks evidence, working progressively back to established facts. Sadly, this paper is far too short to go into details and the interested reader should look at the PKN specification.

An open challenge is how to declaratively model strategies and tactics for reasoning rather than needing to hard code them as part of the reasoner's implementation. Further work is needed to clarify the requirements and to evaluate different ways to fulfil those requirements using an intuitively understandable syntax. The AIF ontology would be a useful source of inspiration.

4 Summary

This paper introduced PKN as a notation and model for defeasible reasoning with knowledge graphs that include knowledge that is uncertain, imprecise, incomplete and inconsistent. Deductive proof is replaced with plausible arguments for, and against, the supposition in question. This builds on thousands of years of study of effective arguments, and more recently work on argumentation theory. Further work is needed on an intuitive syntax for reasoning strategies and tactics.

Large Language Models have demonstrated impressive capabilities in respect to reasoning and explanations. This raises the question of the role of symbolic approaches such as RDF, N3 and PKN. Deep learning over large corpora has totally eclipsed traditional approaches to knowledge engineering in respect to scope and coverage. However, we are likely to continue to need symbolic approaches as the basis for databases which complement neural networks, just as humans use written records rather than relying on human memory.

References

1. De Roo, J., Hochstenback, P.: Notation 3 language. Draft CG Report, 04 July 2023. https://w3c.github.io/N3/reports/20230703
2. Raggett, D.: Plausible Knowledge Notation (PKN). Draft CG Report, 14 July 2023. https://w3c.github.io/cogai/pkn.html
3. Raggett, D.: Web based demo for PKN (2020). https://www.w3.org/Data/demos/chunks/reasoning/
4. Argument and Argumentation, Stanford Encyclopedia of Philosophy. https://plato.stanford.edu/entries/argument/
5. Toulmin, S.: The uses of argument. https://assets.cambridge.org/97805218/27485/sample/9780521827485ws.pdf
6. Walton, D.: Argument schemes for presumptive reasoning (1997). https://www.routledge.com/Argumentation-Schemes-for-Presumptive-Reasoning/Walton/p/book/9780805820720
7. Hahn, U., Oaksford, M.: The rational of informal argumentation (2007). https://psycnet.apa.org/record/2007-10421-007
8. Collins, A., Michalski, R.: The logic of plausible reasoning (1989). https://psycnet.apa.org/record/1989-38822-001
9. Dauphin, J., Cramer, M.: ASPIC-END: Structured argumentation with explanations and natural deduction (2017). https://homepages.abdn.ac.uk/n.oren/pages/TAFA-17/papers/TAFA-17_paper_15.pdf
10. Argument Interchange Format (AIF) (2011). http://www.arg-tech.org/wp-content/uploads/2011/09/aif-spec.pdf
11. Zadeh, L.: Fuzzy sets. Inf. Control **8**, 338–353 (1965)
12. Klyne, G., Carroll, J.: RDF concepts and abstract syntax (2004). https://www.w3.org/TR/rdf-concepts/
13. Hayes, P., Patel-Schneider, P.: RDF 1.1. Semantics (2014). http://www.w3.org/TR/rdf11-mt/
14. Carroll, J., Herman, I., Patel-Schneider, P.: OWL 2 Web Ontology Language RDF-Based Semantics (2012). https://www.w3.org/TR/owl2-rdf-based-semantics/
15. Hartig, O., Champin, P.-A., Kellog, G.: RDF 1.2 Concepts and Abstraction Analysis (2023). https://www.w3.org/TR/rdf12-concepts/

Recommendation of Learning Paths Based on Open Educational Resources

Jonathan Yaguana and Janneth Chicaiza(✉)

Universidad Técnica Particular de Loja, Loja 110107, Ecuador
{jmyaguana2,jachicaiza}@utpl.edu.ec

Abstract. Open Educational Resources include different types of material for learning and teaching. Over time the number of them has been increasing to a great extent. Although the availability of educational material is beneficial for teachers and learners; however, the search for relevant material becomes a complex task due to the limited availability of specialized tools to locate content that meets the learners' level of knowledge. The current research presents a recommendation service that provides a learning path based on Open Educational Resources. The learning path is created according to the topic of interest of users, and the level of understanding that they have about a particular topic. The recommendation method is based on a knowledge graph, that is created based on the metadata of educational resources obtained from an academic repository. Then, the graph is enriched by three methods: 1) keyword reconciliation using Spanish DBPedia as a target, 2) semantic annotation to find semantic resources, and 3) identification of the level of knowledge of each OER associated with a particular topic. The enriched graph is stored in GraphDB, a repository that provides the creation of semantic similarity indexes to generate recommendations. Results are compared with the TF-IDF measure and validated with the precision metric.

Keywords: Learning paths recommendation · OER · RDF graph

1 Introduction

Open Educational Resources (OER) are teaching and learning materials that are ready to use with no access fees and enable innovation in formal or informal education systems [7]. In the last decade, OER have become a key driver for promoting positive transformation in education.

In recent years, due to the large quantity and variety of educational resources, search engines are not always effective in finding the most relevant or potentially useful content for users. To get more precise results, users can make searches on educational repositories; however, this kind of engine does not consider users' attributes to personalize results. Faced with this problem, recommender systems (RS) arise that are capable of selecting the material that best suits the preferences or needs of a user. Especially in the context of the web, RS are popular

F. Ortiz-Rodriguez et al. (Eds.): KGSWC 2023, LNCS 14382, pp. 52–65, 2023.
https://doi.org/10.1007/978-3-031-47745-4_5

because they provide users with the items or resources they are most likely to like or be interested in [13].

Even though RS are key for discovering resources or relevant items in an information-overloaded environment [4]; however, most RS approaches available for e-learning do not consider the user's level of knowledge about a given topic. According to [19], due to e-learners often use time fragments to learn, as a consequence the learning behavior is fragmented and non-consecutive. This increases the difficulty of mastering the knowledge, therefore, a consecutive learning path is preferred.

Therefore, to ensure the top level of knowledge acquisition for students, it is necessary to create a viable transition from the most elemental degree of knowledge to the higher or complete degree [3] using learning paths. To achieve this, it is necessary that RS help the student to move in an orderly manner and sequential for the different degrees of acquisition of knowledge. On the contrary, of the difference in learners' knowledge, background, and preferences, choosing the same learning path for all would unavoidably lead to bad academic performance and low satisfaction [18]. In conclusion, even though there are many educational websites that offer learning material, the order in which the material is delivered cannot be consistent with the level of knowledge and the preferences of the users.

In view of the aforementioned problem, the aim of this paper is to present a recommendation mechanism based on semantic technologies, that takes two user's key features, such as a particular subject or concept of interest and his/her prior knowledge regarding this topic. The system then generates recommendations of sequences of OER ordered by their knowledge levels. In this way, learning paths allow students to drive their learning progressively.

Next, we describe the research background (see Sect. 2). Then, the recommendation method based on semantic technologies is explained (see Sect. 3); and the results of experiments conducted with a specific OER repository are discussed (see Sect. 4). Finally, we present the research conclusion.

2 Background

In this section, we present the background related to the main dimensions of this proposal: OER, semantic technologies and recommendation systems. Also, we identify some proposals to recommend learning paths.

2.1 Open Educational Resources

OER are generally available free of charge on the Web. Their main use is by teachers and educational institutions that support the development of courses and other types of material, but they can also be used directly by students [17].

Digital libraries, universities, and other educational organizations make OER available to users through open-access repositories [2]. Among the best-known repositories are [11,15]: OpenStax[1], MERLOT[2], OER Commons[3], Open Learn[4], NPTEL[5], GalileoX[6], Teaching Commons[7], Universia[8] and RiUTPL[9].

To recommend learning paths based on OER and their levels of knowledge, we must first know how to define or measure those levels. Two options are described below.

According to [10], the assessment of different types of knowledge can be done through taxonomies that define levels of student understanding. Two well-known schemes are the Bloom Taxonomy, which evaluates the level of knowledge acquired in an area or subject, and the SOLO Taxonomy (Structure of the Observed Learning Outcome), which is a means of classifying learning outcomes in terms of their complexity, by enabling teachers to assess students' work in terms of its quality.

Bloom's taxonomy was proposed by Benjamin Bloom in 1956 to create a common language for learning outcomes, thus 1) facilitating communication among the interlocutors involved, 2) defining the objectives of courses and curricula at different levels, and 3) comparing them with the desired learning outcomes. Bloom's taxonomy defines six levels of cognitive learning: Knowledge, Comprehension, Application, Analysis, Synthesis, and Evaluation.

Subsequently, in 2009, Churches in [6] decided to update the version of Bloom's Taxonomy due to the potential that technology has on education. The new digital version of the taxonomy incorporates some updates such as: using verbs instead of nouns for each category, reordering the last levels, changing the name of the first level (knowledge by recall), and expanding the synthesis level. Table 1 summarizes the digital version of Bloom's Taxonomy.

In our recommendation proposal, Bloom's taxonomy and verbs mentioned in the OER abstracts were taken as a reference to define the level of knowledge of each OER.

2.2 Semantic Web and Knowledge Graphs

The idea of the Semantic Web (SW) is that data is available on the Web and interconnected through named relationships. Thanks to SW technologies, we can semantically describe any entity by means of vocabularies or ontologies. Fundamental SW technologies include:

- Resource Description Framework (RDF) is a standard model for data interchange. Using RDF we can describe any specific object (called an individ-

[1] https://openstax.org/.
[2] https://merlot.org/merlot/.
[3] https://www.oercommons.org/.
[4] https://www.open.edu/openlearn/.
[5] https://nptel.ac.in/courses.
[6] https://www.galileo.edu/page/edx-galileox-cursos/.
[7] https://teachingcommons.us/.
[8] https://www.universia.net/es/lifelong-learning/formacion.
[9] https://dspace.utpl.edu.ec/.

Table 1. Digital version of Bloom's Taxonomy.

Bloom level	Description	Verbs	Knowledge level
Remembering	Recall facts or data without understanding	Recognize, list, describe, identify, retrieve, find, highlight and mark favorites	Beginner
Understanding	Demonstrate understanding in finding information from text	Interpret, summarize, infer, paraphrase, classify, compare, explain and categorize	Beginner
Applying	Solve problems by applying knowledge, facts or techniques to a new situation	Implement, carry out, use, run, upload, share and edit	Intermediate
Analysing	Examine and decompose information into parts	Compare, organize, deconstruct, attribute, find, structure, and integrate	Intermediate
Evaluating	Justify opinions by making judgments about information, the validity of ideas, or the quality of a job	Verify, hypothesize, criticize, experiment, judge, test, detect, publish, moderate and collaborate	Intermediate
Creating	Collect information in a different way by combining its elements in a new model or proposing other solutions	Design, invent, program, manage, produce, build or compile mash-ups	Expert

ual), such as physical things (people, locations, organizations, books, etc.) or abstract concepts.
- Vocabulary and ontology are data models that are used as a reference to describe the individuals or entities of a domain, by specifying the concepts and their relationships. The language used to create a vocabulary is RDF(S), and an ontology is OWL. In the Semantic Web, the use of these models is of great importance as they help to reuse information and avoid meaning issues.
- SPARQL is used in SW as a query language. SPARQL allows querying and manipulating RDF data over HTTP or SOAP. SPARQL queries are based on patterns, which have the form of RDF triples, except that one or more references to resources are variables.[10].

In addition to the technologies mentioned above, for the implementation of the learning path recommender, we use the SKOS (Simple Knowledge Organization System) vocabulary. In SW, SKOS is used to describe knowledge organization systems (KOS), which are systems of related concepts of a given knowledge domain such as thesauri or taxonomies.

[10] https://www.w3.org/standards/semanticweb/query.

Table 2. Weaknesses of classic RS approaches.

RS type	Issue
Content-based	*New user issue*: this issue occurs when the profile for the target user can not be built due there is not enough information about she/he *Over-specialization*: the system recommends items similar to those already consumed by the target user, thus creating a "filter bubble" trend, which means that there is no novelty in recommendations
Collaborative	*Lack of transparency*: In general, the filtering algorithms used for recommendation work as a black box and the user does not know how the results were generated [4], i.e., the recommendation process is not transparent to the user, and the recommendations are not explainable *Sparsity*: Not all users rate the items available in the system, this lack of information makes it difficult to compute the similarity between users and items *Cold-start*: This problem occurs when a new user or item (new item issue) enters the system; when there are new items, there is no rating information yet
Demographic	It is the RS that can generate the worst results because it assumed that users whose demographic features are similar, share the same preferences about the items to be recommended. In addition, users tend to be suspicious when giving personal information to RS; making this a problem when making recommendations without having knowledge about the features of a user

2.3 Recommendation Systems

Recommender systems (RS) generate a list of items that a target user may be interested in or like. According to [12], these systems fall into several categories based on the information they use to recommend items. The three fundamental types of RS are:

- Content-based filtering systems: They analyze the content or features of the items to calculate the similarity between them. The systems that implement this filtering technique arise from the idea of recommending based on items that the user liked in the past.
- Collaborative filtering systems: This kind of system, as input, needs a rating matrix that is created with the explicit ratings that, in the past, the users assigned to the items.
- Demographic filtering systems: These are systems that make recommendations using demographic information of the users such as age, gender, profession, geographic location, etc.

Each approach is useful in a given context, however, these methods suffer from some weaknesses [14] as can be seen in Table 2.

In some domains, it may be difficult to adopt the above-mentioned approaches, mainly due to a lack of data. To solve such limitations, there are other approaches such as hybrid recommender systems and knowledge-based systems.

Hybrid filtering systems recommend items by combining two or more methods together, thus they reduce issues and limitations of pure recommender systems and increase the overall system performance [8].

Knowledge-based RS use knowledge about users and items to generate a recommendation by reasoning about which products meet the user's requirements. There are multiple techniques to exploit knowledge and build this type of recommender system. In recent years, graph and semantic web technologies are being adopted to address the problem of the lack of explainability and lack of data of classical systems [5]. In this proposal, we use this type of filtering.

2.4 Recommendation of Learning Paths

Nowadays, RS are applied in several industry sectors, such as e-commerce, tourism, entertainment, health, digital libraries, among others.

Another field of application for RS, which is currently booming, is e-learning. Due to the *massification* and virtualization of learning and the online availability of free material, educational institutions, providers, and educators must face several issues and challenges, mainly, how to manage large classes and make learning more effective. According to [16], in large classes, the instructor provides less attention to individual students, thus depersonalizing the learning experience. Through smart environments, intelligent tutors, and recommendation services, students can receive more effective services for support their learning, and the teacher could reduce their workload. Despite these improvements, as pointed out by [1], e-learning technology does not yet offer the ability to recommend the best learning path for each learner.

Rosenbaum & Williams (2004), cited in [9], state that a learning path is a sequence of learning activities and events that lead to a prescribed level of competence. A learning path that fits the user's profile can help the learner understand more effectively. According to [19], a way to create recommendations of learning paths is leverage semantics relations between knowledge units because they could reveal logical sequences of competency.

A proposal where the recommendation of learning paths is addressed is [18]. Here, the authors propose a framework for learning path discovery based on a differential evolutionary algorithm and a disciplinary knowledge graph. The output of the system is a learning path adapted to the learner's needs and learning resources for learning according to the path.

In this proposal, we try to join the effort of recommending learning sequences to students by suggesting to them a set of OERs ordered according to the level of competence with which they deal with a topic, thus learners can scale up learning to the required level.

3 Description of the Recommendation Approach

The learning path recommendation can be divided into three phases, as described in Fig. 1. The first phase consists of acquiring the OER metadata and transforming them to a RDF-based knowledge graph. The second phase consists of discovering new facts from the keywords and abstracts of OER to enrich the graph. In the final phase, the data consumption components are implemented from the graph to generate recommendations for a target user. The project implementation and documentation are available on GitHub[11].

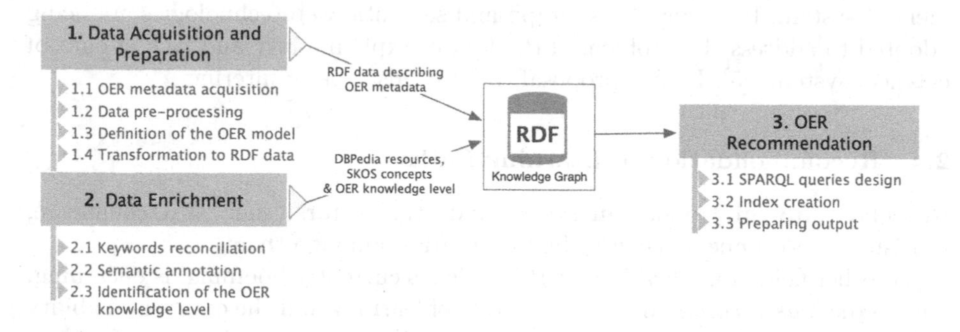

Fig. 1. Process for generating recommendations. To deal with the cold-start problem that faces traditional recommender systems, the proposed process includes a data enrichment phase to identify key contextual information underlying each resource. From the original OER metadata and new data, an RDF knowledge graph is created, that is the source for the OER paths recommendation that is based on SPARQL queries.

3.1 Data Acquisition and Preparation

As can be seen in the Fig. 1, the first step is to obtain OER metadata from an educational repository. After data extraction and cleaning, the semantic model to describe educational resources is defined. Finally, data transformation to RDF format is performed. Further details of implementation of each task are provided below.

OER Metadata Acquisition. In this paper, due to the ease of access to the data, we worked with the data provided by the Institutional Repository (RiUTPL[12]), which is based on DSpace. The repository contains the resources and educational objects generated by the academic community of the Universidad Técnica Particular de Loja (UTPL).

[11] https://github.com/JonathanYaguana/Tesis.
[12] http://dspace.utpl.edu.ec/.

To ensure the appropriate amount of OER in a specific area of knowledge, it was decided to work with the resources generated by students of the Computer Information Systems and Computing career.

The OER metadata was provided by the UTPL Library, through a previously made request. The dataset was provided in JSON format and describes 508 resources. Among the metadata for each resource are: title, abstract, date, author, tutor, keywords, type, format, language, description, and link.

Pre-processing of Data. Based on the analysis of provided metadata, data were cleaned because some quality issues were detected. The data was processed using the OpenRefine tool, which was useful to standardize formats and merge similar terms.

Definition of the OER Model. From the data set obtained in the previous step, the semantic model to describe OER was defined. The definition was based on the reuse well-known vocabularies: schema.org, DCMI Metadata Terms (DCT) and SKOS.

By reusing the three indicated vocabularies we can represent OER metadata such as name, date of publication, keywords, abstract, format, and language, as well as data about the author, contributor, URL, educational level, and citations. Figure 2 presents the model proposed for describing OER.

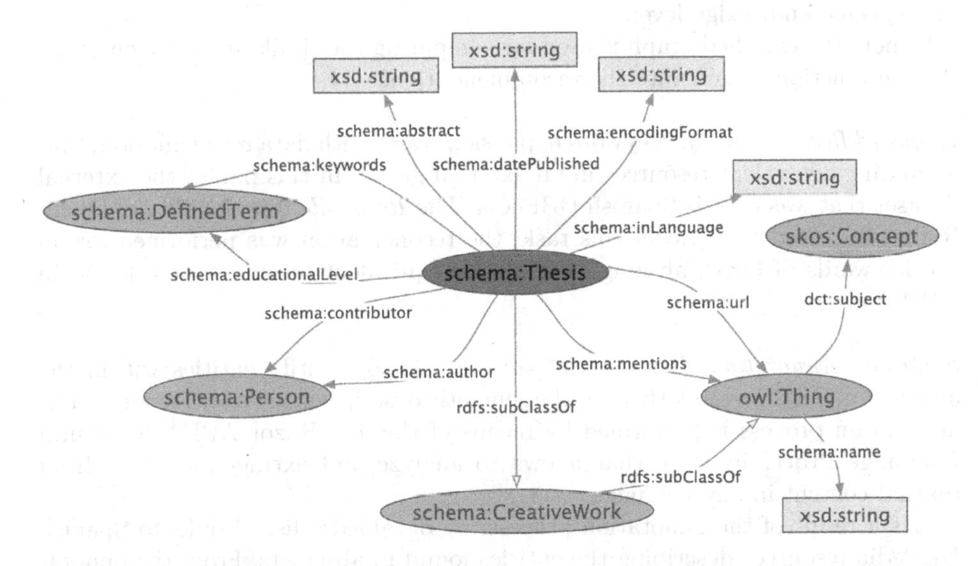

Fig. 2. Semantic model for OER description. The semantic model includes several classes and properties that facilitate the discovery of the more suitable OER, that meet the interests of a target user. The partial view of the model shows how semantic relationships connect OER metadata to concepts or subjects (defined by the class *skos:Concept*) that could be required to support the learning process of a specific learner.

Transformation to RDF data. Once the model has been defined, the objective of this phase is to convert the OER data to RDF. For the transformation, the OpenRefine tool was used, through the RDF Refine extension.

The procedure for the transformation is simple: we edit the RDF Skeleton, which specifies how the RDF data is generated based on the original data structure. Once the mapping is configured, transformed data are exported to an RDF file. Then, we use GraphDB Free to store data due to features to create similarity indexes, a function that was very useful for the implementation of our recommendation method.

3.2 Data Enrichment

The keywords and abstracts of each OER were analyzed to discover new triples and thus extend or enrich the graph created in the previous phase. The objective of analyzing the keywords was to find the equivalent DBPedia resources. DBpedia provides access to extract structured content from the information created in the Wikipedia project.

Likewise, the abstracts were analyzed to discover semantic entities mentioned in the text. From external URIs, we traverse the DBPedia graph to obtain new triples, mainly SKOS concepts related to topics addressed in OER content. Also, from the verbs found in abstracts, we try to classify each educational resource in a specific knowledge level.

Then, the enriched graph is used for computing the similarity between OER, a basic function to generate the recommendations.

Keyword Reconciliation. Keyword expansion starts with data reconciliation, that is, finding equivalent resources in an external graph; in this paper, the external dataset that was used is Spanish DBPedia. The *Reconcile* functionality of OpenRefine was used to perform this task; the reconciliation was performed for all the keywords of OER, although not all the equivalent resources were found in DBPedia.

Semantic Annotation. The goal of this task is to identify entities within the abstract of OER and link them to the semantic descriptions of each resource. The annotation process is performed by means of the TextRazor API[13], a Natural Language Processing tool that allows to analyze and extract metadata from textual content in any language.

As a result of the annotation process, we obtained a list of links to Spanish DBPedia resources describing the entities found in abstracts. From the annotations identified with the API, by means of SPARQL queries, two levels of information were extracted from the semantic entities. The first level uses the relation *dct:subject* to extract SKOS concepts, and the second level uses *skos:broader* to get broader topics. Figure 3 shows an example of the annotation extraction levels.

[13] https://www.textrazor.com/.

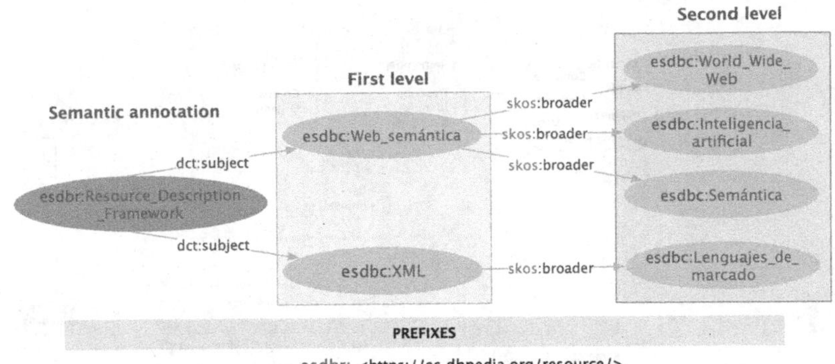

Fig. 3. Concept expansion using the SKOS vocabulary and property paths. The expansion of DBPedia concepts is carried out by using the SKOS vocabulary and SPARQL property paths. Through the property *skos:broader* that connects two *skos:Concept* and property paths like $[*|+|\{\}]$ we can iteratively retrieve the most specific concepts that a target user would like or be interested to know.

According to the OER model, new semantic entities are linked to the semantic representation of each educational resource, through the property *schema:mentions*.

Identification of the level of knowledge of OER. One of the important tasks in generating learning sequences is to determine the level of knowledge of an educational resource, i.e., in addition to identifying the main topic that a resource deals with, we need to know whether the topic is addressed in a superficial way, or in a more advanced way, according to a scale of knowledge level. However, determining that metadata is not easy; according to the study carried out, there is no method with which to do this task.

Nevertheless, in this paper, we propose a simple method to determine the level of knowledge based on Bloom's Taxonomy. Using verbs found in OER abstracts, we get the level of knowledge (Beginner, Intermediate, or Advanced) of each resource (see Table 1 to find mappings between verbs and the knowledge level). To carry out this process, we follow the steps indicated on the Fig. 4.

As Fig. 4 suggests, the first step is to extract the verbs contained in the abstract of OER; the second step is to obtain the Bloom level for the found verb; the third step is to obtain the number of verbs for each Bloom level. In the fourth step, the assignment of the final knowledge level is done according to the highest number of Bloom levels, i.e., the number of verbs of an existing level determines the final level of the educational resource. Finally, the knowledge level was associated with the semantic representation of each educational resource by means of the property *schema:educationalLevel.*

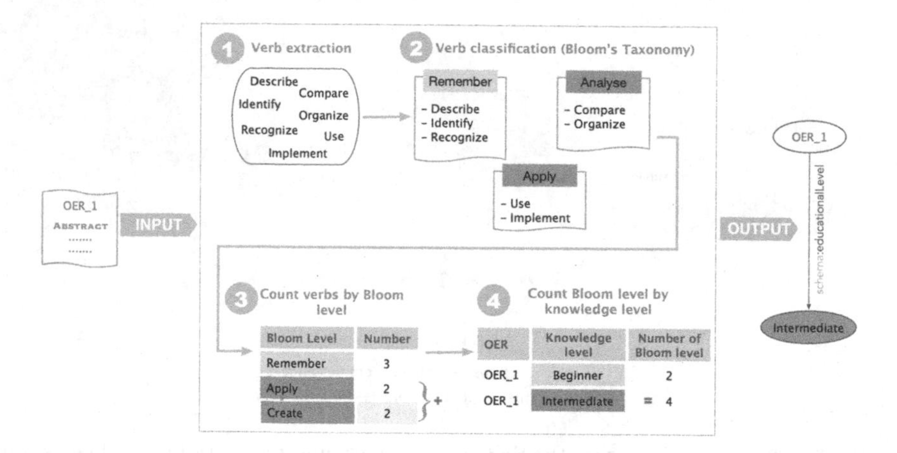

Fig. 4. Process for identifying the knowledge level of OER: from the textual content of an educational resource, such as the abstracts, (1) verbs are extracted and (2) classified according to the knowledge levels defined in Bloom's Taxonomy. Then, (3) by each OER, knowledge levels are counted, and finally 4) we compute which is the predominant level in that OER.

3.3 Learning Path Recommendation

The learning path recommendation is based on a data filtering mechanism based on SPARQL queries. The queries are performed on the OER graph and are the basis for creating an index that determines the similarity between OER.

In this proposal, a text index, named *TextIndex* in GraphDB, was created for selecting OER according to the topic of interest for the target user and a knowledge level equal to or higher than the user. The index computes the score similarity based on textual metadata (such as *schema:keywords, schema:mentions, dct:subject* and *skos:broader*) of resources and their knowledge level.

4 Experimentation

This section describes the test scenario of the proposed method and its results are compared with those of TF-IDF (term frequency-inverse document frequency) which is a statistical measure that evaluates how relevant a term is to a document in a collection of documents.

4.1 Description of the Validation Scenario

The experiment consists of evaluating the results of our recommendation method, i.e., verifying the relevance of the resources that the recommendation method offers to users according to their profiles. As a case study, the base profile of four pre-fabricated users was established. Each profile is based on a topic of interest to the target user and a level of knowledge on that topic.

The topics used for validation were selected considering four of the most popular ones in the OER corpus; the selected topics are neither very general, such as the Internet or the web, nor very specific. In addition, to complete the user profile, a level of knowledge was randomly assigned to each topic, assuming that it is the level the user has. Thus, the four user profiles are as follows: 1) Database → Advanced, 2) Open Educational Resources → Beginner, 3) Data Mining → Intermediate, and 4) Resource Description Framework → Beginner.

To compare recommendation results, the TF-IDF frequency measure was used as a baseline, which was calculated based on the text properties of each educational resource, and without taking into account the level of knowledge.

4.2 Results

To measure the quality of the recommendation, the *relevance* of each suggested resource was determined. Relevance indicates importance of a recommendation (1 = it is relevant, 0 = it is not relevant). The relevance value was assigned by a human, comparing the topic of interest to the user with the metadata of each recommended resource.

With the indicator established, the performance of each method was determined using the precision metric. The precision determines the effectiveness of the recommendation and its value is calculated with the following expression:

$$precision = \frac{Relevant\_items}{Relevant\_items + Irrelevant\_items} \quad (1)$$

To evaluate the recommendation based on the text index, *TextIndex*, we create a Python notebook that reads each user profile, and as output it returns the recommended resources along with a similarity score. Then, the results are evaluated considering the relevance of each recommended item according to the topic of interest and the level of knowledge of the target user.

To compute the precision, for each user's profile, we obtained a list of the first 20 resources with the highest similarity score returned by our method (based on a text index) and by the TF-IDF metric. Table 3 presents the results obtained by each method.

Table 3. Performance of the evaluated methods

Topic of interest	TextIndex Precision	TF-IDF Precision
Database	55.0	50.0
Data mining	75.0	40.0
Open Educational Resources	85.0	30.0
Resource Description Framework	55.0	45.0
Average Precision	67.5	41.25

As can be seen in Table 3, the text-based index obtains a higher precision to generate a list of educational material, although the performance is not excellent as we expected. After analyzing each case evaluated, we were able to detect that one reason that affected the performance was that the OER dataset considered was too small, so there was no variety of material for the recommendations.

5 Conclusion

Throughout a training or study program, the level of knowledge that a learner has on a subject plays a fundamental role when looking for the most appropriate resources that guarantee their evolution until to achieve the proficiency required. The level of knowledge helps to establish a better baseline between the users' understanding and the information they require or want to acquire. For this reason, a learning path based on knowledge levels is a key factor to consider within the educational resources recommendation. In this proposal, each OER was classified in a level of knowledge according to Bloom's taxonomy. This metadata was essential to generate recommendations consistent with the level of understanding of a user on a certain topic.

In addition, the recommendation based on an enriched graph proved to be a fundamental activity for the OER recommendation, since the external graph, DBPedia, helped us to add extra knowledge units to discover related resources, taking advantage of the knowledge systems described by SKOS. Thus, users can find the content they need and understand the relationships between concepts underlying the domain.

For the implementation of the recommendation method, the essential components were built and thus we were able to evaluate its performance. The best results achieved are those obtained by the text index-based recommendation method, which obtained a precision of 67.5% compared to TF-IDF with a precision of 41.25%. Even though the performance of the system was not the best, the result is encouraging, since the user received more relevant resources of interest consistent with their level of knowledge. Currently, we are trying to improve some components of the proposal, in order to release a web prototype that can be used and evaluated for students.

Acknowledgements. The authors thank the Computer Science Department of Universidad Técnica Particular de Loja of Ecuador for sponsoring this academic project.

References

1. Alhasan, K., Chen, L., Chen, F.: Semantic modelling for learning styles and learning material in an e-learning environment. In: International Association for Development of the Information Society, pp. 71–79 (2017)
2. Butcher, N.: A basic guide to open educational resources (OER). UNESCO Biblioteca Digital (2015), https://unesdoc.unesco.org/ark:/48223/pf0000215804
3. Challco, G., Andrade, F., Borges, S., Bittencourt, I., Isotani, S.: Toward a unified modeling of learner's growth process and flow theory. Educ. Technol. Soc. **19**, 215–227 (2016)

4. Chicaiza, J., Valdiviezo-Diaz, P.: Explainable recommender systems: from theory to practice. In: Nagar, A.K., Singh Jat, D., Mishra, D.K., Joshi, A. (eds) Intelligent Sustainable Systems. LNNS, vol. 579, pp. 449–459. Springer, Singapore (2023). https://doi.org/10.1007/978-981-19-7663-6_42
5. Chicaiza, J., Valdiviezo-Diaz, P.: A comprehensive survey of knowledge graph-based recommender systems: technologies, development, and contributions. Information **12**(6) (2021). https://doi.org/10.3390/info12060232
6. Churches, A.: Bloom's taxonomy blooms digitally. Educators' eZine (04 2009)
7. Clinton-Lisell, V., Legerski, E.M., Rhodes, B., Gilpin, S.: Open educational resources as tools to foster equity, pp. 317–337. Teaching and Learning for Social Justice and Equity in Higher Education: Content Areas (2021)
8. Hariyale, I., Raghuwanshi, M.M., Singh, K.R.: Employment of machine learning and data mining technique in hybrid recommender system: a comprehensive survey. In: Kaiser, M.S., Xie, J., Rathore, V.S. (eds.) Information and Communication Technology for Competitive Strategies (ICTCS 2021). LNNS, vol. 401, pp. 505–513. Springer, Singapore (2023). https://doi.org/10.1007/978-981-19-0098-3_49
9. Huang, S.L., Yang, C.W.: Designing a semantic bliki system to support different types of knowledge and adaptive learning. Comput. Educ. **53**(3), 701–712 (2009). https://doi.org/10.1016/j.compedu.2009.04.011
10. Huerta-Palau, M.: Los niveles de van Hiele en relación con la taxonomía SOLO y en los mapas conceptuales. Ph.D. thesis, Universitat de València (1997)
11. Hutson, J., et al.: Open educational resources and institutional repositories: Roles, challenges, and opportunities for libraries. J. High. Educ. Theor. Pract. **22**(18), 100–111 (2022)
12. Jain, S., Grover, A., Thakur, P.S., Choudhary, S.K.: Trends, problems and solutions of recommender system. In: International Conference on Computing, Communication and Automation, ICCCA 2015, pp. 955–958. Institute of Electrical and Electronics Engineers Inc., July 2015. https://doi.org/10.1109/CCAA.2015.7148534
13. Li, X., Wang, T., Wang, H., Tang, J.: Understanding user interests acquisition in personalized online course recommendation. In: U, L.H., Xie, H. (eds.) APWeb-WAIM 2018. LNCS, vol. 11268, pp. 230–242. Springer, Cham (2018). https://doi.org/10.1007/978-3-030-01298-4_20
14. Moya-García, R.: Mapas gráficos para la visualización de relaciones en sistemas de recomendación. Ph.D. thesis, Universidad Politécnica de Madrid (2015)
15. Perifanou, M., Economides, A.: Measuring quality, popularity, demand and usage of repositories of open educational resources (ROER): a study on thirteen popular ROER. Open Learning (2022)
16. Svenningsen, L., Bottomley, S., Pear, J.: Personalized learning and online instruction, May 2018. https://doi.org/10.4018/978-1-5225-3940-7.ch008
17. UNESCO: UNESCO promotes new initiative for free educational resources on the Internet. Technical Report, UNESCO (2002)
18. Wang, F., Zhang, L., Chen, X., Wang, Z., Xu, X.: Research on personalized learning path discovery based on differential evolution algorithm and knowledge graph. In: He, J., et al. (eds.) ICDS 2019. CCIS, vol. 1179, pp. 285–295. Springer, Singapore (2020). https://doi.org/10.1007/978-981-15-2810-1_28
19. Zhu, H., et al.: A multi-constraint learning path recommendation algorithm based on knowledge map. Knowl.-Based Syst. **143**, 102–114 (2018). https://doi.org/10.1016/j.knosys.2017.12.011

An Enterprise Architecture for Interpersonal Activity Knowledge Management

Serge Sonfack Sounchio(✉), Laurent Geneste, Bernard Kamsu-Foguem, Cédrick Béler, Sina Namaki Araghi, and Muhammad Raza Naqvi

Production Engineering Laboratory, UFTMiP 47, Avenue D'Azereix, BP 1629, 65016 Tarbes Cedex, France
sss.sonfack@gmail.com,
{laurent.geneste,bernard.kamsu-foguem,cedrick.beler,sina.namakiaraghi, snaqvi}@enit.fr

Abstract. In today's economy, knowledge is essential for organizations' growth as it allows them to solve problems, be productive, make decisions, and be competitive. Moreover, personal know-how on organizations' activities possessed by individuals, which can benefit other persons within the organization and contribute to the organization's growth, needs to be better managed within information systems generally not designed for these purposes.

Nevertheless, the efficient use of explicit and implicit personal know-how of organizations' activities requires an adequate enterprise architecture to perform tasks such as collecting, transforming, sharing, and using interpersonal activity knowledge. However, existing enterprise architectures that support explicit knowledge do not offer efficient means to capitalize on this humans' interpersonal activity knowledge.

This study provides a holistic view of an enterprise architecture that allows organizations to acquire, transform, share, and use interpersonal activity knowledge for persons and the organization's growth. It describes the proposed architecture through the prism of the Zachman enterprise architecture framework and an information system.

Keywords: Enterprise architecture · Knowledge graph · Zachman framework · Interpersonal activity Knowledge

1 Introduction

In today's economy, knowledge is essential for organizations' growth and competitiveness. To fully use the knowledge produced during activities, organizations should be able to acquire, formalize, apply, and handle outdated knowledge from their information system [1]. Besides explicit knowledge available in documents, a particular knowledge that can contribute to organizations' productivity is the individual know-how persons possess from their working experience within their

F. Ortiz-Rodriguez et al. (Eds.): KGSWC 2023, LNCS 14382, pp. 66–81, 2023.
https://doi.org/10.1007/978-3-031-47745-4_6

fields. This personal know-how is challenging to manage due to its individualistic characteristics and the fact that it is implicit. However, if collected within an organization, it can, on the one hand, improve its performance and be used in decision-making and problem-solving. On the other hand, it can save personal purposes as individuals can learn from each other by querying each other activity know-how [7], internalize it, and use it to produce new knowledge from existing pieces of knowledge.

To effectively manage explicit and implicit know-how produced within organizations, information systems are the main component that contributes to achieving this goal and should strive at four main tasks: (1) knowledge development, (2) codification and storage, (3) transfer and sharing, and (4) utilization [4].

Considering explicit and implicit know-how management by information systems enables organizations to save the entire knowledge produced by persons from their activities, enabling them to avoid losses, dysfunctions in organizations, or knowledge dispersal when faced with factors such as personnel retirements, deaths, staff promotions, or changeovers [3].

A key challenge of activity knowledge management within organizations's information systems is the knowledge representation. This study relies on the work on interpersonal activity knowledge graph [30] to propose an enterprise architecture for interpersonal activity knowledge management.

1.1 Interpersonal Activity Knowledge Graph (IAG)

The interpersonal activity knowledge graph (IAG) is a representation of persons' know-how regarding their activities within an organization. Based on a graph structure representation and its consideration of persons, it is misleading to assume it is a personal knowledge graph. However, interpersonal activity knowledge is an aggregation of personal activity knowledge, which is different from personal knowledge in the sense that it is not centered on the persons' personal information, relationship, social lifestyle, health, or behavior but rather on the activities they carry out to achieve tasks assigned to them. In addition, personal knowledge is generally devoted to services to individuals [30] whereas IAGs represent persons' know-how and can be utilized by individuals and for organizations' decision-making or problem-solving.

Figure 1 shows an overview of personal knowledge graphs and the main groups of concepts they can use. These groups are as follows:

- Personal information includes: This group is made up of concepts such as name, place of birth, town, allergy, or any information about a person.
- Personal social relationships include: This group involves concepts such as father, mother, sisters, friends, or any social relationship connected to a person.
- Personal resources include: This group concerns concepts such as house, guitar, car, shop, or things with which a person is attached.

Unlike personal knowledge graphs, interpersonal activity knowledge graphs are inspired by activity theory studies, used in psychology to study humans in

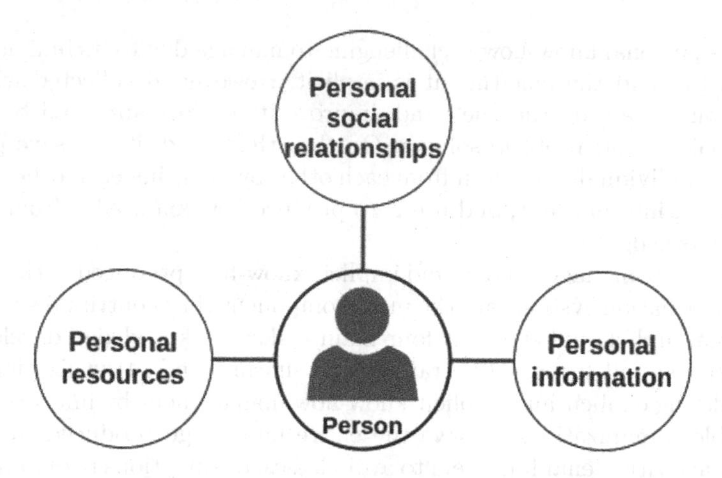

Fig. 1. Primary groups of concepts for personal knowledge representation: The representation is centered on persons

their environment through their activities. In this theory, the concept of activity in general (human or not) is the most fundamental concept [20,21].

In essence, interpersonal activity knowledge representation is centered around activities from which concepts, such as motivation, goal, location, resources, rules, and subject, which characterize these activities, are connected to them. Figure 2 shows a high-level schema of a personal activity knowledge representation, which is the building block of interpersonal activity knowledge graphs (IAGs).

1.2 Enterprise Architecture

An activity is a work carried out within organizations to achieve goals, produce goods and services, and stay competitive. Capturing the knowledge from activities can help organizations study them, learn, and make decisions, but how can the information systems be aware of the activities knowledge as presented above.

This study addresses the capitalization of knowledge derived from organizational activities through enterprise architecture (EA) so that interpersonal activity knowledge benefits organizations for their growth. In other words, how can enterprise architecture describe the management of interpersonal activity knowledge?

Enterprise architecture (EA) is an abstract and holistic description or detail plan of an enterprise information system (IS) and business processes, elaborated as a guideline to achieve a specific goal [25,31]. It describes how the technology and business components are related or work together in order to support organizations' business strategy, plans, activities, business rules, and external relations [22,28].

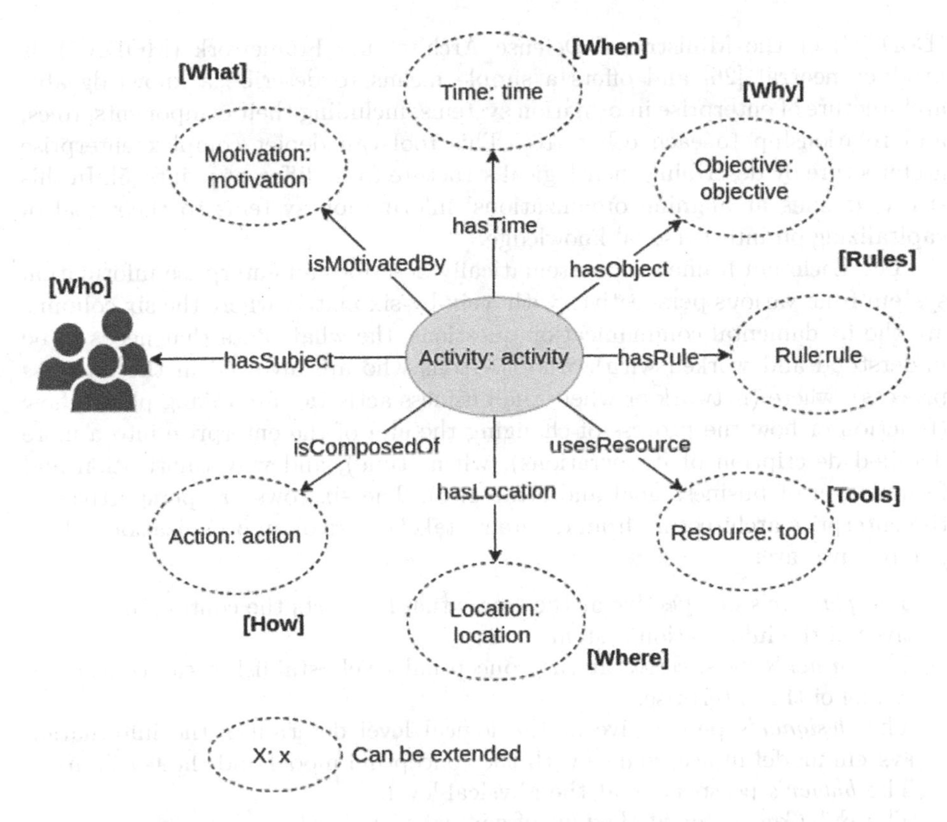

Fig. 2. Activity-centered knowledge representation: A schema for personal activity knowledge graph and a building block for interpersonal activity knowledge graph

EA can be used to analyze the current state of ISs and describe their future directions concerning enterprises' visions. EA frameworks can be classified as follows [27]:

- Military frameworks dedicated to military ISs: the Department of Defense Architecture Framework (DoDAF).
- Company frameworks to achieve ISs for organizations' goals: Zachman Framework (ZF), The Open Group Architecture Framework (TOGAF).
- Government frameworks for government information systems: Federal Enterprise Architecture Framework (FEAF).
- Manufacturing frameworks for manufacturing information systems: Generalize Enterprise Reference Architecture and Methodology (GERAM).

Because this study is intended to describe a generic EA for organizations that want to manage (inter)personal activity knowledge, it will rely on the Zachman framework, which is one of the most used EA frameworks alongside The Open Group Architecture Framework (TOGAF) [23,24]. Zachman framework, unlike other frameworks such as the Department of Defense Architecture Framework

(DoDAF) or the Ministry of Defense Architecture Framework (MODAF), is product neutral [26] and offers a simple means to describe a knowledgeable architecture of enterprise information systems, including their components, roles, and relationship to each other [10]. This tool can depict complex enterprise architecture by describing their logical structure from different points [5]. In this study, it aims at aligning organizations' information systems to their goal of capitalizing on interpersonal knowledge.

The Zachman framework systematically describes an enterprise information system from various perspectives with a six-by-six matrix where the six columns are the fundamental communication questions: the what (data that needs to be understood and worked with), who (persons who are involved in the business process), where (network or where the business activities are taking place), how (function or how the process of changing the aim of the enterprise into a more detailed description of it operations), when (time), and why (motivation and formulation of business goal and strategies). The six rows are perspectives of the enterprise architecture from different stakeholders of an organization. These perspectives are:

- The *planner's* perspective at the contextual level sets the context and objectives of the information system.
- The *owner's* perspective at the conceptual level establishes the conceptual model of the enterprise.
- The *designer's* perspective at the logical level determines the information system model in accordance with the conceptual model and the feasibility.
- The *bulder's* perspective at the physical level.
- The *Sub-Contractor* at the Out-of-context view level.
- The *functioning* level, which represents the actual deployment of the system in the real world from the user's perspective.

In general, the Zachman framework is a language-independent formalism that offers a high level of abstraction for any system development process description [29].

This study elaborates on a generalized enterprise architecture to capture, transform, and share interpersonal activity knowledge for persons' and organizations' benefits. To achieve this goal, it describes these information systems in five (05) layers from the Zachman Enterprise Architecture (ZEA) framework perspective.

The remainder of this study is organized as follows. Section 2 presents relevant studies related to personal knowledge management, Sect. 3 describes the proposed approach of enterprise architecture and knowledge representation, Sect. 4 presents the discussion on the proposed approach, Sect. 5 is the conclusion of this study.

2 Related Work

Knowledge management and enterprise architecture were addressed by [11] in the educational domain. These authors present an academic knowledge management system in three layers: the people layer, made up of academic and non-academic staff who are the primary data sources of the system. The second layer is the technology that serves as a bridge between the people and the process layer. This layer offers an interface for people to access knowledge explicitly and inputs data to the process layer. The last layer is the process layer on which data is processed. Its main data processing tasks include knowledge extraction, data mining, and knowledge collaboration.

To depict the strategy alignment between business and information technology (IT) in enterprises, [14] used The Open Group Architecture Framework (TOGAF) to explain mechanisms enterprise architectures enable alignment between business and IT. From the combination of Archimate language and the business motivation model (BMM), the authors proposed an approach described with five layers, as follows:

- The motivation layer is made up of a hierarchy rooted in the enterprise's missions and spread throughout its strategies.
- The business layer is made up of business capabilities such as lean production, communication, and legislation that are directly connected to concepts in the motivation layer.
- The application layer with application services and the enterprise system suite such as human resources, customer interaction, logistics, and finances.
- The technology layer contains the enterprise system and database servers.

This study provides a generalized framework that combines business strategies and enterprise architecture. It explained how strategy alignment can be achieved between business and IT in enterprises.

In other to allow enterprises to make decisions even during crises or to stay competitive, [18] designed an enterprise architecture to enable knowledge sharing among enterprises working on the same domain and dynamic response to change.

This goal is achieved with blockchain technology and case-based reasoning technique. Each enterprise's case base and reasoning transactions are registered on a knowledge and transaction blockchain for security, trustfulness, and ownership of case sharing among enterprises.

This illustration of dynamic enterprise architecture relies on case-based reasoning and blockchain technology. Its elaboration does not mention any specific framework, and it does not rely on personal know-how.

Even though there are no clear huge differences among enterprise architectural (EA) frameworks, this study uses the Zachman framework for its expressiveness and scalability [5]. This tiny difference among EA frameworks justices the non-dominance of a specific framework.

The study of [32] elaborates on digital enterprise architecture (DEA). It describes technical components that allow this architecture to align business models and digital technologies for digital solutions or transformation. The components of the DEA, such as the operational, technological, and knowledge architecture, relying on the cyber security architecture, integrate digital technologies such as the Internet of Things, mobile systems, big data, cloud computing, and collaboration networks, which are vital tools for digitalization. However, the architecture is not dedicated to collecting and analyzing implicit knowledge, such as activity know-how.

In summary, knowledge management systems in the literature describe with various enterprise architecture frameworks how knowledge can be acquired and processed within organizations or how enterprises can meet their business challenges by relying on a specific architecture. However, no study was dedicated to personal know-how possessed by individuals about their activities. In addition, the frameworks illustrated do not consider the security and semantic layers that offer, on the one hand, data and knowledge protection and, on the other hand, a common understanding and reasoning mechanism of the knowledge shared within these organizations.

3 Proposed Approach

This approach is a direct follow-up of the study on interpersonal activity knowledge representation from which the construction of an interpersonal activity graph (IAG) was described [30]. To illustrate the proposed information system, the Zachman framework is used to describe the EA, and a layer view of how the main components of the system are arranged is presented.

3.1 Interpersonal Activity Knowledge Management Enterprise Architecture

Under the lens of the Zachman framework, this section describes the enterprise architecture for the management of interpersonal activity knowledge within an enterprise. This description is resumed in Tables 1 and 2, which present the six layers and the 5W 1H questions for understanding enterprise architectures [13].

In general, the architecture described in this section allows the acquisition and use of knowledge from activities within an organization in a graph structure. The knowledge acquired integrates information about hard and soft resources used for organizations' activities. This knowledge will be used for learning from activities and decision-making by operators and managers.

People involved in this architecture are operators who carry out tasks of the organization, managers and top managers for tactical and strategic decisions, and finally, the support team in charge of the technological, telecommunication system, and the security of the entire system.

Two principal challenges to overcome in the architecture are (1) the automatical integration of soft resources, such as emails, documents, chats, or videos, and hard resources, such as equipment used to carry out activities, into the activity graph. (2) the alignment of personal activity graphs to form the interpersonal activity graph (IAG) of the organization.

Table 1. Zachman's framework of organizations' interpersonal activity knowledge management with the interrogative abstraction (Why, Who, What) on the columns and the perspectives as rows. This table was split due to space inconveniences

Layer	Why (Motivation)	Who (People)	What (Data)
Scope (Contextual)	Learn and make decisions for the organization's productivity and growth using knowledge about the organization's activities	- Operators - Managers - Top managers	- Activities (manual, automated, virtual) carried out within the organization - Resources (soft, hard) used for activities - Persons carrying out activities and those involved in activities - Information related to activities (location, rules, time, motivation)
Business model (Conceptual)	Infer knowledge from activities carried out within the organization	Organization management team	Information of the organization structure and data generated areas
System model (Logical)	To store interpersonal activity knowledge	Knowledge architect	Interpersonal activity knowledge schema
Technology model (Physical)	- Create the organization's interpersonal knowledge graph (IAG) - Develop knowledge access tools for the IAG	- Knowledge engineer - Domain experts - IT experts	Interpersonal activity graph (IAG)
Detailed Representation (Out-of-Context)	Access data to enrich activity knowledge graph	- Operator - Manager - Top manager	Activities related resources: - Emails - Documents - Audios - Videos - Chats
Functioning Enterprise	- Store personal activity knowledge - Query and learn learn from existing knowledge - Take decisions from activities' knowledge	- Operator - Manager - Top manager	Information related to activities within the organization

3.2 Interpersonal Activity Knowledge Management System

The information system (IS) for the proposed enterprise architecture is described in this section. Figure 3 depicts its five layers (infrastructure, security, semantics, data and information, and knowledge) and shows how knowledge will flow within an organization.

Infrastructure Layer. The infrastructure layer is the first layer of the proposed architecture. Its main goal is to support the communication and storage among humans or devices of the organization.

Table 2. Zachman's framework of organizations' interpersonal activity knowledge management with the interrogative abstraction (When, Where, How) on the columns and the perspective as rows. This table is the follow-up of Table 1, split due to space inconveniences

Layers	When (Time)	Where (Network)	How (Function)
Scope (Contextual)	- When facing challenges on tasks carried out within the organization - When making critical decisions for the organization growth	From the organization's network service	- Query interface for interpersonal activity knowledge (IAG) - Integrate knowledge learned from IAG to decision making systems, such as strategical, tactical and operation decisions
Business model (Conceptual)	When there is a need to utilize IAG	Within the organization	Designing the interpersonal activity knowledge representation and the data collection interface, such as command ligne and graphical interfaces
System model (Logical)	After a clear understanding of activity data components from	From the organization's graph data base management system	Based on an ontology for commun semantics and reasoning
Technology model (Physical)	When a stable data model is available	The organization graph storage infrastructure	Graph-based representation of interpersonal activities knowledge
Detailed Representations (Out-of-Context)	- During interaction - During activities	Organization data storage system	- Digitalize resources (text, image and videos) - Interface for activity knowledge enrichment
Functioning Enterprise	- When lack of knowledge - To take tactical decision - To take strategical decision	The organization network	- Activity reporting module to enter activities - Activity querying module for specific questions - Activity learning module for general knowledge

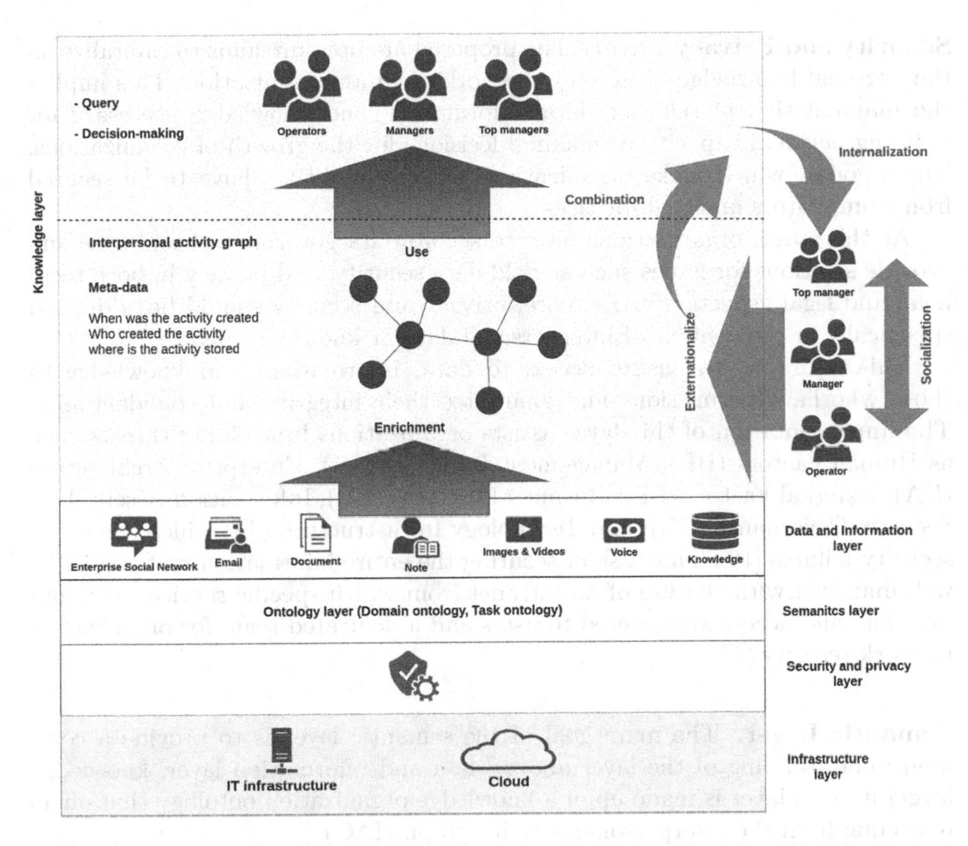

Fig. 3. A layer view and knowledge cycle of interpersonal activity knowledge management within an information system

This layer can be made up of physical equipment such as servers, storage machines, and network connectivity. It can also be a completely virtual online network, storage, and communication service, also known as a cloud. This sort of online infrastructure has three provision type such as Software-as-a-Service (SaaS), Platform-as-a-Service (PaaS), and Infrastructure-as-a-Service (IaaS), which can be consumed on demand while reducing management costs [12].

The choice of either of these infrastructures has to be taken after an evaluation of the total cost ownership (TCO) because these two infrastructures are deployed in completely different approaches. On the one hand, the physical or hardware infrastructure, with servers, networks, and communication tools, needs human operators to configure and maintain the infrastructure. On the other hand, the cloud approach with on-demand network access to online resources offers, among others, agility and scalability [8, 9].

Security and Privacy Layer. The proposed architecture aims to capitalize on the personal knowledge of individuals working in an organization. This implies the manipulation of relevant data, information, and knowledge necessary for building personal capacity or making decisions for the growth of organizations. These goods, which make organizations stay competitive, have to be secured from competitors and malpractices.

At this stage, organizations have to set up data governance to address and provide solutions for issues such as risk, data security, and privacy in both technical and legal aspects. Furthermore, privacy and security should be addressed specifically in the context of interpersonal data or knowledge [34].

This layer has to ensure access to data, information, and knowledge to those who have permissions and guarantee their integrity and confidentiality. The implementation of this layer assists organizations in avoiding threats such as Human Factors (HF), Management Influence (MI), Enterprise Architecture (EA), External Factors (EF), Business Processes (BP), Information Assets (IA), Security Governance (SG), and Technology Infrastructure (TI) which can cause security failures [16]. This task of securing the entire information system can be well managed with the use of an intranet from which specific services, communication, and access are granted to users and a dedicated team for organization network security [17].

Semantic Layer. The main goal of the semantic layer is to provide a common understanding of the layer above (data and information layer, knowledge layer) it. This layer is made up of a knowledge organization ontology that offers reasoning from the interpersonal activity graph (IAG).

The semantic layer also offers a means to integrate interpersonal activity knowledge with other external knowledge. This integration will enhance query possibilities and ameliorate semantics among persons and machines.

Using the semantic web technology to achieve the purpose of this layer, a domain ontology named the personal knowledge organization ontology (PKO-Onto) was designed using a methodology from [19] to represent high-level knowledge of the personal activity and can be extended for specific activities. Figure 4 illustrate the PKO-Onto classes and relations.

Data-Information Layer. This layer faces two challenges: (1) the management of diverse data formats resulting from the diversification of data sources, heterogeneity and (2) the synchronization of all internal and external sources of data and information manipulated or used by employees transparently. For this purpose, tools for managing personal information, such as emails or note-taking software, and collected information, such as videos, documents, and enterprise social networks, are essential for individual data acquisition because they contain helpful information about employees' activities [6]. These sources of data and information enrich the personal activity knowledge graph with context and offer more reasoning possibilities.

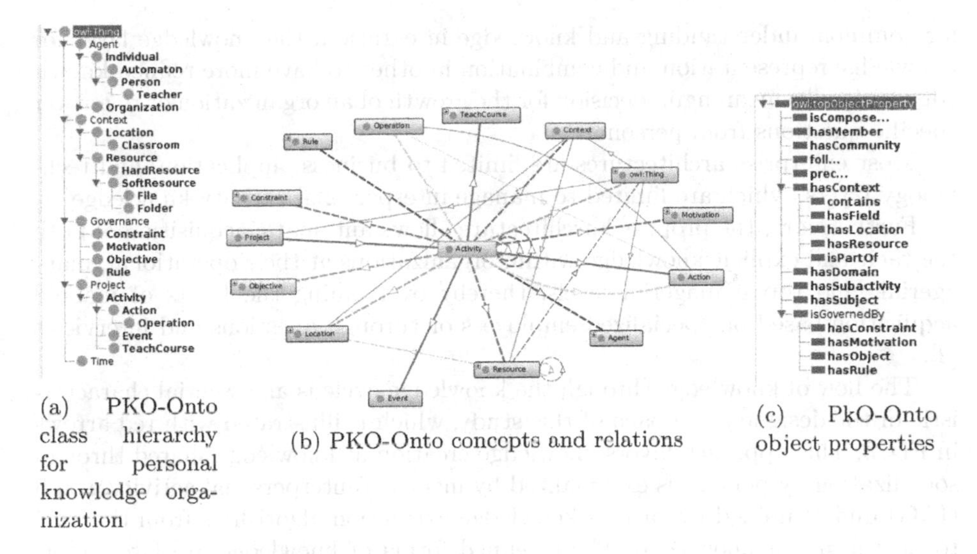

(a) PKO-Onto class hierarchy for personal knowledge organization

(b) PKO-Onto concepts and relations

(c) PkO-Onto object properties

Fig. 4. PKO-Onto Ontology for personal activity knowledge representation from [30]

Knowledge Layer. The knowledge layer is dedicated to knowledge encoding, its transformation, and application for decision-making. This layer comprises two main layers: (1) the encoding of knowledge as a graph, (2) the transformation or creation of a more refined form of knowledge.

- Knowledge representation
  At this level of the enterprise architecture, the activities of an agent working within an organization are encoded as a graph structure to represent how they carry out tasks assigned to them. This graph representation relies on the semantic layer to form an interpersonal activity graph (IAG) which can be exploited from visual search and reasoning using a query language.
- Knowledge combination
  The ability to query and reason from the interpersonal activity graph (IAG) is an excellent way to infer new knowledge from what is available in the knowledge graph to solve problems or make decisions. Some approaches to derive new knowledge that can be used to assist the managerial team in making decisions are tools such as process mining to learn how activities proceed in an organization. Furthermore, learning knowledge maps from IAG provides the distribution of knowledge in the organization, and interpersonal activity knowledge clustering can be used to identify interest groups and how they are related.

4 Discussion

The enterprise architecture proposed in this study particularly distinguishes itself from existing with the following key points: the presence of a semantic layer for common understanding and knowledge integration, the knowledge layer for knowledge representation and combination in other to have more refined knowledge and make meaningful decision for the growth of an organization or to answer specific questions from persons.

Most enterprise architectures are limited to business, application, and technology layers, which are limited to manage interpersonal activity knowledge.

Furthermore, the proposed architecture allows automatic acquisition of both the tacit and explicit knowledge within organizations at their operational, managerial, and top managerial levels, thereby overcoming the limits of manual acquisition based on specialized employees or through questions and interviews [2].

The flow of knowledge through the knowledge cycle is an essential characteristic of the designed approach of this study, which is illustrated with red arrows in Fig. 3. This approach favors knowledge creation as knowledge shared through socialization by persons is externalized by means of interpersonal activity graph (IAG) and combined by various knowledge extraction algorithms from the IAG to new forms of knowledge. These refined forms of knowledge are later internalized by persons by lessons learned from these refined pieces of knowledge [15].

This study does not exclude the use of data governance. Instead, it mentioned the roles it plays, from the security to the knowledge layers. In essence, the goal of data governance will be to define policies, organization, and standards for efficient data access, integration, and representation in the organization [33].

5 Conclusion

This study elaborated on an enterprise (EA) architecture to capture and store organizations' interpersonal activity knowledge in a graph structure called the interpersonal activity knowledge graph (IAG). The proposed EA aims to support persons with knowledge of the organization's activities accessible through queries and the organization's managerial board with knowledge for decision-making.

This interpersonal activity knowledge graph mentioned in this work relies on the activity theory used in psychology and other social sciences to study human activities in their environment. This activity theory driven design of knowledge within organizations permits to capture persons know-how of their activities.

The proposed enterprise architecture presented through the Zachman framework relies on an interpersonal activity knowledge management system made up of: the infrastructure layer to manage communication, the security layer for restriction and access control on data, information, and knowledge in the organization, the data and information layer that supports data and information produced by the organization, the semantic layer for interoperability and common understanding among persons and machines within the organization, the knowledge representation layer to structure knowledge of the organization and offer means to query, learn and make decisions from it.

The proposed enterprise architecture offers a complete knowledge cycle from externalization with the interpersonal activity graph, combination through its utilization, internalization by learning, and socialization by interactions.

Future challenges to this study are, on the one hand, the automated collection and integration of multiple sources of information into an interpersonal activity knowledge graph to enrich it with context. On the other hand, an interest will be granted to identifying and mitigating biases within interpersonal activity knowledge to enhance reasoning and decision-making.

References

1. Kudryavtsev, D., Sadykova, D.: Towards architecting a knowledge management system: requirements for an ISO compliant framework. In: IFIP Working Conference On The Practice Of Enterprise Modeling, pp. 36–50 (2019)
2. Mezahem, F., Salloum, S., Shaalan, K.: Applying knowledge map system for sharing knowledge in an organization. In: International Conference On Emerging Technologies And Intelligent Systems, pp. 1007–1017 (2021)
3. Chergui, W., Zidat, S., Marir, F.: An approach to the acquisition of tacit knowledge based on an ontological model. J. King Saud Univ.-Comput. Inf. Sci. **32**, 818–828 (2020)
4. Zaim, S., Bayyurt, N., Tarim, M., Zaim, H., Guc, Y.: System dynamics modeling of a knowledge management process: a case study in Turkish airlines. Procedia-Soc. Behav. Sci. **99**, 545–552 (2013)
5. Dumitriu, D., Popescu, M.: Enterprise architecture framework design in IT management. Procedia Manuf. **46**, 932–940 (2020)
6. Prilipsky, R., Zaeva, M.: A hybrid system for building a personal Knowledge Base. Procedia Comput. Sci. **169**, 96–99 (2020)
7. Jain, P.: Personal knowledge management: the foundation of organisational knowledge management. South African J. Libr. Inf. Sci. **77**, 1–14 (2011)
8. Zbořil, M., Svatá, V.: Cloud adoption framework. Procedia Comput. Sci. **207**, 483–493 (2022)
9. Agarwal, R., Dhingra, S.: Factors influencing cloud service quality and their relationship with customer satisfaction and loyalty. Heliyon. (2023)
10. Zhang, M., Chen, H., Luo, A.: A systematic review of business-IT alignment research with enterprise architecture. IEEE Access. **6**, 18933–18944 (2018)
11. Maligat, D., Torio, J., Bigueras, R., Arispe, M., Palaoag, T.: Customizing academic knowledge management system architecture framework based on enterprise architecture. In: IOP Conference Series: Materials Science And Engineering, vol. 803, p. 012048 (2020)
12. Farwick, M., Agreiter, B., Breu, R., Häring, M., Voges, K., Hanschke, I.: Towards living landscape models: Automated integration of infrastructure cloud in enterprise architecture management. In: 2010 IEEE 3rd International Conference On Cloud Computing, pp. 35–42 (2010)
13. Gerber, A., le Roux, P., Kearney, C., van der Merwe, A.: The Zachman framework for enterprise architecture: an explanatory is theory. In: Hattingh, M., Matthee, M., Smuts, H., Pappas, I., Dwivedi, Y.K., Mäntymäki, M. (eds.) I3E 2020. LNCS, vol. 12066, pp. 383–396. Springer, Cham (2020). https://doi.org/10.1007/978-3-030-44999-5_32

14. Bhattacharya, P.: Modelling strategic alignment of business and IT through enterprise architecture: augmenting archimate with BMM. Procedia Comput. Sci. **121**, 80–88 (2017)
15. Bandera, C., Keshtkar, F., Bartolacci, M., Neerudu, S., Passerini, K.: Knowledge management and the entrepreneur: insights from Ikujiro nonaka's dynamic knowledge creation model (SECI). Int. J. Innov. Stud. **1**, 163–174 (2017)
16. Loft, P., He, Y., Yevseyeva, I., Wagner, I.: CAESAR8: an agile enterprise architecture approach to managing information security risks. Comput. Secur. **122**, 102877 (2022)
17. Miloslavskaya, N., Stodelov, D., Malakhov, M., Marachyova, A.: Security architecture of network security centers as part of modern intranets. Procedia Comput. Sci. **213**, 58–63 (2022)
18. Ettahiri, I., Doumi, K.: Dynamic enterprise architecture planning using case-based reasoning and blockchain. Procedia Comput. Sci. **204**, 714–721 (2022)
19. Sonfack Sounchio, S., Kamsu-Foguem, B., Geneste, L.: Construction of a base ontology to represent accident expertise knowledge. Cogn. Technol. Work **25**, 1–19 (2023)
20. Kaptelinin, V., Nardi, B.: Activity Theory in a Nutshell. MIT Press, Cambridge (2006)
21. Akhurst, J., Evans, D.: The utility of a model from activity theory for analysing processes in the Nkosinathi community project. Aust. Commun. Psychol. **19**, 83–95 (2007)
22. Tamm, T., Seddon, P., Shanks, G.: How enterprise architecture leads to organisational benefits. Int. J. Inf. Manage. **67**, 102554 (2022)
23. Daoudi, W., Doumi, K., Kjiri, L.: Proposal of a sensing model in an adaptive enterprise architecture. Procedia Comput. Sci. **219**, 462–470 (2023)
24. Kornyshova, E., Deneckère, R.: A proposal of a situational approach for enterprise architecture frameworks: application to TOGAF. Procedia Comput. Sci. **207**, 3499–3506 (2022)
25. Petrov, I., Malysheva, N., Lukmanova, I., Panfilova, E.: Transport enterprise architecture and features of its personnel management. Transp. Res. Procedia **63**, 1462–1472 (2022)
26. Bondar, S., Hsu, J., Pfouga, A., Stjepandić, J.: Agile digital transformation of system-of-systems architecture models using Zachman framework. J. Ind. Inf. Integr. **7**, 33–43 (2017)
27. Gong, Y., Janssen, M.: The value of and myths about enterprise architecture. Int. J. Inf. Manage. **46**, 1–9 (2019)
28. Chen, Z., Pooley, R.: Rediscovering Zachman framework using ontology from a requirement engineering perspective. In: 2009 33rd Annual IEEE International Computer Software And Applications Conference, vol. 2, pp. 3–8 (2009)
29. Ramadan, A.B., Hefnawi, M.: A network security architecture using the Zachman framework. In: Linkov, I., Wenning, R.J., Kiker, G.A. (eds.) Managing Critical Infrastructure Risks. NATO Science for Peace and Security Series C: Environmental Security, pp. 133–143. Springer, Dordrecht (2007). https://doi.org/10.1007/978-1-4020-6385-5_8
30. Sonfack Sounchio, S., Coudert, T., Kamsu Foguem, B., Geneste, L., Beler, C., Namakiaraghi, S.: Organizations' Interpersonal Activity Knowledge Representation. In: HHAI 2023: Augmenting Human Intellect, pp. 254–262 (2023)
31. Syynimaa, N.: The quest for underpinning theory of enterprise architecture: general systems theory. In: International Conference On Enterprise Information Systems (2017)

32. Zimmermann, A., Schmidt, R., Sandkuhl, K., Jugel, D., Bogner, J., Möhring, M.: Evolution of enterprise architecture for digital transformation. In: 2018 IEEE 22nd International Enterprise Distributed Object Computing Workshop (EDOCW), pp. 87–96 (2018)
33. Muñoz, A., Marti, L., Sanchez-Pi, N.: Data Governance, a knowledge model through ontologies. In: International Conference On Technologies And Innovation, pp. 18–32 (2021)
34. Paparova, D., Aanestad, M., Vassilakopoulou, P., Bahus, M.: Data governance spaces: the case of a national digital service for personal health data. Inf. Organ. **33**, 100451 (2023)

Unlocking the Power of Semantic Interoperability in Industry 4.0: A Comprehensive Overview

Fatima Zahra Amara[1(✉)], Meriem Djezzar[1], Mounir Hemam[1], Sanju Tiwari[2], and Mohamed Madani Hafidi[1]

[1] ICOSI Laboratory, Department of Mathematics and Computer Science, University of Abbes Laghrour, Khenchela, Algeria
{f.amara,meriem.djezzar,hemam.mounir, hafidi.mohamedmadani}@univ-khenchela.dz

[2] Universidad Autonoma de Tamaulipas, Tampico, Mexico
tiwarisanju18@ieee.org

Abstract. As Industry 4.0 continues to transform the current manufacturing scene, seamless integration and intelligent data use have emerged as important aspects for increasing efficiency, productivity, and creativity. Semantic interoperability, a critical notion in this disruptive era, enables machines, systems, and humans to comprehend and interpret data from disparate sources, resulting in improved cooperation and informed decision-making. This article presents a thorough overview of semantic interoperability in the context of Industry 4.0, emphasizing its core concepts, problems, and consequences for smart manufacturing. Businesses may unlock the full power of interoperability and promote a new level of data-driven insights and optimizations by investigating the potential of semantic technologies such as ontologies, linked data, and standard data models. The goal of this paper is to provide a full knowledge of the role of semantic interoperability in Industry 4.0, enabling enterprises to embrace the latest advances and propel themselves toward a more intelligent and connected industrial landscape.

Keywords: Internet of Things · Industry 4.0 · Semantic Web Technologies · Ontology · Linked Data

1 Introduction

The Internet of Things (IoT) and Industry 4.0 are two interconnected technology concepts that have shaped how industries operate and develop in recent years. Manufacturing productivity, resource efficiency, and reduced waste may all benefit from digitizing manufacturing and commercial processes and adopting smarter machines and technology [46]. IoT and Industry 4.0 are driving digital transformation in industries however, present issues in terms of data privacy, security, and the need to up skill the workforce in order to operate and maintain these complex technology. With its emphasis on data-driven operations and

F. Ortiz-Rodriguez et al. (Eds.): KGSWC 2023, LNCS 14382, pp. 82–96, 2023.
https://doi.org/10.1007/978-3-031-47745-4_7

digital transformation, Industry 4.0 introduces various data problems that enterprises must address in order to fully reap the benefits of this revolution. Data Volume and data verity represent the primary data problems in Industry 4.0 which lead to semantic interoperability issues.

Semantic interoperability is a fundamental feature of Industry 4.0, guaranteeing that disparate systems, devices, and machines can reliably and meaningfully understand and communicate information with one another. It is a critical enabler of smooth communication and collaboration among numerous components in smart factories and throughout the supply chain. Achieving semantic interoperability is an obstacles in Industry 4.0 as the data generated by IoT devices in various formats impedes the interoperability of applications and platforms that are unable to access the data and act inconsistently on the incoming information [48]. Integrating data from several systems and devices that use multiple data formats and standards can be challenging and time-consuming.

Organizations must invest in comprehensive data management strategies and data analytics technologies. Furthermore, cross-industry collaboration and standardization efforts can be critical in overcoming some of the interoperability problems. The Semantic Web and knowledge representation play essential roles in Industry 4.0 by enabling semantic interoperability, data integration, knowledge management, and smart manufacturing. They are instrumental in transforming raw data into meaningful insights, optimizing processes, and driving innovation across various industrial domains. Semantic modeling involves the generalization of entities along with their descriptions, enabling the representation of relationships among these entities to imbue data with its inherent meaning [44,45]. This paper represents a comprehensive overview of semantic interoperability in the context of Industry 4.0. It emphasizes the significance of seamless data integration and intelligent data utilization in increasing manufacturing efficiency, productivity, and creativity. The article digs into the fundamental ideas of semantic interoperability and emphasizes its importance in facilitating effective interaction among machines, systems, and humans.

The remainder of this paper is structured as follows: Sect. 2 is dedicated to Industry 4.0 fundamentals. Knowledge representation and reasoning are presented in Sect. 3. Section 4 provides a set of research findings on semantic web technologies in the Industry 4.0 Sector. A summary of the paper is given in Sect. 5.

2 Fundamentals of Industry 4.0

Intelligent technologies play a pivotal role in long-term economic growth. They transform homes, offices, factories, and even entire cities into autonomous, self-controlled systems that operate without constant human intervention [26].

2.1 Internet of Things

The IoT paradigm emerged with the aim of collecting and transmitting data autonomously through Internet Protocol-based networks, eliminating the need for constant human intervention. IoT connects billions of objects and people and is recognized as one of the most influential technologies for generating, modifying, and sharing vast amounts of data.

IoT has been defined as follows by [42]: 'Things that possess identities and virtual personalities, operating within smart spaces through intelligent interfaces to connect and communicate within social, environmental, and user contexts'.

Indeed, the use of IoT technologies continues to thrive, as it seeks to connect objects and people, making it one of the most essential technologies in daily life. It finds applications in various devices, industries, and settings.

2.2 Industrial Internet of Things

The Industrial Internet of Things (IIoT), a subset of the broader IoT, specifically focuses on the manufacturing industry [31]. Unlike traditional IoT devices such as smartphones and wireless gadgets, IIoT is designed for more extensive and robust 'things.'

The primary goal of IIoT is to establish connectivity between industrial assets, such as engines, power grids, and sensors, and the cloud through a network [19]. This connectivity empowers the resulting systems and their constituent devices to monitor, collect, exchange, analyze, and respond to data in real-time, autonomously adapting their behavior or environment without human intervention [22]. As a result, IIoT can significantly enhance operational efficiencies and foster the development of entirely new business models.

IIoT plays a crucial role in the transformation of Cyber-Physical Systems and production processes, leveraging big data and analytics within the context of the fourth industrial revolution. Without the connectivity and data provided by IIoT, Industry 4.0 wouldn't exist and wouldn't have the same transformative impact on efficiency. Therefore, IIoT stands as one of the foundational pillars of Industry 4.0. Figure 1 use a Venn diagram to depict the overlaps among IoT, IIoT, CPS, and Industry 4.0.

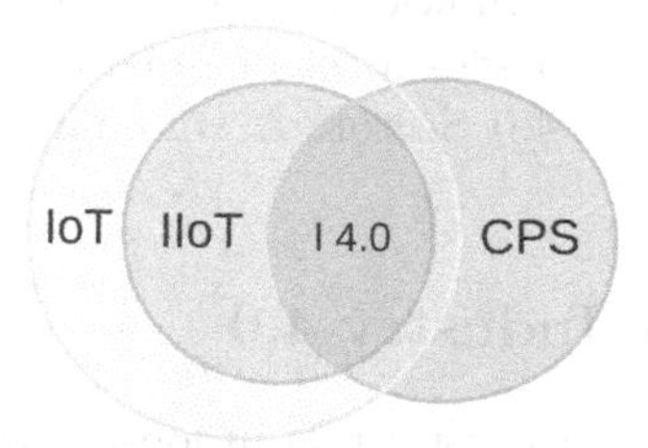

Fig. 1. IoT, IIoT, Industry 4.0, and CPS in Venn Diagram [40].

2.3 Cyber Physical Systems

Cyber Physical Systems (CPS) [1] are an integral part of the service-oriented system architecture. Represents a system of collaborating computer elements to control and command physical entities. In this context, CPS is a synonym for a set of distributed system components implemented as software modules and embedded hardware [28].

The concept of CPS plays a pivotal role in implementing Industry 4.0 principles. CPS represents an emerging class of systems in which physical assets, endowed with computational capabilities, are seamlessly integrated into a network. These systems facilitate the exchange of various types of data, including real-time data from physical assets, a variety of models (such as physics-based and data-driven models), and a range of services such as reconfiguration and monitoring [17].

2.4 Industry 4.0

The rise of the IoT, CPS, and closer collaborations between human-machine and machine-machine systems have transformed the current industrial landscape, resulting in the so-called Industry 4.0 (I4.0) [47]. It refers to the intelligent networking of machines, people, and industrial processes. Industry 4.0 now entails the digital transformation of all industrial and consumer markets, from the introduction of smart manufacturing to the digitization of entire value delivery channels [39] (Fig 2).

Fig. 2. Industry 4.0 Fourth Industrial Revolution.

Many existing industrial systems are not inherently designed for seamless integration into Industry 4.0 frameworks, and coping with the vast amount of data can be a challenge. Industries employ strategies such as real-time monitoring, smart processes, connected devices, paperless operations, automation with minimal human intervention, and big data analysis to digitize their products, manufacturing processes, or create digital value-added services.

Real-time data enables industrial devices to make informed decisions and automate tasks, marking a transformative shift in industrial capabilities, and the Semantic Web holds promise for addressing data interoperability challenges in Industry 4.0 implementations.

3 Knowledge Representation and Reasoning

Knowledge Representation and Reasoning (KRR) is a branch of artificial intelligence (AI) that focuses on how AI agents think and how thinking contributes to their intelligent behavior. It involves representing real-world information and data in a format that machines can use to address real-world problems and demonstrates how this knowledge can be leveraged in a reasoning process to derive new insights from existing information [8]. It is dedicated to the study (identification, modeling, representation and implementation) of the different types of information (knowledge, beliefs, preferences, actions, etc.) and the reasoning necessary for developing such systems.

The evolution of the Web, particularly the Semantic Web [5] perspective, has revitalized the field by introducing the contentious term ontology. Several languages, including the W3C RDF[1], RDFS[2], SKOS[3], and OWL[4] standards, have been developed with this in mind. The term "Semantic Web" refers to the World Wide Web Consortium's vision of a Web of Linked Data. Semantic Web technologies enable people to create Web-based data stores, build vocabularies, and write data-handling rules[5]. semantic web technologies provide a standardized solution to the problem described, such as heterogeneous data and interoperability across manufacturers and devices. As a result, semantic interoperability must evolve from a communication protocol to an ontology coordination challenge, incorporating knowledge representation and artificial intelligence [7].

3.1 Ontology

Ontologies have grown in importance as the use of knowledge graphs, machine learning, natural language processing (NLP), and the amount of data generated on a daily basis has increased [25]. It enables a shared understanding of any domain that is communicated between application systems and people [18]. It includes the entity's name, properties, classes, and relationships. In the case of various systems, ontology consists of the description of the system's components, such as sensors and actuators [13]. Hence an ontology makes data more understandable and usable by providing controlled vocabulary.

[1] https://www.w3.org/RDF/.
[2] https://www.w3.org/TR/rdf-schema/.
[3] https://www.w3.org/2004/02/skos/.
[4] https://www.w3.org/OWL/.
[5] https://www.w3.org/standards/semanticweb/.

Ontologies emerged as an essential tool for representing I4.0 domain knowledge to support integration and interoperability [49]. Ontology-based knowledge representation has aided in the resolution of various issues, including interoperability (between standards and devices) and domain knowledge modeling [52].

3.2 Semantic Reasoning

The reasoning is the mechanism by which an inference engine evaluates logical assertions made in an ontology and related knowledge base [25]. Semantic reasoning refers to a system's ability to infer new facts from existing data using inference rules or ontologies. Rules, in a nutshell, add new information to an existing dataset, providing context, knowledge, and valuable insights. This is an example of Semantic AI. The rule language of the Semantic Web SWRL (Semantic Web Rule Language)[6] is intended to improve inferential power on OWL ontologies by incorporating rules into the language. SWRL allows rules to be combined with OWL ontologies to support deduction on Semantic Web ontologies.

3.3 Metadata, Semantic Annotation, and Linked Data

Metadata. The metadata is defined as *"data that provides information about other data"*[7]. Metadata is data about data, as the name implies, describing the features of a dataset or resource. Semantic metadata can be used to increase search engine traffic as it provides search engines with much more information about the content being searched for [12]. By their role, ontologies aim to facilitate the translation of raw data into semantic metadata to achieve interoperability between physical object services.

Semantic Annotation. Is the process of converting data into knowledge based on predefined concepts in ontologies. The semantic enrichment of data is the foundation of most attempts to overcome the challenge of the dynamic situation of IoT [33,44]. However, manufacturers and developers avoid the extra tasks of including semantic enrichment in their data, services, and modeling techniques through annotation, limiting the interoperability and usability of data and services. Semantic annotation based on Linked Data introduces a new problem in an enormous, complex associated and contextual application scene. These linked and contextual data are critical for intelligent applications [9].

Linked Data. The term "Linked Data" refers to a set of best practices for publishing and interlinking structured data on the Web [6]. Tim Berners-Lee defines it as *"the Semantic Web is not only about putting data on the web. It is*

[6] https://www.w3.org/Submission/SWRL/.
[7] https://www.merriam-webster.com/dictionary/metadata.

about creating links so that a person or a machine can explore the data network. When you have Linked Data, you can find other Linked Data" [4].

The overarching goal of data exchange, discovery, integration, and reuse remain unfulfilled. Every Semantic Web subfield requires more work. Given the variety of methodologies, significant application-oriented consolidation is required, as well as robust tool interoperability and well-documented processes [20].

3.4 Semantic Interoperability

Semantic interoperability came from the need to enable machine-computable logic, inferencing, and knowledge discovery to find better meaningful insights. Different applications cannot share and reuse data without a standard data representation, and they will have difficulty finding the data they require [41].

Interoperability is the capacity of heterogeneous things or systems to connect and share meaningful information with one another [51]. The importance of semantic interoperability grows even more in multidisciplinary contexts, where things are significantly more complex owing to technical and linguistic differences. Interoperability entails accessing real-time data, which leads to a new approach to how businesses can improve their manufacturing operations. It enables manufacturing and their automated systems to exchange information accurately and promptly. As a result, operations are more effective and reliable. Figure 3 shows a general vision of semantic interoperability in an IoT environment.

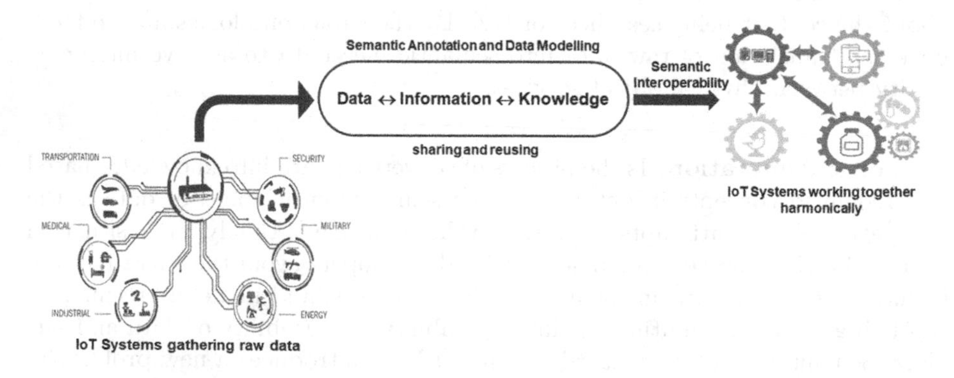

Fig. 3. General vision of Semantic Interoperability in IoT [2].

Incorporating IoT into semantic models enables the reuse of sensor data across various applications and enhances productivity by addressing the challenges of heterogeneity and promoting interoperability [44,45]. The logistics industry faces a significant challenge in achieving semantic interoperability, hindering effective communication between Cyber-Physical Systems, humans, and Smart Factories via the IoT and the Internet of Services.

With transportation and goods handling being integral to businesses, there is a continuous need for efficiency improvement in areas including fuel consumption, CO2 emissions, driver turnover, waiting times, and storage space optimization [15].

4 Semantic Web Technologies in Industry 4.0 Sector

This section presents relevant existing knowledge in the current research area, focusing on semantic interoperability issues in Industry 4.0 and the role of the Semantic Web of Things. It includes a curated selection of research papers.

Ferrer et al. in [14] proposed including human skills and tasks in a manufacturing ontology that uses CPS knowledge repositories. Their work presented a semantic model that allows human operators to model operations. During the production plans, however, they focused more on the orchestration service.

The AutomationML ontology (AMLO) was presented in [27], which covers the AutomationML data exchange standard in the industrial engineering domain. The semantic model serves as a means for data exchange among various Cyber-Physical Systems and improves engineering processes in I4.0.

Patel et al. in [35], they addressed the SWeTI platform, which uses Semantic Web, AI, and data analytics to help developers build smart IoT applications for Industry 4.0. To advocate for the SWeTI platform, they presented a set of realistic use case scenarios.

Teslya et al. proposed an ontology-based approach to describe the industrial components merged from four different scenarios to form an upper-level ontology [43]. Such a union will allow for changes to the created business process, increasing product customization for customers while reducing costs for its producers.

Kaar et al. [23] used the Reference Architectural Model Industrie (RAMI 4.0) to extract context and information, proposing an ontology approach to integrate the industry 4.0 process. This data was gathered from various sources, standards, architectures, and models related to I4.0. The ontology aims to provide an overview of the RAMI4.0 key concepts and their relationships to identify inconsistencies, gaps, and redundancies in its layer descriptions and definitions for I4.0 process development.

SAREF4INMA [21] focuses on extending SAREF for the industry and manufacturing domain to address the lack of interoperability between various types of machinery that produce items in a factory and, once outside the factory, between different organizations in the value chain to uniquely track back the produced items to the related production equipment, batches, material, and precise time of manufacture.

Wan et al. in [50] proposed a resource configuration-based ontology describing domain knowledge of sensible manufacturing resource reconfiguration using web ontology language (OWL). Their work aims to integrate CPS equipment using an ontology-based resource integration architecture. The generated data is stored in a relational database and is associated with and mapped into the manufacturing ontology model instances. The proposed ontology for resource reconfiguration is tested with an intelligent manipulator as a use case, which validates its manufacturing feasibility.

Through semantic technology, Cho et al. [11] proposed an approach providing a novel method for opening up an efficient way to manage/integrate data in Industry 4.0 applications. This approach's most significant contribution is overcoming the limitation of semantic enriched digital twins caused by the lack of mapping standards to address real-time data. It enables digital twins without time delay, and it facilitates constant and instant maintenance decision-making for sustainable manufacturing.

In [36], Ramírez-Durán et al. developed ExtruOnt, that is a development effort to create an ontology for describing a type of manufacturing machine, precisely one that performs an extrusion process (extruder). Although the ontology's scope is limited to a specific domain, it could be used as a model for developing other ontologies for describing manufacturing machines in Industry 4.0 scenarios. The ExtruOnt ontology terms provide various types of information related to an extruder, which is reflected in distinct modules that comprise the ontology.

Da Rocha el al. in [38] proposed a semantic interoperability service that encapsulates information from outside the IEEE 1451 and promotes sensing data and transducer management in the communication process using the JSON-LD data structure. It may also aid in the incorporation of the IEEE 1451 family of standards into Industry 4.0.

Kalayci et al. in [24] demonstrates how the data integration challenge can be addressed using semantic data integration and the Virtual Knowledge Graph approach. They proposed the SIB Framework to semantically integrate Bosch manufacturing data, specifically the data required to analyze the Surface Mounting Process (SMT) pipeline.

Berges et al. in [3] present a proposal materialized in a semantic-based visual query system designed for a real-world Industry 4.0 scenario, allowing domain experts to formulate queries to deal with a customized digital representation of the machine and on-the-fly generated forms. The process is supported by an underlying ontology that describes the machine's main components and sensors.

The issue of standard interoperability across different standardization frameworks is addressed in [16]; researchers developed a knowledge-driven approach that enables the description of standards and standardization frameworks into an Industry 4.0 knowledge graph (I40KG). The STO ontology represents the properties and relationships of standards and standardization frameworks.

To address heterogeneity issues in the IIoT, Ren, et al. [37] proposed a novel concept based on the standardized W3C TD, semantic modeling of artefacts, and KG. They also presented two lightweight semantic model examples. The concept for clogging detection was then demonstrated on a Festo workstation in an industrial use case. Finally, they used three SPARQL queries to demonstrate how to discover and reuse knowledge stored in a KG in order to engineer an on-device IoT application in a distributed network using low-code development.

Table 1. Comparative Table of Semantic Research Papers in Industry 4.0

Paper	Year	Application Domain	Ontology	Platform or Framework
[14]	2017	Manufacturing	Yes	No
[27]	2018	Industrial Engineering	Yes	No
[35]	2018	Industry 4.0	No	Yes
[43]	2018	Business Process	Yes	No
[23]	2018	Industry 4.0	Yes	No
[21]	2018	Manufacturing	Yes	No
[50]	2018	Manufacturing	Yes	No
[11]	2019	Industry 4.0	No	Yes
[36]	2020	Manufacturing	Yes	No
[38]	2020	Industry 4.0	No	Yes
[24]	2020	Manufacturing	No	Yes
[3]	2021	Industry 4.0	No	Yes
[16]	2021	Industry 4.0	Yes	No
[37]	2022	IIoT	No	Yes
[32]	2022	Smart Manufacturing	Yes	No
[10]	2022	Steel Production	Yes	No
[34]	2022	Robotics	Yes	No
[29]	2022	Industry 4.0	No	Yes
[30]	2022	Industry 4.0	No	Yes

May el al. in [32] address interoperability in smart manufacturing and the challenge of efficiently federating diverse data formats using semantic technologies in the context of maintenance in this study, and they present a semantic model in the form of an ontology for mapping pertinent data. An industrial implementation is used to validate and verify the suggested solution.

The authors of this paper [10] present a Common Reference Ontology for Steelmaking (CROS). CROS is a shared steelmaking resource and capability model that aims to simplify knowledge modeling, knowledge sharing, and information management. To address the semantic interoperability issue caused by the data and information needed for supply chain planning and steel production. Process modeling is typically disseminated across organizational boundaries and research communities.

The authors of [34] developed OCRA (Ontology for Collaborative Robotics and Adaptation), an ontology specifically tailored for collaborative robotics. OCRA enables robots to reason about collaboration requirements, plan adaptation, and enhance the reusability of domain knowledge, thereby improving the reliability and flexibility of human-robot collaboration in unstructured scenarios.

In [29], an architectural design and implementation of DIGICOR, a collaborative Industry 4.0 (I4.0) platform, is presented. This platform is intended to facilitate the dynamic formation of supply-chain collaborations for small and medium-sized enterprises (SMEs). The DIGICOR architecture is based on the EDSOA model, enabling collaboration among SMEs, dynamic modeling of their systems and services, and seamless integration into the supply chains of large original equipment manufacturers (OEMs) with the aid of semantic technologies for application integration.

The author of the paper [30] developed and extended the KNOW4I platform into an ontology-based, general-purpose, and Industry 4.0-ready architecture. This architecture is designed to enhance the capabilities of smart operators, with a specific focus on mixed reality applications. The extension involves the creation of a new general ontology using ontology engineering methodology and the adoption of the open-source FIWARE infrastructure to enable interoperability across different systems.

Table 1 categorizes and classifies various proposals related to the Industrial Internet of Things discussed in Sect. 4 of this paper. The table provide an overview of each proposal's key attributes, focusing on whether it primarily centers around ontology development, the scope of the IIoT domain addressed, and whether it includes a complete development framework or platform. The classification provides a structured comparison of these proposals, enhancing the clarity and accessibility of the paper's content for researchers and practitioners in the field.

5 Conclusion

Semantic interoperability plays a crucial role in driving the ongoing Industry 4.0 revolution, enabling seamless data integration and propelling smart manufacturing to unprecedented heights. As technology continues to advance and data becomes increasingly prevalent, the ability to analyze and transmit information across disparate systems becomes paramount for industrial success.

This paper underscores the vital importance of semantic technologies, including ontologies, linked data, and standard data models, in ushering in a new era of data-driven insights and optimizations. Enterprises that grasp the latest technological breakthroughs and fully comprehend the concept of semantic interoperability can position themselves for a more intelligent and interconnected industrial landscape.

The primary objective of this paper is to offer a comprehensive understanding of the significance of semantic interoperability in the context of Industry 4.0. The comparative overview of semantic research papers within the Industry 4.0 domain highlights the diverse and innovative approaches that have emerged to tackle the challenges and opportunities in this transformative field. From the development of ontologies to the creation of advanced platforms and frameworks, these studies collectively underscore the pivotal role of semantic technologies in shaping the future of smart manufacturing, industrial engineering, and collaborative robotics. By exploiting this knowledge, businesses can boldly navigate the changes brought about by Industry 4.0 and position themselves at the forefront of a revolutionized manufacturing landscape.

References

1. Alhafidh, B.M.H., Allen, W.H.: High level design of a home autonomous system based on cyber physical system modeling. In: 2017 IEEE 37th International Conference on Distributed Computing Systems Workshops (ICDCSW), pp. 45–52. IEEE (2017). https://doi.org/10.1109/ICDCSW.2017.14
2. Amara, F.Z., Hemam, M., Djezzar, M., Maimor, M.: Semantic web and internet of things: challenges, applications and perspectives. J. ICT Stand. **10**, 261–292 (2022). https://doi.org/10.13052/jicts2245-800X.1029
3. Berges, I., Ramírez-Durán, V.J., Illarramendi, A.: A semantic approach for big data exploration in industry 4.0. Big Data Res. **25**, 100222 (2021). https://doi.org/10.1016/j.bdr.2021.100222
4. Berners-Lee, T.: Linked data (2006). http://www.w3.org/designissues.LinkedData.html
5. Berners-Lee, T., Hendler, J., Lassila, O.: The semantic web. Sci. Am. **284**(5), 34–43 (2001)
6. Bizer, C., Schultz, A.: The berlin SPARQL benchmark. Int. J. Semant. Web Inf. Syst. (IJSWIS) **5**(2), 1–24 (2009). https://doi.org/10.4018/jswis.2009040101
7. Blobel, B.: Ontologies, knowledge representation, artificial intelligence-hype or prerequisites for international pHealth interoperability? In: e-Health Across Borders Without Boundaries, pp. 11–20. IOS Press (2011). https://doi.org/10.3233/978-1-60750-735-2-11
8. Bouquet, P., Ghidini, C., Giunchiglia, F., Blanzieri, E.: Theories and uses of context in knowledge representation and reasoning. J. Pragmat. **35**(3), 455–484 (2003). https://doi.org/10.1016/S0378-2166(02)00145-5
9. Cai, H., Vasilakos, A.V.: Web of things data storage. In: Managing the Web of Things, pp. 325–354. Elsevier (2017). https://doi.org/10.1016/B978-0-12-809764-9.00015-9
10. Cao, Q., Beden, S., Beckmann, A.: A core reference ontology for steelmaking process knowledge modelling and information management. Comput. Ind. **135**, 103574 (2022). https://doi.org/10.1016/j.compind.2021.103574
11. Cho, S., May, G., Kiritsis, D.: A semantic-driven approach for industry 4.0. In: 2019 15th International Conference on Distributed Computing in Sensor Systems (DCOSS), pp. 347–354 (2019). https://doi.org/10.1109/DCOSS.2019.00076
12. Clarke, M.: The digital revolution. In: Academic and professional publishing, pp. 79–98. Elsevier (2012). https://doi.org/10.1016/B978-1-84334-669-2.50004-4
13. Elmhadhbi, L., Karray, M.-H., Archimède, B.: Toward the use of upper-level ontologies for semantically interoperable systems: an emergency management use case. In: Popplewell, K., Thoben, K.-D., Knothe, T., Poler, R. (eds.) Enterprise Interoperability VIII. PIC, vol. 9, pp. 131–140. Springer, Cham (2019). https://doi.org/10.1007/978-3-030-13693-2_11
14. Ferrer, B.R., Mohammed, W.M., Lobov, A., Galera, A.M., Lastra, J.L.M.: Including human tasks as semantic resources in manufacturing ontology models. In: IECON 2017–43rd Annual Conference of the IEEE Industrial Electronics Society, pp. 3466–3473. IEEE (2017). https://doi.org/10.1109/IECON.2017.8216587

15. Ganzha, M., Paprzycki, M., Pawłowski, W., Szmeja, P., Wasielewska, K.: Semantic interoperability in the internet of things: an overview from the inter-IoT perspective. J. Netw. Comput. Appl. **81**, 111–124 (2017). https://doi.org/10.1016/j.jnca.2016.08.007
16. Grangel-González, I., Vidal, M.E.: Analyzing a knowledge graph of industry 4.0 standards. In: Companion Proceedings of the Web Conference 2021, pp. 16–25 (2021). https://doi.org/10.1145/3442442.3453542
17. Hafidi, M.M., Djezzar, M., Hemam, M., Amara, F.Z., Maimour, M.: Semantic web and machine learning techniques addressing semantic interoperability in industry 4.0. Int. J. Web Inf. Syst. (2023)
18. Hashemi, P., Khadivar, A., Shamizanjani, M.: Developing a domain ontology for knowledge management technologies. Online Inf. Rev. (2018). https://doi.org/10.1108/OIR-07-2016-0177
19. Helmiö, P.: Open source in industrial internet of things: a systematic literature review master's thesis. School of Business and Management, Lappeenranta University of Technology, vol. 21 (2018)
20. Hitzler, P.: A review of the semantic web field. Commun. ACM **64**(2), 76–83 (2021). https://doi.org/10.1145/3397512
21. Jaekel, W., Doumeingts, G., Wollschlaeger, M.: A SAREF extension for semantic interoperability in the industry and manufacturing domain (2018)
22. Jeschke, S., Brecher, C., Meisen, T., Özdemir, D., Eschert, T.: Industrial internet of things and cyber manufacturing systems. In: Jeschke, S., Brecher, C., Song, H., Rawat, D.B. (eds.) Industrial Internet of Things. SSWT, pp. 3–19. Springer, Cham (2017). https://doi.org/10.1007/978-3-319-42559-7_1
23. Kaar, C., Frysak, J., Stary, C., Kannengiesser, U., Müller, H.: Resilient ontology support facilitating multi-perspective process integration in industry 4.0. In: Proceedings of the 10th International Conference on Subject-Oriented Business Process Management, pp. 1–10 (2018). https://doi.org/10.1145/3178248.3178253
24. Kalaycı, E.G., et al.: Semantic integration of Bosch manufacturing data using virtual knowledge graphs. In: Pan, J.Z., et al. (eds.) ISWC 2020. LNCS, vol. 12507, pp. 464–481. Springer, Cham (2020). https://doi.org/10.1007/978-3-030-62466-8_29
25. Kendall, E., McGuinness, D.: Ontology Engineering. Springer, Cham (2022). https://doi.org/10.1007/978-3-031-79486-5
26. Khorov, E., Lyakhov, A., Krotov, A., Guschin, A.: A survey on IEEE 802.11 ah: an enabling networking technology for smart cities. Comput. Commun. **58**, 53–69 (2015). https://doi.org/10.1016/j.comcom.2014.08.008
27. Kovalenko, O., et al.: Automationml ontology: modeling cyber-physical systems for industry 4.0. IOS Press J. **1**, 1–5 (2018)
28. Kunold, I., Wöhrle, H., Kuller, M., Karaoglan, N., Kohlmorgen, F., Bauer, J.: Semantic interoperability in cyber-physical systems. In: 2019 10th IEEE International Conference on Intelligent Data Acquisition and Advanced Computing Systems: Technology and Applications (IDAACS), vol. 2, pp. 797–801 (2019). https://doi.org/10.1109/IDAACS.2019.8924274
29. Liu, Z., et al.: The architectural design and implementation of a digital platform for industry 4.0 SME collaboration. Comput. Ind. **138**, 103623 (2022)
30. Longo, F., Mirabelli, G., Nicoletti, L., Solina, V.: An ontology-based, general-purpose and industry 4.0-ready architecture for supporting the smart operator (part i-mixed reality case). J. Manuf. Syst. **64**, 594–612 (2022)
31. Mahmood, Z. (ed.): The Internet of Things in the Industrial Sector. CCN, Springer, Cham (2019). https://doi.org/10.1007/978-3-030-24892-5

32. May, G., Cho, S., Majidirad, A., Kiritsis, D.: A semantic model in the context of maintenance: a predictive maintenance case study. Appl. Sci. **12**(12), 6065 (2022). https://doi.org/10.3390/app12126065
33. Mishra, S., Jain, S.: Ontologies as a semantic model in IoT. Int. J. Comput. Appl. **42**(3), 233–243 (2020)
34. Olivares-Alarcos, A., Foix, S., Borgo, S., Guillem, A.: OCRA - an ontology for collaborative robotics and adaptation. Comput. Ind. **138**, 103627 (2022). https://doi.org/10.1016/j.compind.2022.103627
35. Patel, P., Ali, M.I., Sheth, A.: From raw data to smart manufacturing: AI and semantic web of things for industry 4.0. IEEE Intell. Syst. **33**(4), 79–86 (2018). https://doi.org/10.1109/MIS.2018.043741325
36. Ramírez-Durán, V.J., Berges, I., Illarramendi, A.: ExtruOnt: an ontology for describing a type of manufacturing machine for industry 4. systems. Semant. Web **11**(6), 887–909 (2020). https://doi.org/10.3233/SW-200376
37. Ren, H., Anicic, D., Runkler, T.A.: Towards semantic management of on-device applications in industrial IoT. ACM Trans. Internet Technol. **22**, 1–30 (2022). https://doi.org/10.1145/3510820
38. da Rocha, H., Espirito-Santo, A., Abrishambaf, R.: Semantic interoperability in the industry 4.0 using the IEEE 1451 standard. In: IECON 2020 The 46th Annual Conference of the IEEE Industrial Electronics Society, pp. 5243–5248. IEEE (2020). https://doi.org/10.1109/IECON43393.2020.9254274
39. Schroeder, A., Ziaee Bigdeli, A., Galera Zarco, C., Baines, T.: Capturing the benefits of industry 4.0: a business network perspective. Prod. Plann. Control **30**(16), 1305–1321 (2019). https://doi.org/10.1080/09537287.2019.1612111
40. Sisinni, E., Saifullah, A., Han, S., Jennehag, U., Gidlund, M.: Industrial internet of things: challenges, opportunities, and directions. IEEE Trans. Industr. Inf. **14**(11), 4724–4734 (2018). https://doi.org/10.1109/TII.2018.2852491
41. Strassner, J., Diab, W.W.: A semantic interoperability architecture for internet of things data sharing and computing. In: 2016 IEEE 3rd World Forum on Internet of Things (WF-IoT), pp. 609–614. IEEE (2016). https://doi.org/10.1109/WF-IoT.2016.7845422
42. Tan, L., Wang, N.: Future internet: the internet of things. In: 2010 3rd International Conference on Advanced Computer Theory and Engineering (ICACTE), vol. 5, pp. V5–376. IEEE (2010). https://doi.org/10.1109/ICACTE.2010.5579543
43. Teslya, N., Ryabchikov, I.: Ontology-driven approach for describing industrial socio-cyberphysical systems' components. In: MATEC Web of Conferences, vol. 161, p. 03027. EDP Sciences (2018). https://doi.org/10.1051/matecconf/201816103027
44. , Tiwari, S., Ortiz-Rodriguez, F., Jabbar, M.: Semantic modeling for healthcare applications: an introduction. Semant. Models IoT eHealth Appl., 1–17 (2022)
45. Tiwari, S., Rodriguez, F.O., Jabbar, M.: Semantic Models in IoT and eHealth Applications. Academic Press, Cambridge (2022)
46. Tortorella, G.L., Fettermann, D.: Implementation of industry 4.0 and lean production in Brazilian manufacturing companies. Int. J. Prod. Res. **56**(8), 2975–2987 (2018)
47. Ustundag, A., Cevikcan, E.: Industry 4.0: Managing The Digital Transformation. SSAM, Springer, Cham (2018). https://doi.org/10.1007/978-3-319-57870-5
48. Venceslau, A., Andrade, R., Vidal, V., Nogueira, T., Pequeno, V.: Iot semantic interoperability: a systematic mapping study. In: International Conference on Enterprise Information Systems, vol. 1, pp. 535–544. SciTePress (2019)

49. Wan, J., et al.: Toward dynamic resources management for IoT-based manufacturing. IEEE Commun. Mag. **56**(2), 52–59 (2018). https://doi.org/10.1109/MCOM.2018.1700629
50. Wan, J., Yin, B., Li, D., Celesti, A., Tao, F., Hua, Q.: An ontology-based resource reconfiguration method for manufacturing cyber-physical systems. IEEE/ASME Trans. Mechatron. **23**(6), 2537–2546 (2018). https://doi.org/10.1109/TMECH.2018.2814784
51. Wang, Y.: Enhancing interoperability for IoT based smart manufacturing: an analytical study of interoperability issues and case study (2020)
52. Yahya, M., Breslin, J.G., Ali, M.I.: Semantic web and knowledge graphs for industry 40. Appl. Sci. **11**(11), 5110 (2021). https://doi.org/10.3390/app11115110

On the Representation of Dynamic BPMN Process Executions in Knowledge Graphs

Franz Krause[1(✉)], Kabul Kurniawan[2,3], Elmar Kiesling[2], Heiko Paulheim[1], and Axel Polleres[2]

[1] Data and Web Science Group, University of Mannheim, Mannheim, Germany
franz.krause@uni-mannheim.de
[2] Vienna University of Economics and Business, Vienna, Austria
[3] Austrian Center for Digital Production, Vienna, Austria

Abstract. Knowledge Graphs (KGs) are a powerful tool for representing domain knowledge in a way that is interpretable for both humans and machines. They have emerged as enablers of semantic integration in various domains, including Business Process Modeling (BPM). However, existing KG-based approaches in BPM lack the ability to capture dynamic process executions. Rather, static components of BPM models, such as Business Process Model and Notation (BPMN) elements, are represented as KG instances and further enriched with static domain knowledge. This poses a challenge as most business processes exhibit inherent degrees of freedom, leading to variations in their executions. To address this limitation, we examine the semantic modeling of BPMN terminology, models, and executions within a shared KG to facilitate the inference of new insights through observations of process executions. We address the issue of representing BPMN models within the concept or instance layer of a KG, comparing potential implementations and outlining their advantages and disadvantages in the context of a human-AI collaboration use case from a European smart manufacturing project.

Keywords: Dynamic Knowledge Graph · Business Process Modeling · Semantic Web · Human-AI Collaboration · Smart Manufacturing

1 Introduction

The representation, standardization, and management of knowledge in KGs have gained significant research attention across various domains, such as biomedicine [24], cybersecurity [15], and manufacturing [5]. However, KGs tend to focus on static concepts while neglecting the dynamic instantiations of entities they are representative of. This issue is particularly obvious in business process management, which is concerned with modeling and capturing the dynamics that arise from actions and interactions within and across organizations. Although standardized modeling languages such as BPMN [18] have been widely adopted to contribute to organizational objectives [10], there is a lack of integration between

F. Ortiz-Rodriguez et al. (Eds.): KGSWC 2023, LNCS 14382, pp. 97–105, 2023.
https://doi.org/10.1007/978-3-031-47745-4_8

(i) semantically explicit process knowledge, (ii) domain knowledge, and (iii) dynamically generated data resulting from process executions. Despite efforts to enhance these languages with ontological foundations for semantic modeling [17,20,22], work on the integration into a shared dynamic KG is scarce.

This represents a significant gap towards enterprise KGs as dynamic and comprehensive representations of domain knowledge within the context of an organization or enterprise, whose benefits extend beyond the standardization and management of domain expertise. In particular, enterprise KGs also hold great potential for various downstream tasks, such as question answering [9], recommendation systems [19], and KG embedding-based applications [25]. Thus, by leveraging organizational knowledge, they can enhance decision-making processes, facilitate efficient information retrieval, and provide valuable recommendations tailored to domain-specific requirements [13]. Consequently, the integration of BPM models and corresponding process executions into dynamic KG structures may offer significant benefits for organizations that employ them.

Therefore, this work analyzes two potential approaches to enable comprehensive representations of process executions along with their underlying process models and BPM terminology in a shared KG, with BPMN as an exemplary BPM formalism. While this paper serves as a disclosure of the significant opportunities that arise from integrating dynamic BPM executions and existing domain knowledge, it also provides valuable insights, guidelines, and potential implementation approaches for future work in this field.

2 Related Work

Numerous approaches have emerged that attempt to capture BPM formalisms within well-defined frameworks [12]. In this context, ontologies such as [2,7,8,17,22] provide uniform representations to capture and analyze BPM concepts unambiguously and effectively. Based on these concepts, ontology-based process modeling (OBPM) aims to enhance the semantic clarity and interoperability of BPM systems [23]. To this end, several approaches exist that intend to model the semantics of BPM components, e.g., by annotating concepts from domain ontologies to process components [8,11,21]. However, OBPM frameworks have yet to be adopted in industrial settings, which has been attributed to the significant manual effort required to develop use case specific BPM ontologies [7].

In contrast to OBPM, graphical business process modeling (GBPM) comprises methods to represent BPM formalisms via graphical notations [12]. These formalisms can be applied to visualize and organize BPM models via symbols and diagrams, including their relationships, sequential flows, and logical policies. Thus, GBPM is particularly useful for understanding and analyzing complex business processes. BPMN [6] has established itself as a de-facto standard formalism that provides a convenient graphical notation, making it easier for stakeholders to collaborate on BPMs. It is supported by various tools and platforms, ensuring compatibility and interoperability across systems.

To embed BPMN elements in corresponding ontologies, various approaches have been proposed [2,22]; the BPMN-Based Ontology (BBO) [2], for instance, is an ontological framework for encoding BPMN models that aim to define a standardized set of concepts and relations to represent business processes. It can be used for sharing and comparing business processes based on BPMN and thus, also facilitates interoperability and automation of BPMN processes. Another notable conceptualization is the BPMN ontology [22], which provides a formal ontological description developed from the BPMN specification. It aims to provide a vocabulary and formal semantics for BPMN models. However, the existing approaches do not cover process instances, which limits their applicability for contextualizing, analyzing, and monitoring process executions.

Whereas OBPM and GBPM formalisms can be utilized to standardize BPM components, the alignment of corresponding business process blueprints with the underlying BPM terminology remains challenging. Accordingly, approaches exist to embed these blueprints in KG structures, such as BPMN2KG [3] which is based on the BBO ontology. However, current conceptualizations typically lack the ability to directly incorporate information related to the actual executions of business processes into formalized representations. Instead, traces generated at execution time are usually captured in separate event logs, ideally generated in a central location by a process engine and stored in standardized formats such as XES [1]. More commonly, however, they are dispersed across the respective systems that execute the process and remain isolated from the process model as well as contextual domain-specific knowledge. Hence, from an ontology engineering perspective, existing methods typically focus exclusively on the formalization of process models as indefinite process instances rather than definite instances, i.e., traces of actual process executions.

3 Industrial Use Case: Human-AI Collaboration in the Smart Manufacturing Domain

This section introduces a motivating use case from the smart manufacturing domain that illustrates the challenge of integrating domain expertise and process knowledge with data generated at execution time. The scenario depicted in Fig. 1 is a BPMN process that has been abstracted from one of several real-world applications that leverage dynamic KGs in manufacturing within the Teaming.AI[1] project. Based on this use case scenario, Sect. 4 will address approaches to represent corresponding process executions in a dynamic KG.

Problem Setting: The task of monitoring the quality of manufactured parts and optimizing product quality through machine setting recommendations poses significant challenges in the smart manufacturing domain. It necessitates the integration of process knowledge, real-time production data, and domain expertise that captures causal patterns for diagnosing and resolving quality issues at runtime. In our motivating scenario, a machine produces parts that are checked regarding qualitative requirements. This monotonic task of determining whether

[1] https://www.teamingai-project.eu/.

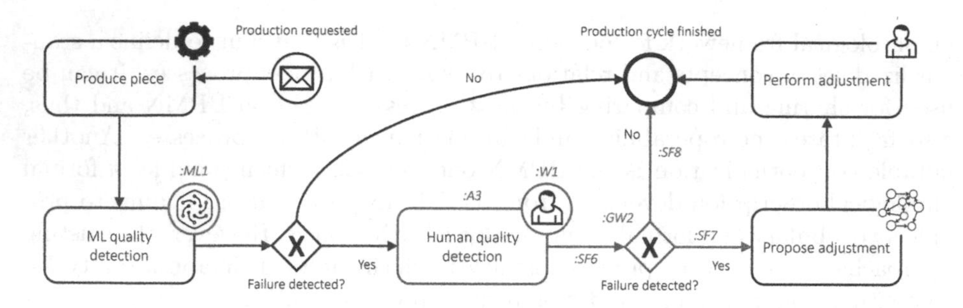

Fig. 1. BPMN model of the human-AI process *:P0* from the Teaming.AI project.

a produced part is OK or Not OK (NOK) can be automated through automated visual quality inspection systems based on Machine Learning (ML) models, which reduces the human workload and only requires human interventions in the case of NOK predictions. Once a failure has been confirmed, the operator is asked to perform an adjustment of the machine. To this end, domain knowledge about product failure types and machine parameter adjustments is queried from the KG with contextual information to provide a potential remedy [4].

Problem Statement: As outlined in Sect. 2, KG encodings of BPMN processes lack the ability to capture dynamic process executions in a semantic manner and link them to dynamic domain knowledge to enable valuable insights and dynamic process improvements. For example, it is of interest which combinations of AI agents *:ML1* and human agents *:W1* have led to an execution of the sequence flow *:SF8*, representing a worker vetoing a precedent ML prediction. Thus, additional background knowledge in the KG could be used to examine whether specific ML methods or training data sets result in differing quality assessments of human and AI agents. However, to allow for such analyses, process executions need to be semantically linked to their roles within the BPMN process model, as well as domain-specific knowledge. In the following, we address two approaches to represent process dynamics, with the main focus on whether process models should be represented as KG concepts or instances.

4 Representing Dynamic BPMN Process Executions in Knowledge Graph Structures

As a framework for representing and standardizing domain knowledge, KGs offer vast potential for advanced data integration, reasoning, and knowledge discovery [13]. They are composed of nodes and edges, where nodes represent entities, concepts, or literal values, and edges represent relationships between these nodes. Typically, KGs are specified by means of some syntax based on the RDF (Resource Description Framework) formalism, including a division into a concept layer and an instance layer, which are also referred to as TBox and ABox, respectively. While the TBox contains a domain ontology, usually based on the Web Ontology Language (OWL), to define concepts and their relationships, the ABox provides and manages the actual data and instances of the KG.

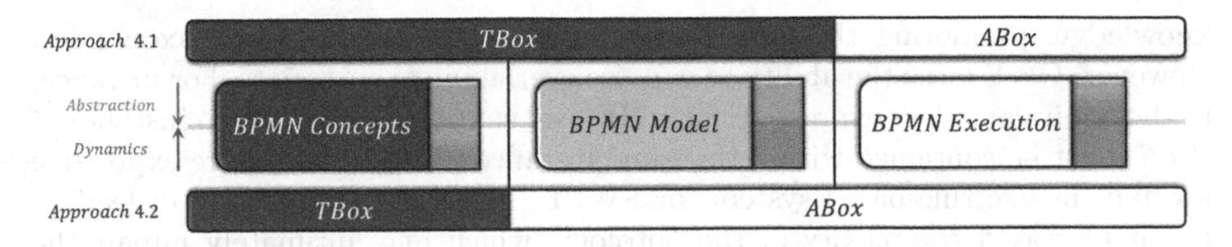

Fig. 2. Abstraction levels pertaining to the three BPMN abstraction layers, along with the potential allocations of KG abstraction levels discussed in the Sects. 4.1 and 4.2.

However, when dealing with BPMN process executions, three abstraction layers are required, as illustrated in Fig. 2. The BBO ontology serves as the BPMN concept layer and is thus included in the TBox of the KG. In contrast, process executions as real-world instantiations of TBox concepts are part of its ABox. Consequently, the positioning of BPMN models within the RDF abstraction layers is ambiguous as they may serve both as instantiations of BPMN concepts and as conceptual models for BPMN executions. In the following, we address approaches to integrate BPMN process models either in (i) the concept layer or (ii) the instance layer of a process-enabled KG by applying the idea of (i) OWL constraints or (ii) definite and indefinite instances, respectively.

4.1 Representing BPMN Models in the Concept Layer of a Knowledge Graph via OWL Constraints

BPMN ontologies such as BBO capture BPMN process models in a KG TBox, allowing for subsequent instantiations in the ABox of the KG based on process executions. For instance, the BPMN process *:P0* in Fig. 1 can be encoded as a subclass of *bbo:Process*, while *:GW2* represents a subclass of *bbo:Gateway*. Moreover, certain constraints must be fulfilled, e.g., each instance of *:GW2* is required to be contained within an execution of *:P0*. In addition, it must be linked to an incoming *bbo:SequenceFlow* of type *:SF6* as well as exactly one outgoing *bbo:SequenceFlow* of type *:SF7* or *:SF8*. OWL class descriptions and axioms can be applied to define such BPMN execution constraints. Accordingly, the class *:GW2* can be defined to be equivalent to

(*bbo:has_container* exactly 1 *:P0*) **and** (*bbo:has_incoming* exactly 1 *:SF6*) **and** (*bbo:has_outgoing* exactly 1 (*:SF7* **or** *:SF8*)).

OWL concepts and properties such as *owl:Restriction*, *owl:qualifiedCardinality*, and *owl:onClass* enable the formulation of these expressions, and domain policies. For example, quality assurance (*:QA*) or quality control (*:QC*) can represent the required qualifications of agents performing *:A3*, as implied by its superclass

inverse *bbo:is_responsibleFor* only (*:has_qualification* min 1 ({*:QA*, *:QC*})).

Thus, TBox encodings of BPMN process models capture valid process flows, as well as constraint-based execution policies that can be enriched with domain knowledge, promoting the definition of reusability and inherent taxonomies. However, OWL lacks the ability to express overlapping constraints. For instance, in the equivalent class expression for *:GW2*, we cannot specify that an instance of *:SF6* must be contained within the same instance of *:P0*. Thus, more expressive formalisms like rule-based systems or SWRL [14] need to be employed, leading to an increased complexity of the ontology, which may ultimately impair the adoption of OBPM (cf. Sect. 2). Moreover, while no restrictions exist regarding the dynamics of BPMN executions as KG instances, this is not the case for BPMN models. Process flows are captured within OWL constraints, i.e., updates typically result in constraint violations of previously performed executions.

4.2 Representing BPMN Models in the Instance Layer of a Knowledge Graph via Indefinite Instances

The desired allocation of BPMN and KG abstraction layers, as illustrated in Fig. 2, can also be achieved by encoding BPMN models and BPMN executions within a shared ABox. This approach aligns with existing work such as BPMN2KG [3], where BPMN models and components are represented as KG instances, maintaining their semantic relations as depicted in the blue box of Fig. 3. Accordingly, process execution rules can be defined via OWL constraints (e.g., to *bbo:Gateway*) or by means of instance-level properties such as node labels or assignments of agent instances that are intended to perform certain activities. For example, to specify the qualifications introduced in Fig. 2 that are required to perform an activity *:A3*, SHACL constraints can be used to design respective node shapes [16]. In contrast to the TBox encodings of BPMN models in Sect. 4.1, process flows are thus represented via explicit facts instead of OWL class restrictions. However, to account for BPMN executions, the ABox needs to be enhanced by an additional abstraction layer.

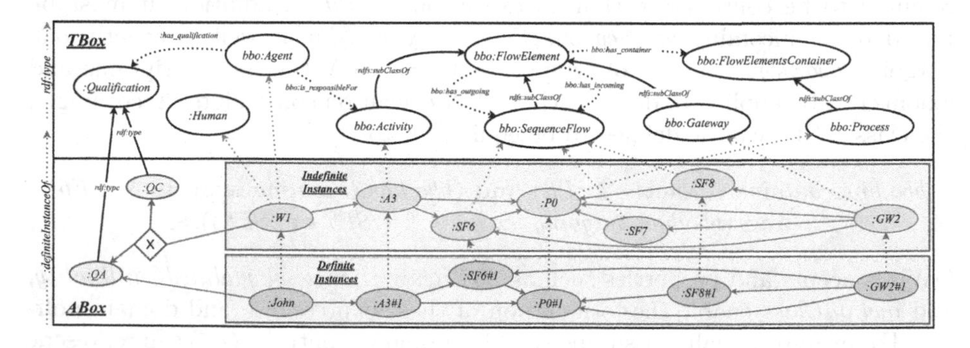

Fig. 3. KG encoding of a process execution and its underlying model.

To this end, we propose to extend the TBox by a class *:IndefiniteInstance* and a property *:definiteInstanceOf* so that each executed BPMN component is a definite instance (red nodes) of an indefinite instance (blue nodes). For example, *:A3#1* is a definite instance (i.e., an execution) of *:A3*. It is executed by the worker *:John* as a human agent who fulfills the required qualification skills implied by the indefinite instance *:W1*. This approach makes it possible to represent dynamic BPMN executions as KG instances by assignments of corresponding indefinite BPMN components. While no explicit constraints are implemented to verify logical correctness, the ABox encodings of process models allow for more dynamic BPMN models compared to the approach discussed in Sect. 4.1. Additionally, representing BPMN process flows as instance-level facts contributes to less complex TBox structures as a key enabler of OBPM implementations [7].

5 Conclusions

To benefit from the potential of knowledge-based approaches in process-aware information systems, it is crucial to effectively establish connections between dynamic process data and static process models. This necessity becomes evident in our real-world use case within a smart manufacturing environment, where process models serve not only as a framework for data integration but also as a foundation for the coordination of human-AI collaborations. To address the challenge of modeling process-centric KGs that capture both BPM models and their actual executions, a comprehensive review of the existing literature was conducted. We identified the key issue of diverging abstraction layers in BPM and KGs, particularly that it is unclear whether process models are to be regarded as KG concepts or instances. To tackle this issue, we reviewed and compared (i) OWL constraints and (ii) definite instances as two approaches for encoding them in a shared KG. Future work will focus on the incorporation of these approaches into process-aware information systems to facilitate human-AI collaboration, alongside a range of additional industrial use cases. Furthermore, we will explore the analytical opportunities offered by these methodologies.

Acknowledgements. This work is part of the TEAMING.AI project which receives funding in the European Commission's Horizon 2020 Research Programme under Grant Agreement Number 957402 (www.teamingai-project.eu).

References

1. Acampora, G., et al.: IEEE 1849TM: the XES standard. IEEE Comput. Intell. Mag., 4–8 (2017)
2. Annane, A., Aussenac-Gilles, N., Kamel, M.: BBO: BPMN 2.0 based ontology for business process representation. In: ECKM, vol. 1, pp. 49–59 (2019)
3. Bachhofner, S., et al.: Automated process knowledge graph construction from BPMN models. In: Strauss, C., Cuzzocrea, A., Kotsis, G., Tjoa, A.M., Khalil, I. (eds.) Database and Expert Systems Applications. Lecture Notes in Computer Science, vol. 13426, pp. 32–47. Springer, Cham (2022). https://doi.org/10.1007/978-3-031-12423-5_3

4. Bachhofner, S., et al.: Knowledge graph supported machine parameterization for the injection moulding industry. In: Villazon-Terrazas, B., Ortiz-Rodriguez, F., Tiwari, S., Sicilia, M.A., Martin-Moncunill, D. (eds.) Knowledge Graphs and Semantic Web. Communications in Computer and Information Science, vol. 1686, pp. 106–120. Springer, Cham (2022). https://doi.org/10.1007/978-3-031-21422-6_8
5. Bader, S.R., Grangel-Gonzalez, I., Nanjappa, P., Vidal, M.-E., Maleshkova, M.: A knowledge graph for industry 4.0. In: Harth, A., et al. (eds.) ESWC 2020. LNCS, vol. 12123, pp. 465–480. Springer, Cham (2020). https://doi.org/10.1007/978-3-030-49461-2_27
6. Chinosi, M., Trombetta, A.: BPMN: an introduction to the standard. Comput. Stand. Interfaces **34**(1), 124–134 (2012)
7. Corea, C., Fellmann, M., Delfmann, P.: Ontology-based process modelling - will we live to see it? In: Ghose, A., Horkoff, J., Silva Souza, V.E., Parsons, J., Evermann, J. (eds.) ER 2021. LNCS, vol. 13011, pp. 36–46. Springer, Cham (2021). https://doi.org/10.1007/978-3-030-89022-3_4
8. Di Martino, B., Esposito, A., Nacchia, S., Maisto, S.A.: Semantic annotation of BPMN: current approaches and new methodologies. In: IIWAS, pp. 1–5 (2015)
9. Diefenbach, D., Giménez-García, J., Both, A., Singh, K., Maret, P.: QAnswer KG: designing a portable question answering system over RDF data. In: Harth, A., et al. (eds.) ESWC 2020. LNCS, vol. 12123, pp. 429–445. Springer, Cham (2020). https://doi.org/10.1007/978-3-030-49461-2_25
10. Dumas, M., et al.: Fundamentals of Business Process Management, vol. 2. Springer, Cham (2018)
11. Fellmann, M., Hogrebe, F., Thomas, O., Nüttgens, M.: Checking the semantic correctness of process models: an ontology-driven approach using domain knowledge and rules. EMISA **6**, 25–35 (2011)
12. Fellmann, M., Koschmider, A., Laue, R., et al.: Business process model patterns: state-of-the-art, research classification and taxonomy. Bus. Process. Manag. J. **25**(5), 972–994 (2019)
13. Hogan, A., et al.: Knowledge graphs. ACM Comput. Surv. **54**(4), 1–37 (2021)
14. Horrocks, I., et al.: SWRL: a semantic web rule language combining OWL and RuleML. W3C Member Submiss. **21**, 1–31 (2004)
15. Kiesling, E., Ekelhart, A., Kurniawan, K., Ekaputra, F.: The SEPSES knowledge graph: an integrated resource for cybersecurity. In: Ghidini, C., et al. (eds.) ISWC 2019. LNCS, vol. 11779, pp. 198–214. Springer, Cham (2019). https://doi.org/10.1007/978-3-030-30796-7_13
16. Knublauch, H., Kontokostas, D.: Shapes constraint language (SHACL). W3C Member Submiss. (2017)
17. Natschläger, C.: Towards a BPMN 2.0 ontology. In: Dijkman, R., Hofstetter, J., Koehler, J. (eds.) BPMN 2011. LNBIP, vol. 95, pp. 1–15. Springer, Heidelberg (2011). https://doi.org/10.1007/978-3-642-25160-3_1
18. Business Process Model and Notation (BPMN) 2.0 specification (2011). https://www.omg.org/spec/BPMN/2.0/PDF. version 2
19. Palumbo, E., Rizzo, G., Troncy, R., Baralis, E., Osella, M., Ferro, E.: Knowledge graph embeddings with node2vec for item recommendation. In: Gangemi, A., et al. (eds.) ESWC 2018. LNCS, vol. 11155, pp. 117–120. Springer, Cham (2018). https://doi.org/10.1007/978-3-319-98192-5_22
20. Pedrinaci, C., et al.: Semantic business process management: scaling up the management of business processes. In: IEEE International Conference on Semantic Computing, pp. 546–553 (2008)

21. Riehle, D.M., Jannaber, S., Delfmann, P., Thomas, O., Becker, J.: Automatically annotating business process models with ontology concepts at design-time. In: de Cesare, S., Frank, U. (eds.) ER 2017. LNCS, vol. 10651, pp. 177–186. Springer, Cham (2017). https://doi.org/10.1007/978-3-319-70625-2_17
22. Rospocher, M., Ghidini, C., Serafini, L.: An ontology for the business process modelling notation. In: Formal Ontology in Information Systems, pp. 133–146 (2014)
23. Thomas, O., Fellmann, M.: Semantic process modeling: design and implementation of an ontology-based representation of business processes. Bus. Inf. Syst. Eng. **1**, 438–451 (2009)
24. Vidal, M.E., et al.: Current Trends in Semantic Web Technologies: Theory and Practice, chap. Semantic Data Integration of Big Biomedical Data for Supporting Personalised Medicine, pp. 25–56. Springer (2019)
25. Wang, Q., Mao, Z., Wang, B., Guo, L.: Knowledge graph embedding: a survey of approaches and applications. IEEE Trans. Knowl. Data Eng. **29**(12), 2724–2743 (2017)

Towards a Framework for Seismic Data

Valadis Mastoras, Alexandros Vassiliades(✉), Maria Rousi, Sotiris Diplaris, Thanassis Mavropoulos, Ilias Gialampoukidis, Stefanos Vrochidis, and Ioannis Kompatsiaris

Information Technologies Institute, Center for Research and Technology Hellas, Thessaloniki, Greece
{mastoras.valadis,valexande,mariarousi,diplaris,mavrathan, heliasgj,stefanos,ikom}@iti.gr

Abstract. Over the past decade, Knowledge Graphs (KGs) gained significant attention as a powerful method for knowledge representation. Driven by increasing interest, a paradigm shift has occurred, where the technology of KGs has transitioned from the research domain to the industry and public sector, with companies and organizations increasingly representing their data as Linked Open Data, gaining in that way significant traction for this technology. This paper focuses on KGs in the context of environmental challenges. More specifically, this work concerns KGs that contain seismic event data, such as location, timestamp, magnitude, depth, target date, as well as images before and after the event occurrence. Moreover, a Natural Language Processing (NLP) module is integrated to enhance the KG. That module enables users to query for seismic events in a free-text manner, before addressing them with a relevant response through a dedicated Information Retrieval (IR) component. The KG was constructed with data retrieved from the *Instituto Nazionale di Geofisica e Vulcanologia*, a rich resource that comes with earthquake-related information, such as magnitude, depth, occurrence location, and timestamp. Additionally, public APIs from the *Copernicus Open Access Data Hub* and *ONDA DIAS* are leveraged to provide access to sentinel data, such as images of the event location before and after its occurrence.

Keywords: Knowledge Graph · Information Retrieval · Natural Language Processing · Semantic Framework

1 Introduction

The usage of KGs as a method for representing data has increased significantly in the last five years. The KGs can provide semantically related knowledge, along with interoperability, interlinking, and re-usability of data, that other knowledge bases cannot, while they can also be applied in a variety of domains, such as the representation of knowledge for environmental challenges. For this reason, we can see that more and more industries, as well as the public sector, translate their data into Linked Open Data (LOD).

F. Ortiz-Rodriguez et al. (Eds.): KGSWC 2023, LNCS 14382, pp. 106–119, 2023.
https://doi.org/10.1007/978-3-031-47745-4_9

In this paper, we present a KG for seismic data. In more detail, we present a KG that can represent information for seismic events across the globe. The KG was built upon the information provided by the Copernicus Data Provider[1] for Sentinels data and the National Institute of Geophysics and Volcanology[2] for earthquake data. It also incorporates a NLP mechanism to aid the users address queries in natural language, and retrieve data about a specific event. Moreover, the KG is equipped with a IR mechanism that retrieves semantically related pairs of pre/post images for an event, among other information.

The motivation behind this paper is to present the first KG that contains, to this extent, knowledge about seismic events and sentinel data, for the state of a target location before and after an event. Moreover, the KG presented in this paper can help in providing knowledge about the state of a target environment and the changes that it goes through, while also provide knowledge for possible disaster management scenarios.

The main objective of our study is to introduce the first, to our knowledge, comprehensive and high-quality knowledge resource, specifically dedicated for seismic events. The KG provides valuable insights about the state of a target environment, including pre-event and post-event changes, while also, it offers guidelines for effective disaster management scenarios. Conclusively, our work provides a high-quality RDF representation of seismic events, covering the biggest possible extent of recorded instances.

The main contribution of this study is the quality of data that are introduced in RDF representation, as it represents information about almost all seismic events ever recorded. The plurality of this information occurs because we extract knowledge from the Copernicus Data Provider and the National Institute of Geophysics and Volcanology, which contain information about all the seismic events ever recorded.

The outline of the paper is as follows, we start with related work (Sect. 2), followed by a presentation of the utilized data and a description of the KG, the NLP module, and the IR mechanism in Sect. 3. Next, in Sect. 4 we provide an evaluation of our framework, where we evaluate the consistency and completeness of the KG. Moreover, we compare our framework against a Large Language Model (LLM). We conclude our paper with Sect. 5, where we summarize the key findings and implications of our study.

2 Related Work

This study delves into three integral components: the KG which represents seismic event data, the NLP mechanism which facilitates user-friendly queries, and the IR mechanism which fetches relevant data from the KG. In this section, we will omit the literature regarding similar NLP components, since the NLP part is mostly a helping component, and we will focus on relevant KG and IR proposals.

[1] https://scihub.copernicus.eu/.

[2] https://github.com/INGV/openapi.

Regarding the KG domain, a variety of research attempts have been proposed in the literature that exhibit an adequate resemblance to the target topic of this work. More precisely, one of the most relevant categories may be the KGs for seismic events, while also, the category of KGs regarding the seismic risk domain.

Starting with the former, some indicative approaches may be those of Li & Li (2013), where a KG for semantic representation of textual information is presented with multi-document summarization, with disaster management being its main aspect, while also the studies of Chou et al. (2010; 2014) may be indicative, where similar KGs were presented. Nevertheless, even though in these works KGs for disaster management may properly be presented, those fall short in representing a specific knowledge for seismic events since they only target information that exists on various disaster management websites, in contrast to the KG proposed in this work.

In a similar sense, regarding the latter category of KGs for the seismic risk domain, an indicative research example may be the work of Murgante et al. (2009). In this work, a promising KG is introduced, which improves semantic interoperability and aims to reduce the economic and social costs associated with seismic events. Nevertheless, the information provided in this work lacks of an extensive depth, in contrast to ours, which encompasses precise details such as location, timestamp, magnitude, and depth of each earthquake.

Continuing with our examination, it would be essential to highlight the IR mechanisms currently utilized in disaster management. To the best of our knowledge, existing IR mechanisms in this domain do not have a direct association with a KG, resulting in a significant oversight of vital semantically related information. For instance, the studies of Islam & Chik (2011); Bouzidi et al. (2019); Shen et al. (2017) can extract data from various websites but lack the reasoning capabilities that a KG can offer. A similar scenario is observed in Pi et al. (2020); Vallejo et al. (2020), where drone footage is analyzed without leveraging a KG. In contrast to such a drawback, our proposed IR mechanism showcases the true potential of a KG, enriching IR and providing an initial direction on how a KG can be used to manage crucial information about a disaster.

Eventually, before concluding our literature overview, a series of research examples regarding the helpfulness of KGs may be useful to be presented. Some example works in that direction are those of Landis et al. (2021), Silva et al. (2013) and Ortmann et al. (2011), where approaches on how a KG could help the perseverance of Cultural and Natural heritage in the case of a disaster (such as an earthquake), are presented. Subsequently, the work of Correia et al. (2023) may be an indicative example of how a KG can help a decision support system to access semantically rich data in order to help users take better decisions during a disaster management scenario. The aforementioned research endeavors may serve as supplementary evidence underlining the potential benefits that KGs could confer.

3 Methodology

In this Chapter, we present a description of the data upon which the KG was constructed, along with the KG itself, the NLP mechanism and the IR mechanism.

Figure 1 provides an overview of the architecture, presenting a user interacting with the system through the NLP mechanism, to query the Seismic KG. Subsequently, the IR mechanism retrieves relevant answers and presents them to the user. The Seismic KG acquires data from Copernicus and the Institute of Volcanology, incorporating it into its internal Knowledge Base (KB).

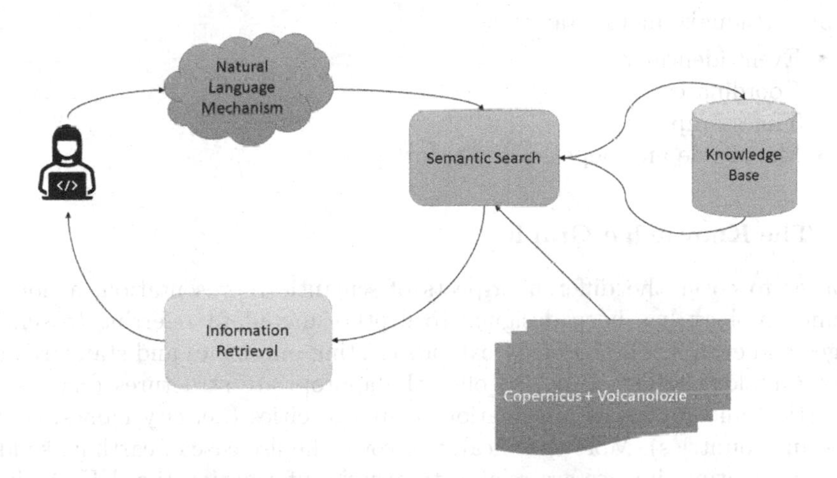

Fig. 1. The Architecture of the Seismic Data Framework

Moreover, it is worth noticing that the NLP mechanism accesses the KG through an endpoint, based on which the KG also plays the role of a client-server application. Whenever a question is received on the KG endpoint, a SPARQL query is automatically produced, containing all the desired information for the KG.

3.1 Nature of the Data

The data and metadata which are represented in the KG are retrieved from two sources: The Copernicus Data Provider (DHuS and/or ONDA DIAS[3]) retrieved via the OData/ OpenSearch APIs[4,5], and the National Institute of Geophysics and Volcanology for earthquake data, retrieved via the fdsnws-event API. The relevant product type of the Copernicus data for earthquake analysis is the

[3] https://www.onda-dias.eu/.
[4] https://www.opensearch.org/.
[5] https://www.odata.org/.

IW SLC for Sentinel-1 Level-1 data. The KG by default will contain the data sources presented below, while also additional data sources related to events or Copernicus data, may also be supported.

- For Sentinel-1 Images:
  - Timestamp (i.e. sensing start)
  - Event ID
  - Image URL
  - Location expressed as polygon or multipolygon
  - Orbit number of the satellite
  - Pass direction of the satellite
- For earthquake metadata:
  - Event identifier
  - Coordinates
  - Timestamp
  - Magnitude and depth of the event

3.2 The Knowledge Graph

In order to cover the different aspects of semantic representation, a domain-specific ontology has been designed to capture metadata referring to sentinel images and events. The ontology extends existing ontologies and standards (i.e. Event Ontology, OWL-Time) to offer the appropriate structures (classes and properties) and represent information related to cities (i.e. city names, coordinates and countries). More specifically, to cover the use case of earthquake identification, sentinel images are related to generic information (i.e. UUID, timestamp, URL), while earthquake events are related to information pertinent to the earthquake (i.e. depth, magnitude, coordinates, timestamp). An example of such a semantic representation is depicted in Fig. 2. For each sentinel image, the same ontological schema is applicable for all types of use cases. In case of events, all types of events will contain coordinates and timestamp. Other information might also be included, based on the needs of each use case. The KG is composed of two main classes **Event** and **Product**. The Event class contains information about the location of the earthquake, such as the country, the city, the coordinates, the timestamp, and the date (i.e., year-date-month). The Product class contains information about the earthquake itself such as the magnitude, the depth, pre/post-seismic photos, a multi polygon for the area that the image represents, the source from which the photos were extracted, the orbit number and pass direction of the satellite, the product URL, and the unique ID of the product. The distinction in two classes emerges as an intuitive way to segregate data, pertaining to the earthquake event itself, from information related to the geographical location where the earthquake transpired.

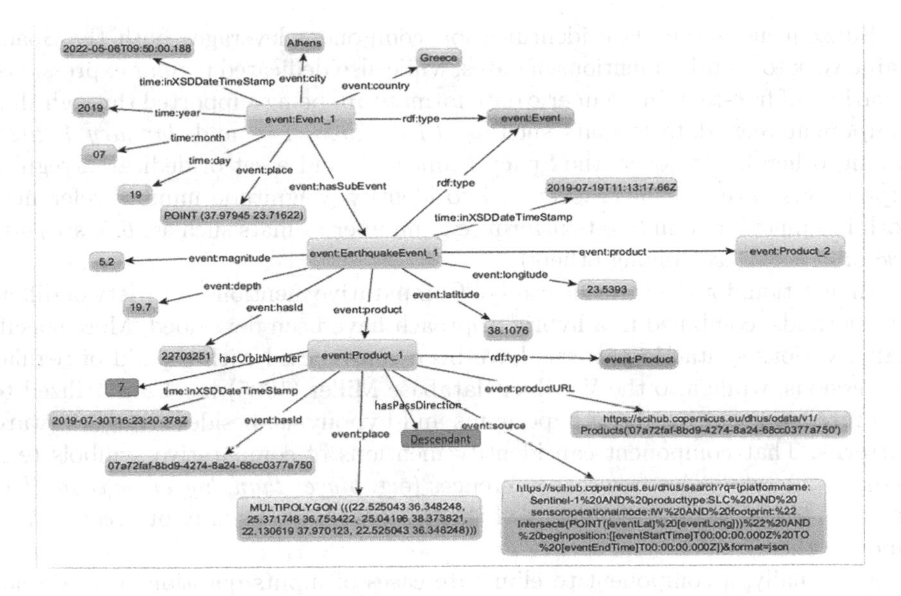

Fig. 2. The Knowledge Graph for Seismic Data

3.3 The Natural Language Mechanism

The NLP mechanism was developed so that the KG would support user inquiries, formed as free-text, and it functions in cooperation with the IR module. More precisely, given an input question in free-text form, the NLP mechanism is responsible for identifying various aspects of information that are required for the IR part. In that sense, the NLP mechanism can identify references of Countries (e.g. Greece), Cities (e.g. Athens), Dates (e.g. 3/5/2023), Magnitude numbers (e.g. 6.5), Comparatives (e.g. bigger than), while it can also verify that a target question is referred on an earthquake event.

Starting with the identification of Countries and Cities, three different approaches have been developed to identify such mentions, so that the biggest possible coverage, even for less-populated and famous cities, may be accomplished. Initially, a parsing of the input sentence is performed with the Spacy framework[6] to identify references of geopolitical entities. Then, a second parsing of the input question is performed with the aid of three different geoparsing libraries (namely geonamescache[7], pycountry[8] and geotext[9]). Eventually, a final parsing with a set of manually constructed regular expressions takes place, based on which country and city mentions may be captured through various syntactic and phrase patterns.

[6] https://spacy.io/.
[7] https://pypi.org/project/geonamescache/.
[8] https://pypi.org/project/pycountry/.
[9] https://pypi.org/project/geotext/.

Subsequently, the Date identification component leverages both the Spacy framework to identify mentions of dates, while also dedicated regular expressions. A variety of free-text and numeric date formats are being supported through that component (e.g. date formats such as, *01-01-2023*, *1/1*, and *January 1 2023*, among others). Moreover, the Spacy framework and a set of dedicated regular expressions have also been leveraged to identify Magnitude number references both in numeric and in free-text form (e.g. number formats such as, *6.5*, *six point five* or *6 point five*, among others).

In addition, for the identification of Comparative mentions, a variety of different methods, combined in a hybrid approach have been developed. More specifically, various syntactic patterns have been constructed with the aid of regular expressions, while also the Wordnet database Miller (1995) has been utilized to also incorporate a variety of hypernyms and hyponyms, besides straightforward patterns. That component can identify mentions of comparative symbols (e.g. $<$ or $=<$), while also free-text references (e.g. *bigger than*, *bigger or equal than* or *equal to* and various synonyms of those, such as *higher than* or *greater than*, among others).

Eventually, a component to eliminate cases of inputs questions that do not refer to earthquake events has also been developed. For that purpose, a list of earthquake-related keywords has been manually constructed, based on which the semantic relatedness of each keyword with the sub parts of the input question is examined, to conclude whether the input question refers on earthquake events.

3.4 The Information Retrieval Mechanism

The IR mechanism initially translates the keywords provided from the NLP mechanism (Subsect. 3.3) from a JSON format into values used by SPARQL queries. The keywords provided are the ones shown below, out of which some are mandatory (M) (i.e., they must exist in order for a SPARQL query to be created), and some optional (O) (i.e., they are not mandatory for a SPARQL query to be formatted).

- **Event:** The type of event (i.e., earthquake (M)).
- **City:** The city of event (M).
- **Country:** The country of event (M).
- **Year:** The year of event (M).
- **Month:** The month of event (O).
- **Day** The day of event (O).
- **Magnitude:** The magnitude of event (O). (If no magnitude is given then the default value is 5.0).
- **Comparative:** If the requested event is greater or smaller than the value of magnitude (O).
- **Point:** Boolean value that indicates if the location is given in the format of coordinates or city-country pair label (M).
- **Latitude-longitude:** The latitude-longitude of the event (O).

For instance, if the user addresses the question *"Give me all the earthquakes that occurred in Montalvo, Ecuador in 2019"*, the resulting JSON, after having been parsed by the IR mechanism, will have the following format.

```
        Information Retrieval Generated JSON
{
  "page": "1",
  "nlp": {
        "event":"earthquake", "city":"Montalvo",
        "country": "Ecuador", "year": 2019,
        "month": "null", "day": "null",
        "magnitude": "null", "comparative": "null",
        }
}
```

Note that the presence of a solitary result signifies that, during that specific year, only a single earthquake event, surpassing a magnitude of 5.0, had occurred.

Based on this JSON a SPARQL query will be automatically generated in order to retrieve the desired information from the KG. One can see the generated SPARQL query below.

```
        Automatically Generated SPARQL Query
prefix event:<http://purl.org/NET/c4dm/event.owl#>
prefix rdf:<http://www.w3.org/1999/02/22-rdf-syntax-ns#>
prefix time:<http://www.w3.org/2006/time#>
prefix xsd:<http://www.w3.org/2001/XMLSchema#>
SELECT * WHERE {
        ?event rdf:type event:Event.
        OPTIONAL { ?event event:city ?city.}
        OPTIONAL {?event event:country ?country.}
        ?event time:year ?year.
        OPTIONAL {?event time:month ?month.}
        OPTIONAL {?event time:day ?day.}
        OPTIONAL {?event event:place ?place.}
        ?event event:hasSubEvent ?e_event.
        OPTIONAL{?e_event event:hasId ?id.}
        OPTIONAL {?e_event event:magnitude ?magnitude.}
        OPTIONAL {?e_event event:depth ?depth.}
        ?e_event event:latitude ?latitude.
        ?e_event event:longitude ?longitude.
        ?e_event time:inXSDDateTimeStamp ?e_timestamp.
        FILTER (?event=<http://purl.org/NET/c4dm
    /event.owl#Event_id>)
}
```

The output returned by the system contains the following information:

- **Magnitude** - The exact magnitude of the earthquake, since the user input might not contain an exact value.
- **Pairs of images** - Pairs of images before and after the event which have the same coordinates and timestamp, with a difference of 12 days (the amount of time needed for the satellite to pass again from the same spot), the orbit number, and the pass direction.
- **Depth** - The depth of the earthquake.
- **Epicenter location** - the exact location of the epicenter of the earthquake.

Returned output


```
{
  "year": "2019",
  "city": "Montalvo",
  "country": "Ecuador",
  "images_before1": {
    "link": "https://colhub.copernicus.eu
    /dhus/search?q=uuid:60f5f27c-9ca5
    -4834-aa0f-0ab9ccb2249d&format=json",
    "sensing_date": "2019-02-13T23:28:27.036Z",
    "location": "POLYGON ((-78.453781 ...
    -1.327742))",
    "orbit_number": "120",
    "pass_direction": "ASCENDING"
  },
  "image_after1": {
    "link": "https://colhub.copernicus.eu
    /dhus/search?q=uuid:db2145a3-2a07-4adf
    -87c0-e02fbfb1c1a6&format=json",
    "sensing_date": "2019-02-25T23:28:27.035Z",
    "location": "POLYGON ((-78.454636 ...
    -1.328015))",
    "orbit_number": "120",
    "pass_direction": "ASCENDING"
  },
  "magnitude": {
    "value": "7.1"
  },
  "depth": {
    "value": "141.4"
  },
  "epicentral_location": {
    "latitude": {
      "value": "-2.13047"
    },
    "longitude": {
      "value": "-76.8867"
    }
  },
  "timestamp": "2019-02-22T10:17:22.50Z"
}
```

Notice that in the output only one pair on images is provided for beautification and space restrictions.

4 Evaluation

To examine the quality of our work, an evaluation of the Seismic KG was conducted on two fronts: completeness and consistency. For completeness assessment, a set of competency questions was formulated to test the KG's ability to answer based on the information it contains (Subsect. 4.1). Additionally, the KG's consistency was evaluated by verifying its adherence to the SHACL constraints (Subsect. 4.2). Moreover, we conducted a comparative evaluation of the framework's performance in contrast to the Large Language Model (LLM), ChatGPT[10]. The aforementioned are presented in more details subsequently.

4.1 Completeness of the Knowledge Graph

The completeness of the KG was evaluated through a set of Competency Questions (CQs) assembled during the formation of the official ontology requirements specification document (ORSD) Suárez-Figueroa et al. (2009). For this reason, before constructing the KG, we asked from a number of users to define a set of questions they would like to be answered by the KG. The users were developers and scientists from ESA[11]. In total a number of 48 CQs was collected. In Fig. 3 we present an indicative sample of 10 CQs.

A series of tests indicated that the completeness of the KG was adequate, since each CQ returned the desired information when translated into a relevant SPARQL counterpart. Indicatively, the translation of the second CQ from Fig. 3 into a SPARQL counterpart, is shown in Subsect. 3.4.

4.2 Consistency of the Knowledge Graph

Additionally to the CQs, we performed a validation procedure in order to inspect the syntactic and structural quality of the metadata in the KB, and to examine the consistency of them. Abiding by the closed-world criteria, custom SHACL consistency check rules and native, ontology consistency checks, such as OWL 2 DL reasoning, were utilised. By using the first, one can discover constraint breaches such as cardinality contradictions or imperfect/missing information. By using the later, the semantics at the terminological level, known as TBox, are taken into consideration as a validation measure like in the occasion of class disjointedness. An exemplary of shapes constraint that unfolds a constraint is shown below, dictating that all targeted instances of the *Event* class will always have exactly one or less string values in their datatype property *event:country*.

[10] https://openai.com/blog/chatgpt.
[11] https://www.esa.int/.

1) Give me all earthquakes located in the country X and Y with magnitude M.

2) Give me all earthquakes located in country X and city Y which occurred in year Z with magnitude M.

3) Give me all earthquakes located in country X and city Y which occurred in year Z and month T with magnitude M.

4) Give me all earthquakes located in country X and city Y which occurred in year Z, month T, and day D with magnitude M.

5) Give me all earthquakes located in the country X with magnitude greater than M.

6) Give me all earthquakes located in country X which occurred in year Z with magnitude greater than M.

7) Give me all earthquakes located in country X which occurred in year Z and month T with magnitude greater than M.

8) Give me all earthquakes located in country X which occurred in year Z, month T, and day D with magnitude greater than M.

9) Give me all earthquakes located in the country X and Y with magnitude greater than M.

10) Give me all earthquakes located in country X and city Y which occurred in year Z with magnitude greater than M.

Fig. 3. Set of Competency Questions

```
event:ev rdf:type sh:NodeShape;
    sh:targetClass event:Event;
    sh:property [
    sh:path event:country;
    sh:datatype xsd:string;
    sh:minInclusive 0;
    sh:maxInclusive 1:
    sh:minCount 1;
    sh:maxCount 1;
    ].
```

4.3 Comparison with Large Language Model

In this subsection we compare how the proposed framework performed against an of-the-self tool for IR on seismic data for disaster management. Since there is no established baseline tool on an extensive event information scale, to compare our framework, we choose to compare its accuracy and quality of information, against the well-known ChatGPT, since the latter stands as a prominent rival in a variety of contemporary contexts.

Concerning the evaluation measures in use, when we refer to *accuracy*, we are evaluating the capacity of the framework and the LLM, to effectively respond to a given question. On the other hand, the *quality of information* pertains to the nature and relevance of the information contained in the answer. In particular, with regard to quality, we assessed whether our framework could furnish information related to the earthquake's epicenter, depth, and location, presented as coordinates or multipolygon data. Additionally, we evaluated whether the framework could supply links to pertinent images associated with a target earthquake in question. Similarly for the LLM.

We tested both tools in a set of 600 questions, out of which our framework could answer 480, and the LLM could answer 200. One can check the accuracy of both tools in Table 1. All the evaluation questions were retrieved from the aforementioned CQ templates.

Table 1. Accuracy of the framework and the LLM.

Seismic Framework	80%
ChatGPT	33.3%

Conclusively, of utmost significance for our evaluation was to discern the nature of information that ChatGPT would present on its responses and compare that with the information our framework can deliver. It appeared that ChatGPT could solely furnish information on the earthquake's depth and magnitude across the limited 200 questions it could address. On the contrary, and as it was already mentioned, our framework can deliver an extensive array of essential details, such as the earthquake's epicenter, depth, precise timestamp, multipolygon representation of the location, and pre/post-images containing vital information about the image's orbit number, pass direction, sensing timestamp, and multipolygon data.

5 Conclusion

In this paper, we have presented a KG specifically designed for the seismic domain which enables the representation of seismic events worldwide. The KG incorporates a NLP mechanism to facilitate the user experience, which allows users to pose queries in free-text for data retrieval. Additionally, the KG incorporates an IR mechanism, capable of retrieving semantically related pairs of pre/post images for a target seismic event, among other relevant information.

The key innovation of this study lies in the high-quality RDF representation of the data, which encompasses information about all ever-recorded seismic events. This breadth of information is made possible by extracting knowledge from the comprehensive datasets, that are provided by the Copernicus Data Provider and the National Institute of Geophysics and Volcanology.

To evaluate the current state of our work, a two fold goal was set. On the one hand, the completeness and the consistency of the KG were evaluated, while on the other hand the quantity and quality of the responses retrieved by the target framework against those of a famous LLM, were compared. The completeness of the KG (Subsect. 4.1) was evaluated with a series of CQs, which were collected by domain experts. More specifically, each CQ was converted into a SPARQL query and examined whether each one of them could return appropriate results. All the evaluated CQs returned appropriate results, indicating in that way that the proposed KG is capable of providing essential information in a disaster management scenario. Moreover, the consistency of the KG (Subsect. 4.2) was evaluated with a set of 21 SHACL validation expressions (shapes), where none of them returned any invalidation of the rule. Eventually, a validation to identify whether instances that belong to an intersection of classes, had also been applied. No such intersection exists in the KG, which indicates that the KG does not carry any noise or any sort of conflicting information.

Regarding the evaluation against other baseline methods (in our case ChatGPT), our initial expectation was indeed for our framework to outperform ChatGPT in terms of accuracy. This anticipation was based on our framework's access to a diverse set of knowledge bases containing domain-specific information on earthquakes and satellite data (e.g. Onda-Dias[12], Colhub[13], ColHub2[14], ColHub3[15], SciHub[16], and ApiHub[17]). However, the extent that our framework outperformed ChatGPT was a quite unexpected, though pleasant outcome. To elaborate further, the only questions that our proposed framework could not answer, were those that contained a combination of city-country pairs which was not valid or did not exist. Additionally, there were some cases where the framework couldn't response due to a misspelled city or country name.

Looking ahead, our future objectives would be to align the framework presented in this paper with other disaster management frameworks that utilize KGs as their underlying knowledge representation. Achieving such an alignment, would make a possible extension of our KG with information obtained from another KG, to enhance the overall knowledge pool (e.g. the XR4DRAMA disaster management KG Vassiliades et al. (2023), or add information in the existing KG from other vocabularies such as the Management of Crisis vocabulary Shih et al. (2013). Furthermore, we aim to extend the representation of information to encompass other catastrophic natural events beyond earthquakes, broadening in that way the scope and the utility of the KG.

Acknowledgments. This work has been supported by the EC-funded projects H2020-101004152 CALLISTO and research carried out under a program funded by the European Space Agency. The view expressed herein can in no way be taken to reflect the official opinion of the European Space Agency.

References

Bouzidi, Z., Amad, M., Boudries, A.: Intelligent and real-time alert model for disaster management based on information retrieval from multiple sources. Int. J. Adv. Media Commun. **7**(4), 309–330 (2019)

[12] https://www.onda-dias.eu/cms/.
[13] https://colhub.copernicus.eu/dhus/.
[14] https://colhub2.copernicus.eu/dhus/.
[15] https://colhub3.copernicus.eu/dhus/.
[16] https://scihub.copernicus.eu/.
[17] https://apihub.copernicus.eu/.

Chou, C.-H., Zahedi, F.M., Zhao, H.: Ontology for developing web sites for natural disaster management: methodology and implementation. IEEE Trans. Syst., Man, Cybernet.-Part A: Syst. Hum. **41**(1), 50–62 (2010)

Chou, C.-H., Zahedi, F.M., Zhao, H.: Ontology-based evaluation of natural disaster management websites. MIS Q. **38**(4), 997–1016 (2014)

Correia, A., Água, P.B., Simões-Marques, M.: The role of ontologies and linked open data in support of disaster management. In: Scholl, H.J., Holdeman, E.E., Boersma, F.K. (eds.) Disaster Management and Information Technology. Public Administration and Information Technology, vol. 40, pp. 393–407. Springer, Cham. Springer (2023). https://doi.org/10.1007/978-3-031-20939-0_18

Islam, S.T., Chik, Z.: Disaster in Bangladesh and management with advanced information system. Disast. Prevent. Manage. Int. J. **20**(5), 521–530 (2011)

Landis, C., Wiseman, C., Smith, A.F., Stephens, M.: GaNCH: using linked open data for Georgia's natural, cultural and historic organizations' disaster response. Code4Lib J. (50) (2021)

Li, L., Li, T.: An empirical study of ontology-based multi-document summarization in disaster management. IEEE Trans. Syst., Man, Cybernet.: Syst. **44**(2), 162–171 (2013)

Miller, G.A.: Wordnet: a lexical database for English. Commun. ACM **38**(11), 39–41 (1995)

Murgante, B., Scardaccione, G., Casas, G.: Building ontologies for disaster management: seismic risk domain. In: Urban and Regional Data Management, pp. 271–280. CRC Press (2009)

Ortmann, J., Limbu, M., Wang, D., Kauppinen, T.: Crowdsourcing linked open data for disaster management. In: Proceedings of the terra Cognita workshop on foundations, technologies and applications of the geospatial web in conjunction with the ISWC, pp. 11–22 (2011)

Pi, Y., Nath, N.D., Behzadan, A.H.: Disaster impact information retrieval using deep learning object detection in crowdsourced drone footage. In: EG-ICE 2020 Workshop on Intelligent Computing in Engineering, Proceedings, pp. 134–143 (2020)

Shen, S., Murzintcev, N., Song, C., Cheng, C.: Information retrieval of a disaster event from cross-platform social media. Inf. Discovery Deliv. **45**, 220–226 (2017)

Shih, F., Seneviratne, O., Liccardi, I., Patton, E., Meier, P., Castillo, C.: Democratizing mobile app development for disaster management. In: Joint Proceedings of the Workshop on AI Problems and Approaches for Intelligent Environments and Workshop on Semantic Cities, pp. 39–42 (2013)

Silva, T., Wuwongse, V., Sharma, H.N.: Disaster mitigation and preparedness using linked open data. J. Ambient. Intell. Humaniz. Comput. **4**, 591–602 (2013)

Suárez-Figueroa, M.C., Gómez-Pérez, A., Villazón-Terrazas, B.: How to write and use the ontology requirements specification document. In: OTM Confederated International Conferences" on the Move to Meaningful Internet Systems", pp. 966–982 (2009)

Vallejo, D., Castro-Schez, J.J., Glez-Morcillo, C., Albusac, J.: Multi-agent architecture for information retrieval and intelligent monitoring by UAVs in known environments affected by catastrophes. Eng. Appl. Artif. Intell. **87**, 103243 (2020)

Vassiliades, A., et al.: Xr4drama knowledge graph: a knowledge graph for disaster management. In: 2023 IEEE 17th International Conference on Semantic Computing (ICSC), pp. 262–265 (2023)

Using Pregel to Create Knowledge Graphs Subsets Described by Non-recursive Shape Expressions

Ángel Iglesias Préstamo(✉) and Jose Emilio Labra Gayo

WESO Lab, University of Oviedo, Oviedo, Spain
angel.iglesias.prestamo@gmail.com

Abstract. Knowledge Graphs have been successfully adopted in recent years, existing general-purpose ones, like Wikidata, as well as domain-specific ones, like UniProt. Their increasing size poses new challenges to their practical usage. As an example, Wikidata has been growing the size of its contents and their data since its inception making it difficult to download and process its data. Although the structure of Wikidata items is flexible, it tends to be heterogeneous: the shape of an entity representing a human is distinct from that of a mountain. Recently, Wikidata adopted Entity Schemas to facilitate the definition of different schemas using Shape Expressions, a language that can be used to describe and validate RDF data. In this paper, we present an approach to obtain subsets of knowledge graphs based on Shape Expressions that use an implementation of the Pregel algorithm implemented in Rust. We have applied our approach to obtain subsets of Wikidata and UniProt and present some of these experiments' results.

Keywords: Knowledge Graphs · Graph algorithms · RDF · Linked Data · RDF Validation · Shape Expressions · Subsets · Pregel

1 Introduction

Knowledge graphs have emerged as powerful tools for representing and organizing vast amounts of information in a structured manner. As their applications continue to expand across various domains, the need for an efficient and scalable processing of these graphs becomes increasingly critical.

Creating subsets of knowledge graphs is a common approach for tackling the challenges posed by their size and complexity. Such subsets are essential not only to reduce computational overhead but also to focus on specific aspects of the data.

In this paper, we explore the synergy between two essential concepts in the field of graph processing: Shape Expressions (ShEx) [12] and the Pregel model [10]. Shape Expressions allow to describe and validate knowledge graphs based on the Resource Description Framework (RDF). These expressions have gained significant adoption in prominent projects like Wikidata. On the other

F. Ortiz-Rodriguez et al. (Eds.): KGSWC 2023, LNCS 14382, pp. 120–134, 2023.
https://doi.org/10.1007/978-3-031-47745-4_10

hand, Pregel is a distributed graph processing model designed for efficiently handling large-scale graphs across multiple machines.

Motivated by the need for handling massive graphs in a scalable manner, we propose the concept of creating subsets of knowledge graphs using Shape Expressions. By selecting relevant portions of the graph, we can focus computational efforts on specific areas of interest, leading to enhanced efficiency and reduced processing times.

Furthermore, we delve into the capabilities of the Pregel algorithm and its potential for distributed graph processing. We emphasize that the scalability of graph computation can be achieved not only by increasing the number of machines but also by optimizing the use of multi-threading solutions to leverage a single machine's capabilities. Hence, our solution aims to distribute the problem across multiple threads of a single-node machine. This is, a multi-threaded Pregel. The idea is not only to provide a solution that can run on any hardware efficiently but also to explore the capabilities of Rust for enabling some performance gains regarding single-node computation. The main contributions of this paper are the following:

1. We present an approach for subset generation of Knowledge Graphs based on Shape Expressions using the Pregel Framework.
2. We have implemented it in Rust.
3. We have applied it to generate subsets of Wikidata and UniProt and presented some optimizations and results.

Section 2 establishes the alternatives and work related to what is presented in the document. Section 3 presents the key concepts required for describing the foundations of the problem to be solved. Section 4 explains the most important algorithms for creating Knowledge Graph subsets. Section 5, the novel approach introduced by this paper is described. Section 6 depicts the experiment for analyzing how the Pregel-based Schema validating algorithm behaves. Section 7 contains the conclusions and future work.

2 Related Work

2.1 Knowledge Graph Descriptions

Several Knowledge Graph descriptions have been proposed, with many outlined in [4,6]. Notably, this paper focuses on Property, RDF, and Wikibase graphs.

Shape Expressions, which are used to create the subsets, were first introduced in 2014. While SHACL (Shapes Constraint Language) is the W3C recommendation[1], the Wikidata community has been using Shape Expressions [14] since 2017. The preference for Shape Expressions arises from their superior adaptability in describing data models when compared to SHACL. A comparison between both can be found in the book [9].

[1] https://www.w3.org/TR/2017/REC-shacl-20170720/.

2.2 Knowledge Graph Subsets

Although it is possible to create subsets of the RDF Knowledge Graph through SPARQL construct queries, there are limitations to this approach. Notably, the lack of support for recursion. While proposals to extend SPARQL with recursion have been made [13], such extensions are not widely supported by existing processors. In light of these limitations, a new method using Shape Expressions for creating Knowledge Graph subsets is described in [4]. `PSchema` follows a similar approach to that presented in [16]. However, `SP-Tree` uses a SPARQL to query the Knowledge Graph, while `PSchema` uses Shape Expressions and Rust. As such, the `PSchema` algorithm is more flexible, leaving room for optimizations.

The creation of Knowledge graph subsets has gained attention, starting from the 12th International SWAT4HCLS Conference[2]. It has since been selected as a topic of interest in the Elixir Europe Biohackathon 2020[3] and the SWAT4HCLS 2021 Hackathon, which resulted in a preprint collecting various approaches [7].

It has been discussed that the Wikidata Knowledge Graph is not feasible to be processed in a single domestic computer using the existing techniques. To address this issue, a novel method to split the Wikidata graph into smaller subsets using Shape Expressions was introduced in [4].

A comparison between several approaches and tools for creating Wikidata Knowledge graph subsets has been discussed [2], where they evaluated the performance of different approaches and tools. Their methodology for measuring performance and conducting experiments served as the primary inspiration for designing the experiments in Sect. 6.

3 Background

3.1 Knowledge Graphs

Definition 1 (Knowledge Graph [4,6]**).** *A Knowledge Graph is a graph-structured data model that captures knowledge in a specific domain, having nodes that represent entities and edges modeling relationships between those.*

Definition 2 is a general and open description of a Knowledge Graph. There are several data models for representing Knowledge Graphs, including Directed edge-labeled and Property Graphs [6], to name a few. In this paper, we will focus on RDF-based Knowledge Graphs, a standardized data model based on Directed edge-labeled graphs [6].

RDF-Based Knowledge Graphs. The Resource Description Framework (RDF) is a standard model for data interchange on the Web. It is a W3C Recommendation for representing information based on a directed edge-labeled graph,

---

[2] https://www.wikidata.org/wiki/Wikidata:WikiProject_Schemas/Subsetting.

[3] https://github.com/elixir-europe/BioHackathon-projects-2020/tree/master/projects/35.

where labels are the resource identifiers. The idea behind the RDF model is to make statements about things in the form of subject-predicate-object triples. The subject denotes the resource itself, while the predicate expresses traits or aspects of it and expresses a relationship between the subject and the object, another resource. This linking system forms a graph data structure, which is the core of the RDF model. If the dataset represents Knowledge of a specific domain, the Graph will be an RDF-based Knowledge Graph. There are several serialization formats for RDF-based Knowledge Graphs, including Turtle, N-Triples, and JSON-LD. Its formal definition is as follows:

Definition 2 (RDF-based Knowledge Graph [4]**).** *Given a set of IRIs* $\mathcal{I}$*, a set of blank nodes* $\mathcal{B}$*, and a set of literals* l*. An RDF-based Knowledge Graph is defined as a triple-based graph* $\mathcal{G} = \langle \mathcal{S}, \mathcal{P}, \mathcal{O}, \rho \rangle$ *where* $\mathcal{S} \subseteq \mathcal{I} \cup \mathcal{B}$*,* $\mathcal{P} \subseteq \mathcal{I}$*,* $\mathcal{O} \subseteq \mathcal{I} \cup \mathcal{B} \cup l$*, and* $\rho \subseteq \mathcal{S} \times \mathcal{P} \times \mathcal{O}$*.*

Example 1 (RDF-based Knowledge Graph of Alan Turing).[4] Alan Turing (23 June 1912 – 7 June 1954) was employed by the government of the United Kingdom in the course of WWII. During that time he developed the computer for deciphering Enigma-machine-encrypted secret messages, namely, the Bombe machine. Additional information about relevant places where he lived is also annotated, including his birthplace, and the place where he died.

$$
\begin{aligned}
\mathcal{I} = \{ & \ alanTuring,\ wilmslow,\ GCHQ,\ unitedKingdom,\ warringtonLodge,\ bombe \\
& \ town,\ computer,\ dateOfBirth,\ placeOfBirth,\ employer,\ placeOfDeath, \\
& \ country,\ manufacturer,\ instanceOf \ \} \\
\mathcal{B} = \{ & \ \emptyset \ \} \\
l = \{ & \ 23 \text{ June } 1912 \ \} \\
\rho = \{ & \ (alanTuring,\ instanceOf,\ Human), \\
& \ (alanTuring,\ dateOfBirth,\ 23 \text{ June } 1912), \\
& \ (alanTuring,\ placeOfBirth,\ warringtonLodge), \\
& \ (alanTuring,\ placeOfDeath,\ wilmslow), \\
& \ (alanTuring,\ employer,\ GCHQ), \\
& \ (bombe,\ discoverer,\ alanTuring), \\
& \ (bombe,\ manufacturer,\ GCHQ), \\
& \ (bombe,\ instanceOf,\ computer), \\
& \ (wilmslow,\ country,\ unitedKingdom) \\
& \ (wilmslow,\ instanceOf,\ town) \\
& \ (warringtonLodge,\ country,\ unitedKingdom) \ \}
\end{aligned}
$$

URIs in Wikidata follow a linked-data pattern called *opaque URIs* representing them as unique sequences of characters that are language-independent. As an example, Alan Turing's identifier is serialized as `Q7251`. Furthermore, within Wikidata, there is a designated property known as `instanceOf` that serves to describe the type of entity it is associated with, which resembles the `rdf:type` constraint. This can be employed to perform an early evaluation of the nodes.

[4] https://rdfshape.weso.es/link/16902825958.

3.2 ShEx

Shape Expressions (ShEx) were designed as a high-level, domain-specific language for describing RDF graph structures. The syntax of ShEx is inspired by Turtle and SPARQL, while the semantics are motivated by RelaxNG and XML Schema. In this manner, Shape Expressions specify the requirements that RDF data graphs must fulfill to be considered conformant, they allow systems to establish contracts for sharing information; through a common schema, systems agree that a certain resource should be part of the graph. This pattern behaves similarly to interfaces in the object-oriented paradigm. Shapes can be specified using a JSON-LD syntax or a human-friendly concise one called ShExC.

Example 2. The schema below describes the `Person` Shape Expression, which is used to validate the RDF-based Knowledge Graph of Alan Turing (see Example 1). Recall, `Person` is just the label of the resource and does not relate to the Human entity. The ShExC-serialized schema can be found in RDFShape.

$$
\begin{aligned}
\mathcal{L} = \{ & \, Person, Place, Country, Organization, Date \, \} \\
\delta(Person) = \{ & \, \_ \xrightarrow{placeOfBirth} \text{@Place}, \\
& \, \_ \xrightarrow{dateOfBirth} \text{@Date}, \\
& \, \_ \xrightarrow{employer} \text{@Organization} \, \} \\
\delta(Place) = \{ & \, \_ \xrightarrow{country} \text{@Country} \, \} \\
\delta(Country) = \{ & \, \} \\
\delta(Organization) = \{ & \, \} \\
\delta(Date) \in & \;\; \texttt{xsd:date}
\end{aligned}
$$

3.3 Pregel

Pregel (*Parallel, Graph, and Google*) is a data flow paradigm created by Google to handle large-scale graphs. Even if the original instance remains proprietary at Google, it was adopted by many graph-processing systems, including Apache Spark. For a better understanding of Pregel, the idea is to *think like a vertex* [11]; this is, the state of a given node will only depend on that of its neighbors, the nodes linked to it by an outgoing edge (see Definition 4). Hence, by *thinking like a vertex*, the problem is divided into smaller ones. Instead of dealing with huge graphs, smaller ones are processed: a vertex and its neighbors.

The series of steps that Pregel follows to process a graph are depicted in Fig. 1. The execution starts by sending the initial messages to the vertices at iteration 0. Then, the first – actual – *superstep* begins. In our current implementation, this loop will last until the current iteration is greater than the threshold set at the creation of the Pregel instance. At each iteration, the vertices will send messages to their neighbors, provided the given direction, and subsequently, they may receive messages sent from other nodes. Moving forward, an aggregation function is applied, and the vertices are updated accordingly. Finally, the iteration counter is incremented and the next iteration starts.

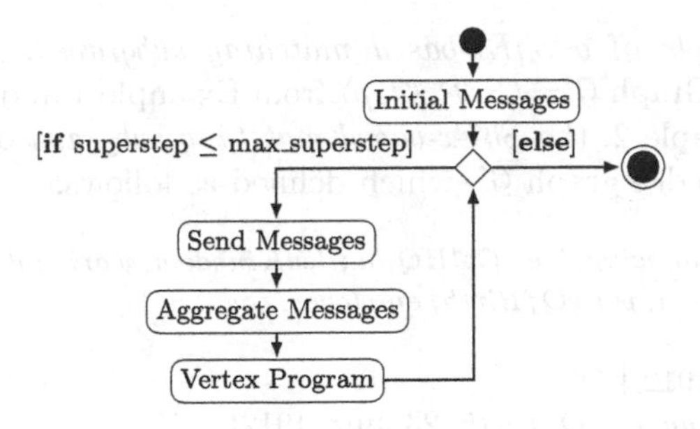

Fig. 1. Pregel model as implemented in pregel-rs

4 Subsetting Approaches

4.1 Knowledge Graph Subsets, a Formal Definition

Definition 3 (RDF-based Knowledge Graph subset [4]**).** *Given a Knowledge Graph $\mathcal{G} = \langle \mathcal{S}, \mathcal{P}, \mathcal{O}, \rho \rangle$, as defined in Definition 2, a RDF sub-graph is defined as $\mathcal{G}' = \langle \mathcal{S}', \mathcal{P}', \mathcal{O}', \rho' \rangle$ such that: $\mathcal{S}' \subseteq \mathcal{S}$, $\mathcal{P}' \subseteq \mathcal{P}$, $\mathcal{O}' \subseteq \mathcal{O}$ and $\rho' \subseteq \rho$.*

Example 3 (Example of an RDF-based Knowledge Graph subset). Given the RDF-based Knowledge Graph $\mathcal{G} = \langle \mathcal{S}, \mathcal{P}, \mathcal{O}, \rho \rangle$ from Example 1, the subset $\mathcal{G}'$ that only contains information about Alan's birthplace is as follows:

$$\begin{aligned}
\mathcal{I}' &= \{\ alanTuring,\ warringtonLodge,\ dateOfBirth,\ placeOfBirth\ \} \\
\mathcal{B}' &= \{\ \emptyset\ \} \\
\mathcal{L}' &= \{\ \text{23 June 1912}\ \} \\
\rho' &= \{\ (alanTuring,\ dateOfBirth,\ \text{23 June 1912}), \\
&\qquad (alanTuring,\ placeOfBirth,\ warringtonLodge)\ \}
\end{aligned}$$

4.2 ShEx-Based Matching Generated Subsets

ShEx-based matching comprises using a ShEx schema *se* as input, including any nodes whose neighborhood matches any of the shapes from *se* in the produced subset [4]. This approach is used by the `PSchema` algorithm.

Definition 4 (Neighborhood of a node in a Knowledge graph). *The neighbors of an item $s \in \mathcal{S}$ in a RDF-based Knowledge graph $\mathcal{G} = \langle \mathcal{S}, \mathcal{P}, \mathcal{O}, \rho \rangle$ are defined as $neighbors(s) = \{(s, p, o) : \exists v \in \mathcal{S}$ such that $\rho(v) = (p, o)\}$.*

Example 4 (Neighborhood of Alan Turing (Q7251)*).* Given the RDF-based Knowledge graph $\mathcal{G} = \langle \mathcal{S}, \mathcal{P}, \mathcal{O}, \rho \rangle$ from Example 1, the neighborhood of Alan Turing (Q7251) $\in \mathcal{S}$ is defined as follows:

$$\begin{aligned}
neighbors(alanTuring) = \{\ &(alanTuring,\ instanceOf,\ Human), \\
&(alanTuring,\ dateOfBirth,\ \text{23 June 1912}), \\
&(alanTuring,\ placeOfBirth,\ warringtonLodge), \\
&(alanTuring,\ placeOfDeath,\ wilmslow), \\
&(alanTuring,\ employer,\ GCHQ)\ \}
\end{aligned}$$

Example 5 (Example of a ShEx-based matching subgraph). Given the RDF-based Knowledge Graph $\mathcal{G} = \langle \mathcal{S}, \mathcal{P}, \mathcal{O}, \rho \rangle$ from Example 1 and the ShEx schema *se* defined in Example 2, the *ShEx-based matching subgraph* of $\mathcal{G}$ from *se* is the RDF-based Knowledge graph $\mathcal{G}'$, which defined as follows:

$$
\begin{aligned}
\mathcal{I}' &= \{\ alanTuring,\ wilmslow,\ GCHQ,\ unitedKingdom,\ warringtonLodge,\\
&\qquad dateOfBirth,\ placeOfBirth,\ employer,\ country\ \}\\
\mathcal{B}' &= \{\ \emptyset\ \}\\
\mathcal{l}' &= \{\ \text{23 June 1912}\ \}\\
\rho &= \{\ (alanTuring,\ dateOfBirth,\ \text{23 June 1912}),\\
&\qquad (alanTuring,\ placeOfBirth,\ warringtonLodge),\\
&\qquad (alanTuring,\ employer,\ GCHQ),\\
&\qquad (wilmslow,\ country,\ unitedKingdom)\\
&\qquad (warringtonLodge,\ country,\ unitedKingdom)\ \}
\end{aligned}
$$

5 Pregel-Based Schema Validating Algorithm

In this section, both the support data structure and the subsetting algorithm are described, including the different steps followed in the Pregel implementation.

5.1 Shape Expression Tree

The Shape Expression tree is a hierarchical data structure representing Shapes in a tree format. Each node in the tree corresponds to a Shape Expression, with the root node being the one subject of study. Nodes can reference other Shape Expressions, which become its children in the tree.

Definition 5 (Shape Expression tree). *Given a Shape Expression se, the Shape Expression tree $\mathcal{T}$ is defined as follows:*

- *If se does not reference any other* `Shape`*, then $\mathcal{T}$ is a leaf node.*
- *If se references other* `Shapes`*, then $\mathcal{T}$ is an internal node, and its children are the* `Shapes` *referenced by se. Which will be the root nodes of their respective Shape Expression trees.*

Example 6. Given the Shape Expression *se* defined in Example 2, the Shape Expression tree $\mathcal{T}$ obtained from *se* was created using the RDFShape and is depicted in Fig. 2. `Person` is the root node of $\mathcal{T}$, a non-terminal `Shape` that references `Organization`, `Date`, and `Place`. Thus, the children of the root are the `Shapes` referenced by `Person`, which are the root nodes of their respective Shape Expression trees. In the case of the first child, `Organization` is a terminal `Shape`, and thus, it is a leaf node. The same applies to `Date`. However, `Place` is a non-terminal `Shape`, and thus, it is an internal node. Its children are the `Shapes` referenced by it. This representation is recursive, and thus, the `Shapes` referenced by `Place` are the root nodes of their respective ShEx trees.

The currently supported ShEx language does not support recursion; however, it is planned to implement a solution based on the idea of *unfolding* the recursive schema.

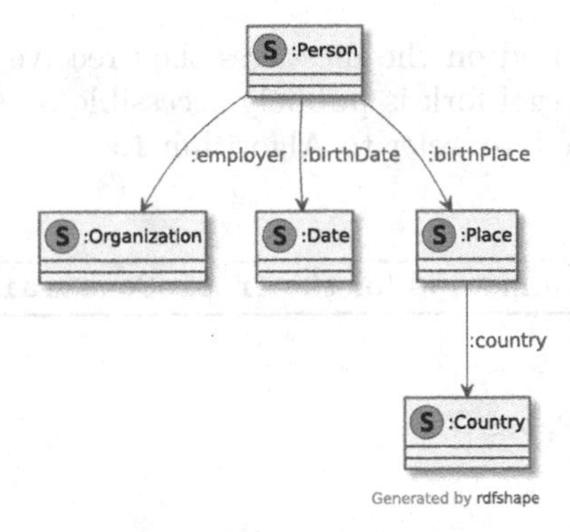

Fig. 2. Example of a Shape Expression tree for the `Person` Shape Expression

5.2 Subsetting Algorithm Using Pregel and ShEx

Algorithm 1: The PSchema algorithm as implemented in Rust

Input parameters:

g: Graph$[\mathcal{V}, \mathcal{E}]$
l: $\mathcal{L}$

Output:

sub: Graph$[\mathcal{V}, \mathcal{E}]$

`maxIters` = see Lemma 1
`initialMsgs` = None
return *Pregel(g,maxIters,initialMsgs,sendMsg,aggMsgs,vProg)*

def `sendMsg`(l: $\mathcal{L}$, g: Graph$[\mathcal{V}, \mathcal{E}]$) = $msgs$ where **foreach** $l \in \mathcal{L}$

$$msgs.insert\left(\begin{cases} \text{validate}(l,\, g) \text{ if } l = \textit{TripleConstraint} & \textit{see Algorithm } 2 \\ \text{validate}(l,\, g) \text{ if } l = \textit{ShapeReference} & \textit{see Algorithm } 3 \\ \text{validate}(l,\, g) \text{ if } l = \textit{ShapeAnd} & \textit{see Algorithm } 4 \\ \text{validate}(l,\, g) \text{ if } l = \textit{ShapeOr} & \textit{see Algorithm } 5 \\ \text{validate}(l,\, g) \text{ if } l = \textit{Cardinality} & \textit{see Algorithm } 6 \\ \text{None} \quad \text{otherwise} & \end{cases}\right)$$

def `aggMsgs`($msgs$: $\mathcal{M}$) = $msgs$ where

$$msgs.insert\left(\begin{cases} msg & \text{if } msg \neq \text{None} \\ \emptyset & \text{otherwise} \end{cases}\right)$$

def `vProg`(l: $\mathcal{L}$, g: Graph$[\mathcal{V}, \mathcal{E}]$, $msgs$: $\mathcal{M}$) = $labels$.concatenate($msgs$)

`PSchema` is a Pregel-based algorithm that creates subsets of RDF-based Knowledge Graphs using Shape Expressions. The algorithm's core idea is to validate the nodes of the Shape Expression tree $\mathcal{T}$ in a bottom-up manner, proceeding from the leaves to the root. The validation is performed in a *reverse level-order traversal* of the tree. The algorithm comprises two main phases: initialization and validation. During the initialization phase, the initial messages are generated and sent to the vertices, while also setting up the *superstep* counter and threshold. This phase establishes the baseline for subsequent steps. In the validation phase, referred to as the *local computation*, the `Shapes` of the tree $\mathcal{T}$ are validated. The vertices are updated based on the messages they receive from their neighbors. The aforementioned Pregel fork is publicly accessible on Github[5]. For a formal description of the procedure, refer to Algorithm 1.

Algorithm 2: Validate method for the `TripleConstraint` Shape

Input parameters:
- l: $\mathcal{L}$
- (_, p, o): $(\_, p \in \mathcal{P}, o \in \mathcal{O})$

Output:
- msg: $\mathcal{M}$

match $l.object$
- **case** $Value(v)$
  - **if** $p == l.predicate \land o == v$ **then**
    - **return** l
- **case** Any
  - **if** $p == l.predicate$ **then**
    - **return** l

A formal description of the *validating* algorithms for each of the currently implemented Shapes is provided. Note that the input parameters are simplified, as in the actual implementation of the algorithm it can access the whole Graph. Having said that, the first method that is described is the validating algorithm for the `TripleConstraint` Shape, as seen in Algorithm 2. This Shape corresponds to the most basic representation of a Node Constraint. In that manner, it is verified if a node satisfies the predicate and object constraints. In other words, if it exists an outgoing edge from a certain node to another determined by the Shape that is being validated currently. Note that the Object may be an actual value or any.

Algorithm 3: Validate method for the `ShapeReference` Shape

Input parameters:
- l: $\mathcal{L}$
- (_, p, o): $(\_, p \in \mathcal{P}, o \in \mathcal{O})$

Output:
- msg: $\mathcal{M}$

if $p == l.predicate \land o == l.reference.object$ **then**
- **return** l

[5] https://github.com/weso/pregel-rs.

The `ShapeReference` is in charge of evaluating the cases in which the object of a Triple Constraint is an IRI; that is, a reference to another Shape. Even if the algorithm behaves similarly to the description before, the implementation details vary as the value referenced has to be retrieved. Refer to Algorithm 3.

Algorithm 4: Validate method for the `ShapeAnd` Shape

Input parameters:
- l: $\mathcal{L}$
- $(\_, \mathrm{p}, \mathrm{o})$: $(\_, p \in \mathcal{P}, o \in \mathcal{O})$

Output:
- msg: $\mathcal{M}$

$ans \leftarrow$ `true`
forall $l \in l.shapes$ **do**
 $ans \leftarrow ans \wedge$ `validate`(l, g)
if ans **then**
 return l

The `ShapeAnd` constraint checks whether *all* the Shapes wrapped by it are valid. Having all the children already evaluated, it is going to be checked if all of them hold for every node in the graph. This is, the `ShapeAnd` acts as a *logical and* for a grouping of Shapes. Refer to Algorithm 4.

Algorithm 5: Validate method for the `ShapeOr` Shape

Input parameters:
- l: $\mathcal{L}$
- $(\_, \mathrm{p}, \mathrm{o})$: $(\_, p \in \mathcal{P}, o \in \mathcal{O})$

Output:
- msg: $\mathcal{M}$

$ans \leftarrow$ `false`
forall $l \in l.shapes$ **do**
 $ans \leftarrow ans \vee$ `validate`(l, g)
if ans **then**
 return l

The `ShapeOr` constraint checks whether *any* Shape wrapped is valid. Having all the children already evaluated, it is going to be checked if any of them holds for every node in the graph. This is, the `ShapeOr` acts as a *logical or* for a grouping of Shapes. Refer to Algorithm 5.

Algorithm 6: Validate method for the `Cardinality` Shape

```
Input parameters:
    l: 𝓛
    (_, p, o): (_, p ∈ 𝓟, o ∈ 𝓞)
    prevMsg: 𝓜
Output:
    msg: 𝓜

count ← prevMsg.count()
match l.min
    case Inclusive(min)
        if count ≤ min then
            min ← true
    case Exclusive(min)
        if count < min then
            min ← true

match l.max
    case Inclusive(max)
        if count ≥ max then
            max ← true
    case Exclusive(max)
        if count > max then
            max ← true

if min ∧ max then
    return l
```

The `Cardinality` constraint is in charge of checking whether the Shape referenced is valid a certain number of times in the neighborhood of every node in the graph. That is, the number of times the neighboring nodes are valid for a certain Shape. The concept of inclusivity or exclusivity allows for the range to be closed or open, respectively. Refer to Algorithm 6.

Lemma 1 (Convergence of the PSchema algorithm). *Given a* **`Shape Expression`** *tree $\mathcal{T}$ and a Knowledge Graph $\mathcal{G}$, let h denote the height of $\mathcal{T}$; then the* **`PSchema`** *algorithm is going to converge in h supersteps. This is, the algorithm is going to validate all the* **`Shapes`** *of $\mathcal{T}$ in h supersteps.*

Example 7 (Example of the subset generated by the **`PSchema`** *algorithm).* Given the RDF-based Knowledge Graph $\mathcal{G} = \langle \mathcal{S}, \mathcal{P}, \mathcal{O}, \rho \rangle$ from Example 1 and the ShEx schema *se* defined in Example 2, the *ShEx-based matching subgraph* of $\mathcal{G}$ from *se* is the RDF-based Knowledge graph $\mathcal{G}'$ from Example 5, which is represented in `Turtle` syntax as follows, refer to RDFShape for more information:

```
1  PREFIX :    <http://example.org/>
2  PREFIX xsd: <http://www.w3.org/2001/XMLSchema#>
3
4  :alan       :placeOfBirth   :warrington            ;
5              :dateOfBirth    "1912-06-23"^^xsd:date ;
6              :employer       :GCHQ                  .
7
8  :warrington :country        :uk                    .
9
10 :wilmslow   :country        :uk                    .
```

5.3 Optimizations

Columnar storage [1] is a data warehousing technique used in databases and data processing systems. Unlike traditional row-based storage, where data is stored in rows, columnar storage organizes data in columns, grouping values of the same attribute. Columnar storage enables better data compression, as similar data types are stored together, reducing the storage footprint. This leads to faster data retrieval and reduced I/O operations when querying specific columns. It is advantageous for analytical workloads that involve aggregations or filtering on specific attributes, as they only need to read the relevant columns. It also enhances query performance by leveraging vectorized processing, as modern CPUs can perform operations on entire sets of data (vectors) more efficiently than on individual elements, further improving query speeds.

Caching. Dictionary encoding [15] is a data compression technique where unique values in a column are assigned numerical identifiers (dictionary indices) and stored in a separate data structure. The actual data in the column is replaced with these compact numerical representations. This technique significantly reduces the storage footprint, especially when columns contain repetitive or categorical data with limited distinct values, as predicates in real-life scenarios.

In data processing scenarios with repeated patterns and aggregations, caching with columnar storage and dictionary encoding can lead to performance improvements. The reduced data size allows more data to be cached within the same memory capacity, maximizing cache utilization. Additionally, the focused access ensures that only the necessary data is retrieved, further enhancing cache efficiency. The cache can hold a large amount of relevant data, minimizing the need for costly disk accesses and accelerating query response times, ultimately resulting in a more efficient and responsive data processing system.

6 Experiments and Results

Two different Knowledge Graphs are going to be used to test the algorithm, namely, Uniprot[6] and Wikidata. The former is a database that contains information about proteins [3], while the latter is a general-purpose knowledge base

[6] https://ftp.uniprot.org/pub/databases/uniprot/current_release/rdf/.

having a dump created the 21st August 2017. As the serialization format of *Uniprot* is `RDF/XML`, the `riot` utility from *Apache Jena* is used to convert from `RDF/XML` to `N-Triples`. Refer to the examples in the GitHub repository[7].

As it is seen, the results obtained are similar regarding the size of the subsets. That is, the optimizations have no impact on the *validity* of the tool; this is, the subsets are correct for the `Shape` defined. For this, two Shapes were created during the Japan BioHackathon 2023 [8], namely, `protein` and `subcellular_location`. When comparing the optimized version against its counterpart, the time consumption is reduced by 38% and 35%; while the memory consumption is decreased by 43% and 38%, respectively. Hence, the optimizations are effective both time and memory-wise. The next experiment will focus on how the algorithm behaves when its parameters are modified (Fig. 3).

Shape Expression	Initial triples	Resulting triples	Time (s)	Memory (GB)
`protein`	7,346,129	226,241	23.35	6.74
`subcellular_location`	7,346,129	1,084,151	57.56	6.04

(a) Execution of the **PSchema** algorithm with no optimization enabled

Shape Expression	Initial triples	Resulting triples	Time (s)	Memory (GB)
`protein`	7,346,129	226,241	14.58	3.87
`subcellular_location`	7,346,129	1,084,151	37.76	3.75

(b) Execution of the **PSchema** algorithm with all the optimizations enabled

Fig. 3. Time and memory consumption to create `Uniprot`'s *subsets*

In the second experiment, the number of `Wikidata` entities, the depth, and the width of the `ShEx` tree were modified. The results are depicted in Fig. 4. It was observed that the execution time followed a linear trend in all scenarios. This indicates that as the number of `Wikidata` entities increased, the execution time increased at a consistent rate. Additionally, the depth and width of the `ShEx` tree influenced the execution time similarly, displaying a linear correlation.

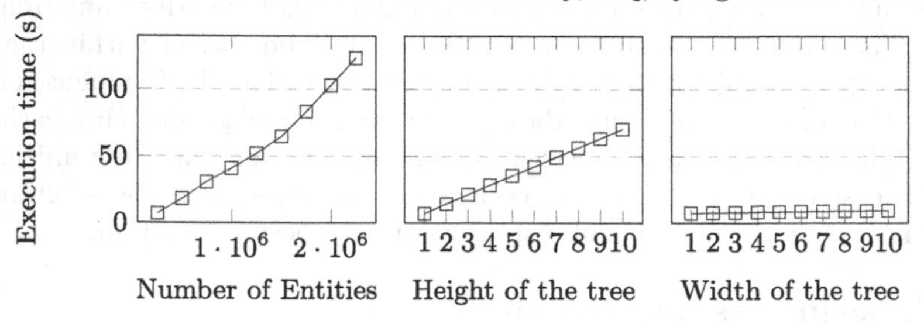

Fig. 4. Time to create the *subsets* of Wikidata with `pschema-rs`

[7] https://github.com/angelip2303/pschema-rs/tree/main/examples/from_uniprot.

7 Conclusions and Future Work

A novel approach for creating subsets of RDF-based Knowledge Graphs using Shape Expressions and the Pregel framework is presented. The `PSchema` algorithm is described, including the different steps followed in the Pregel implementation. Moreover, the support data structure and the optimizations applied are also described. Two Shapes were created during the Japan BioHackathon 2023 [8] for testing the tool and its validity regarding the optimizations applied. The subsets resulting subsets have the same size in both scenarios. When comparing the optimized version against its counterpart, the time consumption is reduced by 38% and 35%; while the memory consumption is decreased by 43% and 38%, respectively. Hence, the optimizations are effective. The next experiment focused on how the algorithm behaves when its parameters are modified. It was observed that the execution time followed a linear trend in all cases.

`PSchema` could be extended to support more complex ShEx features like recursive Shapes, and an early-prune strategy to reduce the cost of the local computation. The algorithm should receive the `ShEx` schema as an input, rather than programmatically creating desired Shape instance. It is planned to give support for WShEx [5], a ShEx-inspired language for describing Wikidata entities, where qualifiers about statements and references can be used for validating purposes. This would allow the algorithm to be used in a wider range of scenarios.

To conclude, `PSchema`, being a Pregel-based Knowledge Graph validating algorithm, allows the processing of large-scale Knowledge Graphs. This is especially relevant in the *Bioinformatics*, where the integration of data from multiple sources is needed. What's more, inference algorithms can be applied to the subsets generated, which is not possible in larger Graphs due to their sizes.

Acknowledgements. This project has received funding from NumFOCUS, a nonprofit organization promoting open-source scientific projects, and has been supported by the ANGLIRU project, funded by the Spanish Agency for Research. The opinions and arguments employed herein do not reflect the official views of these organizations.

References

1. Abadi, D.J., Madden, S.R., Hachem, N.: Column-stores vs. row-stores: how different are they really? In: Proceedings of the 2008 ACM SIGMOD International Conference on Management of Data, SIGMOD 2008, pp. 967–980. Association for Computing Machinery, New York (2008). https://doi.org/10.1145/1376616.1376712
2. Beghaeiraveri, S.A.H., et al.: Wikidata subsetting: approaches, tools, and evaluation (2023). https://www.semantic-web-journal.net/system/files/swj3491.pdf
3. The UniProt Consortium: UniProt: the Universal Protein Knowledgebase in 2023. Nucleic Acids Res. **51**(D1), D523–D531 (2022). https://doi.org/10.1093/nar/gkac1052
4. Gayo, J.E.L.: Creating knowledge graphs subsets using shape expressions (2021). https://doi.org/10.z8550/ARXIV.2110.11709. https://arxiv.org/abs/2110.11709
5. Gayo, J.E.L.: Wshex: a language to describe and validate wikibase entities (2022). https://arxiv.org/abs/2208.02697

6. Hogan, A., et al.: Knowledge graphs. CoRR abs/2003.02320 (2020). https://arxiv.org/abs/2003.02320
7. Labra-Gayo, J.E., et al.: Knowledge graphs and wikidata subsetting (2021). https://doi.org/10.37044/osf.io/wu9et. http://biohackrxiv.org/wu9et
8. Labra-Gayo, J.E., et al.: RDF Data integration using Shape Expressions (2023). https://biohackrxiv.org/md73k
9. Labra Gayo, J.E., Prud'hommeaux, E., Boneva, I., Kontokostas, D.: Validating RDF Data. No. 1 in Synthesis Lectures on the Semantic Web: Theory and Technology, Morgan & Claypool Publishers LLC (2017). https://doi.org/10.2200/s00786ed1v01y201707wbe016
10. Malewicz, G., et al.: Pregel: a system for large-scale graph processing. In: Proceedings of the 2010 International Conference on Management of Data, New York, NY, USA, pp. 135–146 (2010). https://doi.org/10.1145/1807167.1807184
11. McCune, R.R., Weninger, T., Madey, G.: Thinking like a vertex: a survey of vertex-centric frameworks for large-scale distributed graph processing. ACM Comput. Surv. **48**(2) (2015). https://doi.org/10.1145/2818185
12. Prud'hommeaux, E., Labra Gayo, J.E., Solbrig, H.: Shape expressions: an RDF validation and transformation language. In: Proceedings of the 10th International Conference on Semantic Systems, SEMANTICS 2014, pp. 32–40. ACM (2014)
13. Reutter, J.L., Soto, A., Vrgoč, D.: Recursion in SPARQL. In: Arenas, M., et al. (eds.) ISWC 2015. LNCS, vol. 9366, pp. 19–35. Springer, Cham (2015). https://doi.org/10.1007/978-3-319-25007-6_2
14. Thornton, K., Solbrig, H., Stupp, G.S., Labra Gayo, J.E., Mietchen, D., Prud'hommeaux, E., Waagmeester, A.: Using shape expressions (ShEx) to share RDF data models and to guide curation with rigorous validation. In: Hitzler, P., et al. (eds.) ESWC 2019. LNCS, vol. 11503, pp. 606–620. Springer, Cham (2019). https://doi.org/10.1007/978-3-030-21348-0_39
15. Witten, I.H., Moffat, A., Bell, T.C.: Managing Gigabytes: Compressing and Indexing Documents and Images, 2nd edn. Morgan Kaufmann Series in Multimedia Information and Systems. Morgan Kaufmann, San Francisco (1999)
16. Xu, Q., Wang, X., Li, J., Zhang, Q., Chai, L.: Distributed subgraph matching on big knowledge graphs using pregel. IEEE Access **7**, 116453–116464 (2019). https://doi.org/10.1109/ACCESS.2019.2936465

ITAQ: Image Tag Recommendation Framework for Aquatic Species Integrating Semantic Intelligence via Knowledge Graphs

S. S. Nitin Hariharan[1], Gerard Deepak[2](✉), Fernando Ortiz-Rodríguez[3], and Ronak Panchal[4]

[1] Department of Computer Science and Engineering, Alagappa Chettiar Government College of Engineering and Technology, Karaikudi, India

[2] Department of Computer Science and Engineering, Manipal Institute of Technology Bengaluru, Manipal Academy of Higher Education, Manipal, India
gerard.deepak.christuni@gmail.com

[3] Universidad Autonoma de Tamaulipas, Ciudad Victoria, Mexico

[4] Cognizant Technology Solutions, Pune, India

Abstract. In the era of Web 3.0, there is an increasing demand for social image tagging that incorporates knowledge-centric paradigms and adheres to semantic web standards. This paper introduces the ITAQ framework, a recommendation framework specifically designed for tagging images of aquatic species. The framework continuously integrates strategic knowledge curation and addition at various levels, encompassing topic modelling, metadata generation, metadata classification, ontology integration, and enrichment using knowledge graphs and sub graphs. The ITAQ framework calculates context trees from the enriched knowledge dataset using AdaBoost classifier which is a lightweight machine learning classifier. The CNN classifier handles the metadata, ensuring a well-balanced fusion of learning paradigms while maintaining computational feasibility. The intermediate derivation of context trees, computation of KL divergence, and Second Order Co-occurrence PMI contribute to semantic-oriented reasoning by leveraging semantic relatedness. The Ant Lion optimization is utilized to compute the most optimal solution by building upon the initial intermediate solution. Finally, the optimal solution is correlated with image tags and categories, leading to the finalization of labels and annotations. An overall precision of 94.07% with the lowest value of FDR of 0.06% and accuracy of 95.315 % has been achieved by the proposed work.

Keywords: CNN · AdaBoost · KL Divergence · SOC-PMI

1 Introduction

Image tag recommendations are keywords or phrases associated with images to categorize and describe their content. They aid in enhancing discover-ability

F. Ortiz-Rodriguez et al. (Eds.): KGSWC 2023, LNCS 14382, pp. 135–150, 2023.
https://doi.org/10.1007/978-3-031-47745-4_11

and engagement. Tags are employed in social media, blogs, and other platforms. The selection of precise tags is essential for representing the subject or theme of the image. It enables users interested in specific topics to easily locate the content. Popular or trending tags can enhance visibility and reach. The addition of relevant and descriptive tags augments the probability of appearing in search results. Image tags play a crucial role in organizing and optimizing image-based content.

Motivation: There is a clear need for a strategic model framework that specifically focuses on generating tags and labelling images for the aquatic species. The aquatic domain often receives less attention in tagging efforts due to the unique nature of these underwater animals. However, it is of utmost importance to tag and label such species, as there exist several unknown entities within this domain. As the web's data explodes and we move from Web 2.0 to Web 3.0, a semantically driven model gains vital importance. It helps understand data meaning, making web navigation and information utilization easier. Hence, there is an urgent requirement for a model that can effectively tag and label aquatic species datasets using a semantic approach.

Contribution: The proposed framework introduces several innovative approaches to strategically enhance the knowledge density of entities derived from the dataset. This is accomplished using techniques such as structural topic modelling, the generation of RDF entities using resource description frames, and the incorporation of upper ontologies, knowledge graphs, and subgraphs. Another key strategy involves metadata generation through PyMarc, alongside other methods employed to encompass and consolidate knowledge within the model. We use a powerful deep learning classifier, specifically a CNN, to handle and refine the metadata, resulting in improved accessibility. This contributes to the formalization of context trees utilizing an agent. For an Effective initial classification, we use a lightweight machine learning classifier, the AdaBoost classifier. Semantic reasoning is achieved by utilizing computation techniques such as KL divergence and SOC-PMI measures to generate the initial solution, which is further optimized through the Ant Lion optimizer to yield the most optimal set of solutions.

Organization: Meaning of the paper is organized as follows. Related Work is depicted in section two. Proposed System Architecture is depicted as Section Three. Section Four depicts the Performance Evaluation and Results. Atlast the Performance is concluded in Section Five.

2 Related Works

Lui et al. [1] present experimental findings that highlight the effectiveness of their proposed algorithm in enhancing the accuracy of tag recommendation. In a separate study, Lin et al. [2] focus on object recognition and propose a novel method that combines sparse coding and spatial pyramid matching to improve the accuracy of object recognition systems. They provide evidence of the improved performance achieved through their approach. Mamat et al. [3] propose a technique to enhance the image annotation process specifically for fruit classification. Their method utilizes deep learning algorithms to automatically classify and annotate fruit images, resulting in improved accuracy and efficiency.

Deepak et al. [4] proposes a personalized and enhanced hybridized semantic algorithm for web image retrieval. It incorporates ontology classification, strategic query expansion, and content-based analysis to improve the accuracy and efficiency of image search on the web. Kannan et al. [5] introduces a technique for suggesting web images that combines sophisticated approaches in deep learning and machine intelligence. The suggested method aims to improve the recommendation of web images by offering more precise and personalized suggestions. Sejal et al. [6] proposes an image recommendation method based on ANOVA cosine similarity. The approach utilizes ANOVA and cosine similarity measures to provide effective image recommendations.

Chen et al. [7] introduces a method called Attentive Collaborative Filtering for multimedia recommendation. This approach incorporates item- and component-level attention mechanisms to improve the accuracy of recommendations. Lin et al. [8] discusses how image processing techniques can be harnessed to extract valuable information from images, enabling a more comprehensive understanding of user preferences. By integrating image data into the recommendation process, the proposed system aims to provide more personalized and relevant recommendations to users, enhancing their overall experience. Shankar et al. [9] introduces SIMWIR, a semantically inclined model for web image recommendation. It utilizes integrative intelligence to provide more effective and con-textually relevant recommendations for web images.

Kong et al. [10] presents a construction method for an automatic matching recommendation system for web page image packaging design. The system utilizes a constrained clustering algorithm to generate relevant and visually appealing recommendations. It is published in Mobile Information Systems. Niu et al. [11] introduces a neural personalized ranking method for image recommendation. This approach aims to enhance the accuracy and relevance of image recommendations by leveraging personalized ranking algorithms within a neural network framework. By combining neural networks and personalized ranking algorithms, the proposed approach in the paper addresses the challenge of effectively recommending images based on individual preferences and user feedback. Bobde et al. [12] presents SemWIRet, a semantically inclined strategy for web image recommendation that leverages hybrid intelligence. The approach incorporates semantic analysis and hybrid intelligence techniques to improve the accuracy and relevance of image recommendations.

In [4,13–16], and [17] semantic models for inclusive recommen-dations have been put forth where the focus is on knowledge based modeling and recommendations through facts on the basis of semantics oriented reasoning us-ing benchmark knowledge is discussed.

3 Proposed System Architecture

The system architecture of the ITAQ framework, as shown in Fig. 1, has been designed with the objective of improving the assessment of information associated with categories and labels within an image dataset. This improvement is achieved by implementing techniques from structural topic modelling (STM). STM is used to uncover underlying topics and structures within the dataset, making categories and labels more meaningful and informative. By considering the World Wide Web as a reference corpus, STM incorporates topic modelling to reveal previously undiscovered and relevant entities as topics, thereby enriching the informative content of categories and labels. However, it's important to acknowledge that STM has certain limitations in identifying topics, which high-lights the need for strategic knowledge accumulation. In the context of the aquatic species domain, relying on a single source of knowledge is insufficient. There-fore, the framework generates RDF (Resource Description Framework) to incorporate interdisciplinary heterogeneous multi-source knowledge. The RDF enables the integration of diverse knowledge from various sources, allowing for a comprehensive understanding of aquatic species. This integration of knowledge enhances the information measure of categories and labels by incorporating insights from different disciplines.

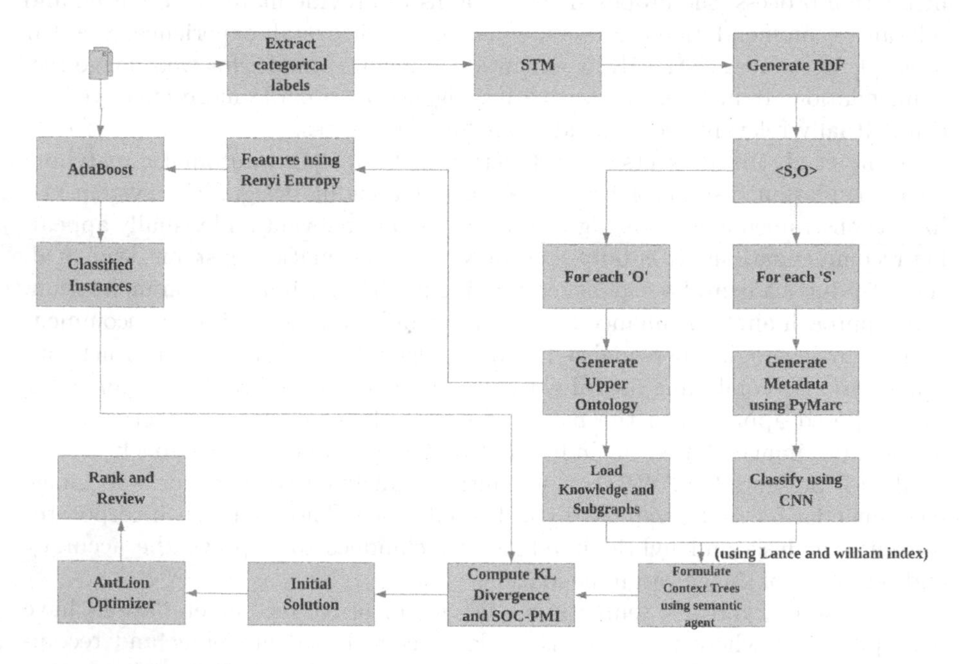

Fig. 1. System Architecture for the Proposed **ITAQ** Framework

RDF, also known as Resource Description Framework, is a framework that utilizes subject, predicate, and object (S, P, O) to express any information. It follows a triadic structure where the predicate acts as the bridge between the subject and object. The predicate can take the form of a URL, a term, or a link. However, for simplicity, the predicate can be omitted, and the RDF structure can be represented as (S, O). The reason for omitting the predicate is because it complicates the entire process computationally as. Since predicate is placed between subject and object, predicate can either be a term, sentence or a URL. When it is a term or sentence, it can be easily parsed and handled. Rather, when it is a URL, the URL must be handled by inputting. By looking onto the actual contents of keywords in the URL which definitely needs more computational cycles and it can be omitted due to the reason and the fact that the subject and the object of the ontology due to the reason, the fact that the subject and the object of an ontology are specific entities which itself is sufficient to provide lateral co-occurrence semantics. The predicate will simply add to the semantics of a subject and object which is always not which is not required. The subject and object itself will be correlated and they will be able to establish shift with each other and henceforth predicate is dropped. This representation preserves the fundamental elements of RDF.

The subject terms within the RDF structure are used to generate metadata using PyMarc, a strategic tool. PyMarc leverages the existing structure of the World Wide Web to produce metadata, which leads to significant increase in the volume of metadata. This substantial growth enhances the density of knowledge. To transform this knowledge into atomic units and enable its integration into the model, a powerful deep learning classifier called Convolutional Neural Networks (CNN) is used for classification. In relation to each object term in the generated RDF, upper ontologies are constructed using a tool named OntoCollab. OntoCollab simplifies the creation of upper ontologies, which offer a higher-level categorization and organization of concepts and entities within the RDF structure.

To generate upper ontologies, OntoCollab is chosen as the preferred tool. The decision to prioritize upper ontologies from the objective of achieving a high density of relevant entities. The upper ontology is designed with a maximum of seven levels, excluding individuals, while preserving core concepts and sub-concepts. For each upper ontology created, nodes are connected to the Google Knowledge Graph API, which facilitates the loading of knowledge graphs and subgraphs. The reason for using onto collab is because it integrates domain experts for which belongs to who belong to the domain of the ontology. In this case, 23 domain experts 24 domain experts were shortlisted and were asked to log into. So, based on the domain of aquatics species, several web documents and e-books comprising of aquatic and marine sciences were given as link for which ontology was generation. The upper ontology was restricted up to seven levels, excluding the individuals, and it was ensured that the seven levels were directly related to the domain of choice and had very little deviants from the immediate superseding. The immediate concept above the features of the generated ontology comprised of all the domain related specifics in terms of concepts and sub concepts of aquatic and marine sciences.

The Google Knowledge Graph API acts as a community-contributed, community-verified, and crowd-sourced knowledge repository. Developers gain access to an extensive collection of structured data through Google's Knowledge Graph API. It enables seamless retrieval of comprehensive information about entities such as individuals, locations, and organizations. By harnessing this API, developers can enrich their applications with intelligent and context-aware functionalities. The Knowledge Graph API enables developers to build powerful and informative applications with abundant and detailed information.

Using the Google Knowledge Graph API, highly relevant knowledge graphs and subgraphs are loaded for each node. These knowledge graphs and subgraphs are then aggregated into their respective nodes, resulting in the creation of a comprehensive formalized knowledge graph that encompasses instances. Alongside the entities classified by the CNN classifier, this knowledge graph is utilized to build a context tree. The context tree follows a hierarchical structure, extending up to 18 levels. Unlike a graph, a context tree at each level consists of a single sub-node or child, forming a linear context chain. This ensures a well-structured and organized representation of the information. To facilitate this process, a semantic agent is employed, developed using AgentSpeak. The semantic agent plays a crucial role in managing and manipulating the knowledge graph, entities, and the construction of the context tree, enabling efficient and effective processing of information.

AgentSpeak is utilized as the strategic tool for designing the agent, enabling effective communication and reasoning. The Lance and Williams Index is used to compute strategic relevance. The context trees are formed with a stringent step deviation of 0.15. This is necessary due to the large number of entities obtained from the Google graph and subgraph, as well as the metadata acquired from the PyMarc tool. Subsequently, the upper ontology is generated to provide a higher-level categorization and organization of concepts and entities within the system. The combination of AgentSpeak, strategic relevance computation, context tree formation, and upper ontology generation enhances the overall representation of knowledge and improves the reasoning capabilities of the system.

The generated upper ontologies undergo feature selection using Renyi entropy with a step deviation of 0.10. The selected features are then inputted into the AdaBoost classifier to classify the image dataset, resulting in classified instances that enhance the atomicity and permeability of the dataset. AdaBoost is chosen for its lightweight nature and strong classification capabilities. Furthermore, the classified instances and context trees undergo further processing utilizing KL divergence and SOC-PMI (Second-Order Co-occurrence Pointwise Mutual Information) measures. KL divergence with a step deviation of 0.15, while SOC-PMI uses a median threshold of 0.50. The median threshold is selected for SOC-PMI to capitalize on its inherent strength, while KL divergence is applied with strictness. This ensures a balanced approach. Despite the robust step deviation employed in the Lance and Williams Index, the formulated context trees yield a significantly large number of entities, forming the initial solution set. This set's size is attributed to the multitude of entities derived from Ontology Knowledge,

Google Knowledge Graph subgraph, and the PyMarc tool. To obtain an optimal solution set, the Ant Lion Optimizer, a metaheuristic algorithm, is deployed. It utilizes SOC-PMI as the intermediate objective function and iterates until all entities in the initial solution set are exhaustively considered. The resulting entities are subsequently ranked and subjected to review by domain experts. After the review process, these entities are integrated with the category annotations in the image dataset using categories or labels, and the tags are finalized.

The finalized tags do not require any further review as the review process already incorporates human cognition. The consumption of tags is entirely automated in nature, eliminating the need for manual intervention.

A convolutional neural community (CNN) is designed to technique structured grid-like data, like pictures or movies. It excels in computer vision tasks by using autonomously studying and extracting widespread capabilities from the raw input information. The key element of a CNN is the convolutional layer, which employs small filters or kernels to carry out convolution operations, producing function maps. The values of those characteristic maps are computed the usage of Eq. 1, involving the input image represented as f and the kernel denoted as h. The resulting matrix's row and column indexes are labeled as m and n respectively.

$$G[m,n] = (f * h)[m,n] = \sum_{j}\sum_{k} h[j,k] \cdot f[m-j, n-k] \tag{1}$$

In CNNs, we add extra layers like pooling and fully connected layers to improve their performance. Pooling layers help shrink the feature maps, keeping important details intact. This reduces the computational workload and makes the network more adaptable to different inputs. Fully connected layers, placed at the network's end, deal with more complex features and ultimately determine the classification or regression results. These additional layers enhance the CNN's ability to recognize patterns and make accurate predictions.

The dimensions of the input tensor, referred to as the 3D matrix, follow Eq. 2: n represents the image size, f is the filter size, n_c denotes the number of channels in the image, p is the used padding, s is the used stride, and n_f stands for the number of filters.

$$[n,n,n_c] * [f,f,n_c] = \left[\left\lfloor \frac{n+2p-f}{s} + 1 \right\rfloor, \left\lfloor \frac{n+2p-f}{s} + 1 \right\rfloor, n_f\right] \tag{2}$$

CNNs have a special ability to learn and understand data in a hierarchical way. As they dig deeper into the information, they can grasp more complex and abstract features. This means they can detect simple things like edges and textures, but also recognize more advanced concepts like objects or scenes. To train a CNN, we need a large set of labeled examples. The network adjusts its internal settings using a process called backpropagation, which helps it get closer to the correct answers. Through this repetitive learning, the CNN becomes skilled at making accurate predictions even on new, unseen data.

AdaBoost, also known as Adaptive Boosting, is a type of machine learning method used for classification tasks. AdaBoost is a clever algorithm that makes use of multiple weak classifiers to do a good job in classification tasks. It does this by adjusting the importance of training examples. In the beginning, all examples have the same importance. Then, AdaBoost trains a weak classifier, which is not perfect but better than random guessing. After each round of training, it pays more attention to the examples it got wrong, giving them higher importance in the next round.

This process is repeated for a set number of iterations. Finally, the weak classifiers are combined to create a strong classifier, and their alpha values, which indicate their importance, are used as weights. During classification, each weak classifier makes a prediction, and the final prediction is determined by considering the weighted combination of these individual predictions. AdaBoost focuses more on challenging examples, giving them more weight in the learning process. By combining multiple weak classifiers, it achieves high accuracy even when using simple classifiers. The key to its success lies in selecting the right weak classifiers. AdaBoost works well for various classification problems, especially when dealing with lots of features or noisy data.

1. For a dataset with N number of samples, we initialize the weight of each data point with w_i =1/N
2. For m = 1 to M:
   (a) Sample the dataset using the weights w_i to obtain training samples x_i
   (b) Fit a classifier K_m using all the training samples x_i
   (c) Compute $\epsilon = \frac{\sum_{y_i \neq K_m(x_i)} w_i^{(m)}}{\sum_{y_i} w_i^{(m)}}$
   where y_i is the ground truth value of the target variable, $w_i^{(m)}$ is weight of the sample i at iteration m
   (d) Compute $\alpha_m = \frac{1}{2} \ln \frac{1-\epsilon}{\epsilon}$
   (e) Update all the weights $w_i^{(m+1)} = w_i^{(m)} e^{-\alpha_m y K_m(x)}$
3. New predictions computed by $K(x) = \text{sign}\left[\sum_{m=1}^{M} \alpha_m K_m(x)\right]$

Adaboost is a helpful classification algorithm, but it has some challenges. If the weak classifiers are too complex or if we use too many iterations, it might overfit the data, meaning it focuses too much on the training examples and doesn't generalize well to new data. Additionally, Adaboost is sensitive to outliers, which are unusual data points that can be misclassified and given too much importance. Despite these issues, Adaboost has many practical applications. It's used in various fields, such as face detection and object recognition. The final classifier created by Adaboost is good at handling difficult examples, which reduces errors in classification. Its strength lies in combining simple classifiers to deal with complex tasks effectively.

The Ant Lion Optimization Algorithm is a clever method inspired by how ant lions behave in nature. The algorithm starts by randomly creating a population of Ant Lions, and each Ant Lion represents a potential solution to a specific optimization problem. As the algorithm runs, it goes through two main behaviors: waiting and pouncing. In the waiting phase, Ant Lions stay in their current positions and build traps to attract ants. In this context, the Ant Lion's position represents a potential solution, and the number of ants attracted to the trap reflects how good that solution is. Ant Lions continuously adjust their traps based on the quality of the solutions they've caught so far. During the pouncing phase, Ant Lions assess nearby traps and explore other potential solutions. Based on what they find, they decide whether to move to a new position in the search space. The movement rules determine where the Ant Lion goes, helping it explore areas that haven't been tried yet. After moving, the Ant Lion evaluates the quality of its new position and adjusts its trap accordingly.

This process of waiting and pouncing continues for a set number of iterations or until a specific condition is met. As time goes on, the Ant Lions get better at finding superior solutions because their traps become more enticing, capturing higher-quality solutions. The algorithm maintains a balance between exploring new possibilities and exploiting promising areas. It uses randomness to achieve this balance, which helps it explore different parts of the solution space while also making the most of valuable areas. This is how the Ant Lion Optimization Algorithm efficiently solves optimization problems.

$$\vec{Ant_i} = [A_{i,1}, A_{i,1}, \ldots, A_{i,d}] \tag{3}$$

Ant_i represents the ith ant, while $A_{i,d}$ represents the position of the ith ant in the dth dimension.

$$x(t) = [0, c(2t(t_1) - 1), c(2t(t_2) - 1), \ldots, c(2t(t_T) - 1)] \tag{4}$$

T represents the maximum number of iterations. t_i^{th} refers to the ith iteration, and t_T^{th} represents the Tth iteration. c denotes the cumulative summation, and $r(t)$ is a random function calculated as specified in Eq. 5.

$$r(t) = \begin{cases} 1 & \text{if rand} \geq 0.5 \\ 0 & \text{if rand} < 0.5 \end{cases} \tag{5}$$

t represents the iteration index, and rand is a randomly generated number in the range [0, 1].

The effectiveness of the Ant Lion Optimization Algorithm depends on its settings, like trap size, trap updating rules, and movement rules. By adjusting these parameters, the algorithm can be customized to work well for different types of problems. The great thing about this algorithm is that it can be used for a wide range of optimization problems. It works for continuous, discrete, and combinatorial problems. It has been successfully applied in various fields, including engineering, data mining, and many others. This flexibility and proven success make the Ant Lion Optimization Algorithm a valuable tool for solving different kinds of real-world challenges.

The RDF is a framework for representing information about resources in a structured and machine-readable manner. It uses a subject-predicate-object (S-P-O) triple format to express relationships between entities. RDF enables the integration and exchange of data across different domains and applications. It provides a flexible and extensible means to model and link diverse sources of information on the web, facilitating efficient data integration and interoperability.

Let $p(x)$ and $q(x)$ be two probability distributions of a discrete random variable x. That is, both $p(x)$ and $q(x)$ sum up to 1, and $p(x) > 0$ and $q(x) > 0$ for any x in χ is described in Eq. 6.

$$D_{KL}\left(p(x)||q(x)\right) = \int_{\chi} p\left(x\right) \log\left(\frac{p\left(x\right)}{q\left(x\right)}\right) \mu\left(dx\right) \tag{6}$$

$$f^{pmi}\left(t_i\ , w\right) = \log_2 \frac{f^b\left(t_i\ , w\right) \times m}{f^t\left(t_i\ \right) f^t\left(w\right)} \tag{7}$$

Equation 7 describes the process of calculating relative semantic similarity using PMI (Pointwise Mutual Information) values from the opposite list, where $f^t\left(t_i\ \right)$ represents the frequency of type t_i appearing in the entire corpus, $f^b\left(t_i\ , w\right)$ denotes the frequency of word t_i appearing with word w in a context window, and m is the total number of tokens in the corpus. For a specific word w a set of words, X^w is defined and sorted in descending order based on their PMI values with w The top-most β words are then selected from this set, subject to the condition that their $f^{pmi}\left(t_i\ , w\right)$ values are greater than zero.

A rule of thumb is used to choose the value of β. The β -PMI summation function of a word is defined with respect to another word. For word w_1 with respect to word w_2 is mentioned in Eq. 9.

$$f\left(w_1, w_2, \beta\right) = \sum_{i=1}^{\beta} \left(f^{pmi}\left(X_i^{w_1}\ \ , w_2\right)\right)^{\gamma} \tag{8}$$

where $f^{\text{pmi}}(X_i^{w_1}, w_2) > 0$, which sums all the positive PMI values of words in the set $X_i^{w_1}$ also common to the words in the set $X_i^{w_2}$.

The Semantic PMI similarity function between the two words, w_1 and w_2, is defined in Eq. 9.

$$\text{Sim}\left(w_1, w_2\right) = \frac{f\left(w_1, w_2, \beta_1\right)}{\beta_1} + \frac{f\left(w_2, w_1, \beta_2\right)}{\beta_2} \tag{9}$$

The semantic word similarity is normalized, so that it provides a similarity score between 0 and 1 inclusively.

The Renyi entropy of order α, where $\alpha \geq 0$ and $\alpha \neq 1$, is defined in Eq. 10.

$$H_{\alpha}(X) = \frac{1}{1-\alpha} log\left(\sum_{i=1}^{n} p_i^{\alpha}\right) \tag{10}$$

Here, X is a discrete random variable with possible outcomes in the set $A = \{x_1, x_2, x_3, \ldots, x_n\}$ and corresponding probabilities $p_i = \Pr(X = x_i)$ for $i = 1, 2, 3, \ldots, n$.

The Eq. 11 represents the Lance and William index. This index is used to calculate the semantic similarity between two groups of entities in a specific sample. The equation considers the total number of entities in the sample (denoted by N), and it measures the semantic similarity between the entities in the two groups. By computing the total semantic information measure (X) for all entities and the individual semantic information measure (Y) for each entity, this index provides insights into the overall similarity and information content of the two groups.

$$1 - \left(\frac{1}{n} \sum_{i} \frac{|x_i - y_i|}{(x_i + y_i)} \right) \tag{11}$$

4 Performance Evaluation and Results

The ITAQ framework proposed, which focuses on recommending tags for aquatic species, has undergone evaluation using various metrics to evaluate its performance. The selected metrics for evaluation in this context includes precision, recall, accuracy, F-measure percentages to measure the relevance rate and false discovery rate (FDR) to measure the error rate. The ITAQ framework under consideration has achieved the highest of all values which includes precision of 94.07%, recall of 96.56%, accuracy of 95.315%, and F measure of 95.2987379% (all in their average form). Additionally, it has obtained the lowest FDR value, which is 0.6. In order to assess, compare, and benchmark the performance of the proposed ITAQ framework, a comparison is made with similar frameworks for image tag recommendation, namely PTD, ISR, and EIA. The PTD framework has yielded precision of 87.45%, recall of 89.08%, accuracy of 88.265%, F measure of 88.2574747% and FDR of 0.13 (all in their average form). The ISR framework has achieved precision of 89.74%, recall of 90.12%, accuracy of 89.93%, F measure of 89.9295986% and FDR of 0.11 (all in their average form). The EIA framework has resulted precision of 91.74%, recall of 92.09%, accuracy of 91.915%, F measure of 91.9146668% and FDR of 0.09 (all in their average form). It is evident that the proposed ITAQ framework has exhibited the highest precision, accuracy, F measure percentages, and the lowest FDR value among the compared frameworks.

The information which is handled in the knowledge graphs comprises of comprise the conditions under which the comparison of other models were made is for the exact same data set for the exact configuration onto which the experimentations were conducted onto which the experimentations were conducted, the exact conditions and exact configurations onto which the experimentation were conducted for the exact same data set and the exacts same experimental setup. The comparison was made by implementing the baseline models in the exact same environment of the proposed model. The baseline models, particularly the PTD fail to perform well because it utilizes tensor decomposition for tag recommendation and achieves specialization and personalization within the model and incorporates social awareness. A major drawback of the PTD model is the limited presence of auxiliary knowledge. While some auxiliary knowledge exists in the PTD model through personalization and the through some social structure, it is considerably less compared to the proposed ITAQ framework. The PTD model lacks strategic relevance and computational mechanisms for semantic-oriented reasoning (Table 1).

Table 1. Comparison of Performance of the proposed **ITAQ** with other approaches

Model	Average Precision (%)	Average Recall(%)	Average Accuracy (%)	Average F-Measure (%)	FDR
PTD [1]	87.45	89.08	88.265	88.2574747	0.13
ISR [2]	89.74	90.12	89.93	89.9295986	0.11
EIA [3]	91.74	92.09	91.915	91.9146668	0.09
Proposed ITAQ	94.07	96.56	95.315	95.2987379	0.06

The Proposed framework surpasses the ISR model because the ISR model falls behind in terms of comprehensive integration of diverse knowledge. It lacks the inclusion of a broad spectrum of features, resulting in a limited understanding of the data. The ISR model has sparse knowledge. The existing tags are not effectively synthesized to enhance contextual information, leading to an incomplete grasp of the data's meaning. Furthermore, the learning models employed in the ISR model are simplistic and lack sophistication. They struggle to capture and depict the relationships between data points. This deficiency contributes to the model's inability to accurately calculate tag-tag correlations and image-tag correlations.

The EIA framework is also inferior to the proposed model because it uses tag-specific linear sparse reconstruction for image tag completion. It incorporates linear sparse recommendations and utilizes image similarity and item-tag associations to establish tag-tag concurrence, indicating a strong correlation between images and tags. This approach results in strong semantic relevance computation and reasoning, but it lacks enough substrate data for all semantics. The knowledge integrated into the model is limited to the available data, and there is a need for improvement in terms of strategic relevance computation mechanisms. The permeability of knowledge into the localized framework is restricted, indicating that the knowledge remains dense but fails to effectively integrate into the framework.

In the experiments, researchers used three different datasets of fish and aquatic animals from Lake Biwa in Japan, observed through carp-mounted video cameras. The dataset employs carp-mounted video loggers for underwater observations, capturing various aquatic species, particularly fishes, in their natural habitat. It contains both video and still-image data, offering a comprehensive view of aquatic behaviors and characteristics. The Marionette Sea Animals Image Dataset provides a comprehensive collection of images featuring various sea creatures and aquatic organisms. It serves as a rich source of visual data, allowing researchers and enthusiasts to explore and study the diverse marine fauna. They also included data from the Lake Erie, Western Basin Aquatic Vegetation dataset from 2018. This dataset, compiled in 2018, provides valuable information and data related to the various species of aquatic plants and vegetation present in this region. Researchers and environmentalists can use this dataset to gain insights into the distribution, composition, and ecological significance of aquatic vegetation in Lake Erie's Western Basin. Instead of using these datasets separately, the researchers combined them into one big dataset using a custom annotator. To do this, they annotated the images based on their tags and categories. If an image didn't have categorical information, they transformed it by removing at least one label and adding it as a category. This process improved and enriched the dataset. Although the datasets didn't have strong correlations, the researchers partially merged them based on their categories. The remaining data points were given priority and integrated into the unified dataset. By combining and integrating these datasets, the researchers created a larger and more diverse dataset. This allowed them to conduct more comprehensive and reliable experiments in their study.

The researchers used Google Collaboratory and Python 3 for their implementation. They used specific tools like Onto-Collab to create the upper ontology, a Meta Tag RDF generator to generate RDF, and PyMarc to create metadata. For the Convolutional Neural Network (CNN) configuration, they used the Keras platform. To develop the semantic agents, they utilized AgentSpeak, a framework designed for building intelligent agents. The formulation of formulas involved context trees, KL divergence, and SOC-PMI, which they integrated into the agent using AgentSpeak. Additionally, they incorporated the Ant Lion Optimizer within the AgentSpeak framework. The effectiveness of the proposed ITAQ framework over the baseline models can be attributed to various factors. Firstly, it uses various methods to generate auxiliary knowledge, assimilating categorical labels from the dataset. Furthermore, the framework leverages structural topic modeling to enhance this auxiliary knowledge and provide contextualization.

Additionally, the framework generates RDF to represent the acquired knowledge. To handle the complexity of the triadic structure, the framework retains the subject-object relationship and strengthens lateral co-occurrence and semantic knowledge by keeping them integrated. For every object, an upper ontology is created, and both the knowledge graph and subgraphs are obtained using Google's Knowledge Graph API. Metada-ta generation is achieved through the implementation of PyMarc. Furthermore, a Convolutional Neural Network (CNN) is utilized as a robust deep learning classifier for tasks involving classification and automation. Moreover, the ITAQ framework incorporates context-free formalization computing to compute semantic relevance through techniques such as KL divergences, SOC-PMI, and Ant Lion optimization. Feature selection is performed using Renyi entropy, and the dataset is classified using the Adaboost classifier. These techniques contribute to a robust ecosystem of learning and reasoning, with the strategic inclusion of knowledge at each stage of the model.

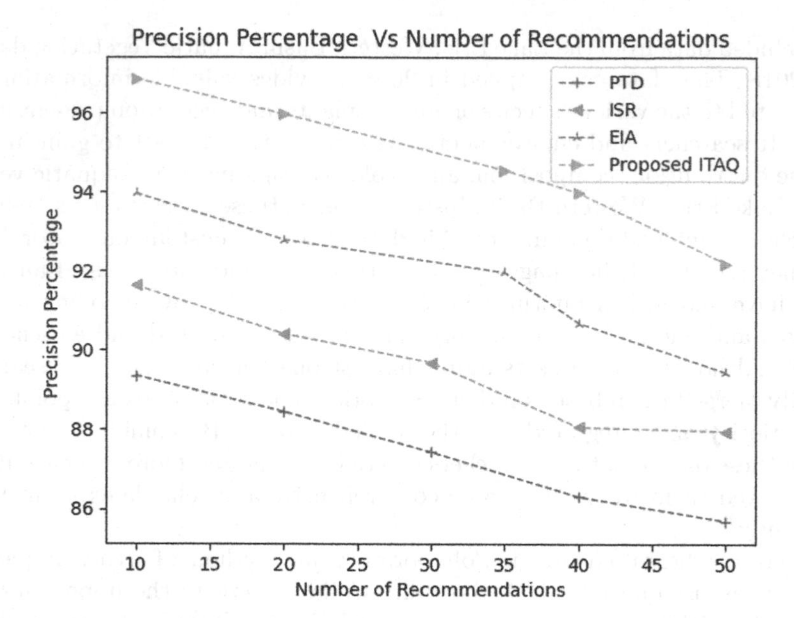

Fig. 2. Precision Vs Number of Recommendations Graph

Figure 2 illustrates the recommendation precision of different models, including the ITAQ framework and baseline models (PTD, ISR, EIA). The ITAQ frame-work shows the highest precision and is positioned at the top of the distribution curve followed by EIA, ISR and PTD. ITAQ outperforms PTD by incorporating a wider array of supplementary information, enhancing the precision of tag recommendations, and utilizing a more powerful classification model for improved reasoning abilities. ITAQ's comprehensive integration of knowledge enables a more thorough comprehension of the data, while PTD's limited utilization of auxiliary knowledge and reliance on tensor decomposition methods hinder its performance.

ITAQ outperforms ISR by achieving a more holistic understanding of the data and generating accurate reconstructions. ISR's limited feature integration, simplistic learning models, and weak relevance computation limit its ability to produce meaningful and accurate results. ITAQ also surpasses EIA by integrating a broader range of knowledge sources, resulting in a more comprehensive understanding of the data. On the other hand, EIA lacks enough substrate data for all semantics, which hampers its performance. Although EIA incorporates robust semantic relevance computation and reasoning, its lack of effective knowledge integration and strategic relevance computation mechanisms limit its overall performance.

5 Conclusion

This paper introduces the ITAQ model, which serves as a domain-specific framework for labelling and tagging aquatic plants. Notably, this model stands out as one of the pioneering efforts aimed at specifically addressing the tagging of images containing aquatic species. The process of knowledge enrichment is accomplished through a range of techniques, including structural topic modelling, RDF generation, and the utilization of the Google Knowledge Graph API to create an upper ontology and enhance knowledge graphs. The RDF object is employed to generate metadata, which is subsequently classified using a CNN classifier. The formalization of context trees is achieved through the implementation of semantic agents, while the computation of KL divergence and SOC-PMI measure aids in semantics-oriented reasoning, resulting in the formulation of an initial solution. Optimization techniques utilizing an Ant Lion-based optimizer are employed to obtain the optimal set of solutions, while the Adaboost classifier plays a crucial role in the initial classification of the dataset. These contributions demonstrate novelty and uniqueness within the proposed model and overall recall of 96.56% and F measure of 95.2987379% with the lowest value of FDR 0.06% in the proposed model.

References

1. Liu, S., Liu, B.: Personalized social image tag recommendation algorithm based on tensor decomposition. In: 2021 2nd International Conference on Smart Electronics and Communication (ICOSEC), Trichy, India, pp. 1025–1028 (2021). https://doi.org/10.1109/ICOSEC51865.2021.9591909
2. Lin, Z., Ding, G., Hu, M., Wang, J., Ye, X.: Proceedings of the IEEE Conference on Computer Vision and Pattern Recognition (CVPR), pp. 1618–1625 (2013)
3. Mamat, N., Othman, M.F., Abdulghafor, R., Alwan, A.A., Gulzar, Y.: Enhancing image annotation technique of fruit classification using a deep learning approach. Sustainability **15**(2), 901 (2023)
4. Deepak, G., Priyadarshini, J.S.: Personalized and enhanced hybridized semantic algorithm for web image retrieval incorporating ontology classification, strategic query expansion, and content-based analysis. Comput. Electr. Eng. **72**, 14–25 (2018)
5. Kannan, B.D., Deepak, G.: I-DLMI: web image recommendation using deep learning and machine intelligence. In: Abraham, A., Hong, T.P., Kotecha, K., Ma, K., Manghirmalani Mishra, P., Gandhi, N. (eds.) HIS 2022. LNNS, vol. 647, pp. 270–280. Springer, Cham (2022). https://doi.org/10.1007/978-3-031-27409-1_24
6. Sejal, D., Ganeshsingh, T., Venugopal, K.R., Iyengar, S.S., Patnaik, L.M.: Image recommendation based on ANOVA cosine similarity. Procedia Comput. Sci. **89**, 562–567 (2016)
7. Chen, J., Zhang, H., He, X., Nie, L., Liu, W., Chua, T.S.: Attentive collaborative filtering: multimedia recommendation with item-and component-level attention. In: Proceedings of the 40th International ACM SIGIR Conference on Research and Development in Information Retrieval, pp. 335–344 (2017)
8. Lin, Q.: Intelligent recommendation system based on image processing. In: Journal of Physics: Conference Series, vol. 1449, no. 1, p. 012131. IOP Publishing (2020)

9. Shankar, A., Deepak, G.: SIMWIR: a semantically inclined model for annotations based web image recommendation encompassing integrative intelligence. In: Abraham, A., Hanne, T., Gandhi, N., Manghirmalani Mishra, P., Bajaj, A., Siarry, P. (eds.) International Conference on Soft Computing and Pattern Recognition. LNNS, vol. 648, pp. 899–910. Springer, Cham (2022). https://doi.org/10.1007/978-3-031-27524-1_88
10. Kong, X.: Construction of automatic matching recommendation system for web page image packaging design based on constrained clustering algorithm. Mob. Inf. Syst. **2022** (2022)
11. Niu, W., Caverlee, J., Lu, H.: Neural personalized ranking for image recommendation. In: Proceedings of the Eleventh ACM International Conference on Web Search and Data Mining, pp. 423–431 (2018)
12. Bobde, M.Y., Deepak, G., Santhanavijayan, A.: SemWIRet: a semantically inclined strategy for web image recommendation using hybrid intelligence. In: Singh, Y., Verma, C., Zoltán, I., Chhabra, J.K., Singh, P.K. (eds.) ICRIC 2022. LNEE, vol. 1011, pp. 467–478. Springer, Singapore (2022). https://doi.org/10.1007/978-981-99-0601-7_36
13. Deepak, G., Santhanavijayan, A.: OntoBestFit: a best-fit occurrence estimation strategy for RDF driven faceted semantic search. Comput. Commun. **160**, 284–298 (2020)
14. Tiwari, S., Rodriguez, F.O., Abbes, S.B., Usip, P.U., Hantach, R. (eds.): Semantic AI in Knowledge Graphs. CRC Press, Boca Raton (2023)
15. Mihindukulasooriya, N., Tiwari, S., Enguix, C.F., Lata, K.: Text2KGBench: A Benchmark for Ontology-Driven Knowledge Graph Generation from Text. arXiv pre-print (2023)
16. Dogan, O., Tiwari, S., Jabbar, M.A., Guggari, S.: A systematic review on AI/ML approaches against COVID-19 outbreak. Complex Intell. Syst. **7**(5), 2655–2678 (2021). https://doi.org/10.1007/s40747-021-00424-8
17. Rai, C., Sivastava, A., Tiwari, S., Abhishek, K.: Towards a conceptual modelling of ontologies. In: Hassanien, A.E., Bhattacharyya, S., Chakrabati, S., Bhattacharya, A., Dutta, S. (eds.) Emerging Technologies in Data Mining and Information Security. AISC, vol. 1286, pp. 39–45. Springer, Singapore (2021). https://doi.org/10.1007/978-981-15-9927-9_4

Automatic Semantic Typing of Pet E-commerce Products Using Crowdsourced Reviews: An Experimental Study

Xinyu Liu, Tiancheng Sun, Diantian Fu, Zijue Li, Sheng Qian, Ruyue Meng, and Mayank Kejriwal(✉)

University of Southern California, Marina del Rey, CA 90292, USA
kejriwal@isi.edu

Abstract. This paper considers the problem of semantically typing pet products using only independent and crowdsourced reviews provided for them on e-commerce websites by customers purchasing the product, rather than detailed product descriptions. Instead of proposing new methods, we consider the feasibility of established text classification algorithms in support of this goal. We conduct a detailed series of experiments, using three different methodologies and a two-level pet product taxonomy. Our results show that classic methods can serve as robust solutions to this problem, and that, while promising when more data is available, language models and word embeddings tend both to be more computationally intensive, as well as being susceptible to degraded performance in the long tail.

Keywords: pet products · e-commerce taxonomy · text classification · semantic typing

1 Introduction

Semantic typing and ontology alignment have both been considered as critical methods in domain-specific knowledge graph (KG) construction pipelines [1,2]. The latter aims to match concepts between two ontologies, while the former can take on several forms in practice. One application of semantic type is to take a *text* description or context of a KG instance (e.g., a product for sale) and to assign it automatically to one or more concepts in a given ontology. In e-commerce, ontologies often tend to look more like taxonomies, and these taxonomies can have significant impact on such aspects as the website-layout of the e-commerce provider [3–5], as well as recommendations served to consumers visiting the e-commerce website.

These authors contributed equally to this work.

F. Ortiz-Rodriguez et al. (Eds.): KGSWC 2023, LNCS 14382, pp. 151–167, 2023.
https://doi.org/10.1007/978-3-031-47745-4_12

However, while consumer items such as books, clothes, and movies have been widely studied in the recommendation and machine learning literature, *pet products* (despite their economic footprint and their relevance to many modern pet-rearing consumers [6,7]) have been less studied. As with other e-commerce products, consumers often leave reviews for the pet products that they purchase. Our guiding hypothesis in this paper is that such reviews may help us to semantically type the product into fine-grained categories (e.g., *Cat Accessories*), without requiring additional information such as a description of the product.

At the same time, the problem is not trivial because many reviews contain a lot of distracting or irrelevant information (or may not contain enough relevant information that would point us toward the semantic type of the product), and independent reviewers tend to exhibit a variety of writing styles. Fortunately, because of research conducted in natural language processing and other similar communities, it is possible to apply existing and promising text classification methods even with a small amount of training data. These methods are also widely used in industry. Because these methods are already established, the focus of this paper is not to propose a 'new' method. Rather, we consider whether, and to what extent, such methods can be feasibly adapted for the stated problem of semantically typing products using only their reviews. With this hypothesis-driven goal in mind, we organize the paper as a experimental study investigating the following research questions (RQs):

1. **RQ 1:** Can ordinary text classification methods be feasibly applied to publicly contributed reviews of pet products to provide a non-trivial semantic type for the product using a pre-defined pet product taxonomy?
2. **RQ 2:** Are reviews describing pet products in some fine-grained, relatively uncommon semantic types in the taxonomy (e.g., *Reptile Decor & Accessories*) more difficult to classify than other more common semantic types in the taxonomy (e.g., *Dog Food & Treats*)?

The first research question is seeking to understand the merits and demerits of various existing text classification methods, and to understand whether these are indeed suitable for semantic typing on the basis of reviews alone. To the best of our knowledge, there has been no study evaluating this question, despite the enormous economic impact of pet product sales, advertisements and online recommendations. As part of the first research question, we also investigate whether some semantic typing *methodologies* work better than others. For example, is it better to semantically type each review of a product (since a product can have multiple independently provided reviews) independently, followed by aggregating the semantic types using some type of ensemble technique, or is it better to 'concatenate' or combine all reviews of a product into a single document and simply classify that one document (per product)? Furthermore, when evaluating each text classification method, should we compute accuracy measures on a per-review basis, or on a per-product basis? We discuss these methodologies and their experimental implications in detail as part of our results.

The second research question is investigating whether there is a 'long-tail' problem in this domain, which may preclude the use of more data-intensive methods (like fine-tuned language models) for categories like *Reptile Decor & Accessories* leading to higher probability of errors than for common categories like *Dog Food & Treats*.

The rest of this paper is structured as follows. We begin with a discussion of relevant related work in Sect. 2. Because we have organized this paper as an experimental study of the enumerated research questions above, rather than as proposing *new* approaches or algorithms, we follow Sect. 2 with a detailed section on the materials and methods for the proposed study in Sect. 3. The results of the study are discussed in Sect. 4, followed by a brief discussion (Sect. 5) before concluding the paper (Sect. 6).

2 Related Work

There has been increasing focus on e-commerce applications in recent years due to the digital footprint of many businesses on the Web, and the clear application of KGs and other Semantic Web technologies to this domain. Beyond e-commerce, semantic typing itself has been recognized as an important problem in recent years [8–11], especially in the Semantic Web, where the use of text data is still a relatively new development considered to an earlier era when text data was not so extensive. Aligning two ontologies or taxonomies has instead witnessed far more research, but is orthogonal to the problem being addressed in this paper.

Although our work focuses on semantic typing, rather than KG construction as traditionally designed, we note that, within certain *ontologically heavy* domains like biology and e-commerce, ontologies and KGs are largely intertwined. In biology, the Gene Ontology, while technically an ontology as its name suggests, has been argued to serve the purpose of a KG in practice [12,13], and similar arguments have been presented for e-commerce ontologies such as the Amazon Product Graph [14].

Text classification, including the methods used in this paper, have a long history and many applications, extending far beyond the KG community. We cite [15,16] as representative papers and surveys. Given this tradition of research, as noted earlier, our focus in this paper is not on developing novel text classification methods and algorithms, but on evaluating their application to a different problem where they have not been applied specifically in support of semantic typing. We note finally that semantic typing, as modeled in this paper, is a *multi-class* classification problem, which separates it from more ordinary binary classification. The former has been recognized to be more challenging than the latter in a number of papers; see, for example, [17].

3 Materials and Methods

3.1 Construction of Pet Product Taxonomy

We constructed the pet product taxonomy for our study by drawing upon pet product concepts in the widely used Google Product Taxonomy taxonomy[1] as a guide. To determine the most relevant categories to use, we sampled the product reviews and aligned them with the general taxonomy, limiting the taxonomy to two levels for a more concise approach. For the first level of the taxonomy, we limited our focus to the five most common pet types in the dataset: *Dogs, Cats, Birds, Fish* and *Reptiles*, which were designated as the (Level 1) product category. To further distinguish between different product sub-types (e.g., *Dog Supply* versus *Dog Pharmacy*), we created another, more precise level, based on our manual assessment of a small set of sampled product descriptions and reviews. The resulting structure of the final taxonomy used for the study is shown in Fig. 1:

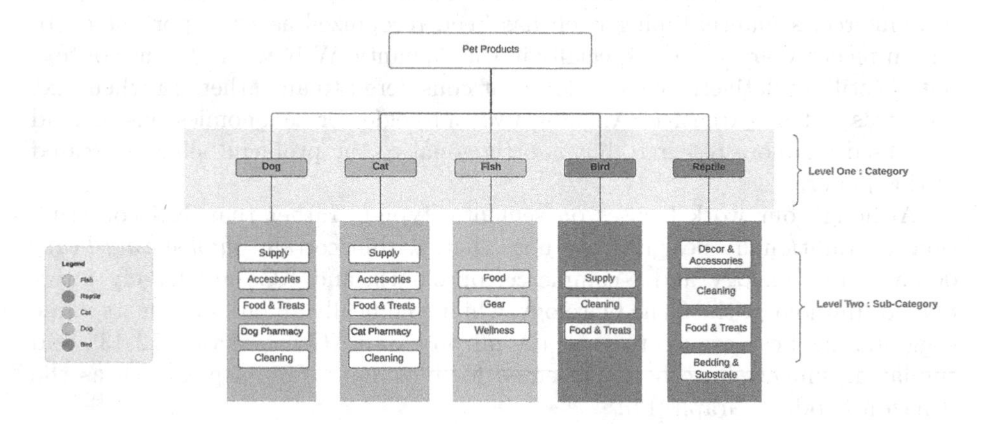

Fig. 1. The two-level taxonomy for pet products used in this paper for the experimental study.

3.2 Data

We used public data on pet product reviews available in the HuggingFace repository[2]. The dataset contains 2,643,619 reviews (each representing a purchase by a consumer) of Amazon pet products in the US, featuring 15 different e-commerce related attributes, including *product title, product id, review body, review headline,* and *review date.*

---

[1] https://www.google.com/basepages/producttype/taxonomy-with-ids.en-US.txt.

[2] https://huggingface.co/datasets/amazon_us_reviews/viewer/Pet_Products_v1_00/train; Accessed: Jan 25, 2022.

Table 1. Keywords(case insensitive) used to categorize products in the datasets in top-level categories.

Top-level category	Keywords
Dog	dog, puppy, doggie
Cat	cat, kitten, kitty, kittie
Fish	fish, betta
Bird	bird, budgies
Reptile	reptile, bearded dragon, pogona

To determine the top-level concepts in our taxonomy, we used a set of manually determined animal-specific keywords on the product title (Table 1). Based on keyword matching, we determined that approximately 64% were related to dogs, 30% were about cats, and the remainder were related to birds, fish, and reptiles.

Each pet's numerous product types, such as toys, bowls, cages, and carriers, can make classifying the second-level categories a challenging task. To simplify this process, we used our own judgment and referred to the general taxonomy to group the product types into a more limited number of common categories. The outcome of this classification is presented in Table 2, where the simplified second-level subcategories can be found.

To address data imbalance in training the text classifiers, we used stratified sampling with manual labeling of 1,000 data points. Stratification was applied to the Level 1 categories (Product Level) by examining the *product title* field to identify the associated animal category. In cases where the product title was ambiguous, product reviews were used for further identification. We then manually labeled each datapoint with the Level 2 subcategory as the label.

The labeled dataset enabled us to retrieve a total of 24,804 product transaction *reviews* (since there can be multiple reviews per product) to train, and then test, the text classifiers. We divided the manually labeled sample per Level 1 category into a training set (80% of sample) and test set (20% of sample). The division was conducted at the level of the category to ensure adequate representation of Level 1 categories in both sets[3].

Frequency statistics of the training dataset are tabulated in Table 2. One caveat to note is that the percentages in the table are computed over products, not reviews. Although the dataset contains over 2 million reviews, the number of unique products is much smaller (124,487 unique products).

[3] We could have also done a 'global' 80-20 split of the entire manually labeled sample of 1000 products, but this would not have guaranteed that all Level 1 categories were, in fact, represented. This is especially the case due to the 'long-tail' nature of the distribution: Level 1 categories like *Reptile* are significantly less prevalent in the sample and in the overall dataset than *Dog* or *Cat*.

3.3 Text Classification Methods and Metrics

We consider five established text classification systems in our experiments:

1. **Bag-of-words:** The bag-of-words, or TF-IDF, model is a classic approach for doing text classification. TF-IDF (Term Frequency-Inverse Document Frequency) is used to measure the importance of a term in a document within a collection of documents. It evaluates the significance of a term by considering both its frequency in the document and its rarity across the entire document collection. We used the TfidfVecorizer[4] available in the Python Sk-learn package for the vector transformation, removal of punctuation[5] and stop words[6]. We set the dimensionality to 100, by only retaining the 100 most

Table 2. Frequency statistics, with a product title example, at the sub-category level (top-level category is indicated by the first word in the sub-category) in the training partition. The frequency across all sub-categories for a top-level category sums to 100%.

Sub-category	Frequency (%)	Example of product title (in sub-category)
Dog Supply	34.23%	Lentek Bark Free Dog Training Device
Dog Accessories	33.71%	PetSafe Gentle Spray Bark Collar for Dogs
Dog Food, Treats	19.56%	Wellness Health Natural Wet Canned Dog Food
Dog Cleaning	8.85%	Mutt Mitt Dog Waste Pick Up Bag
Dog Pharmacy	3.65%	Dancing Paws Canine Multi Vitamin, 90-Counts

Sub-category	Frequency (%)	Example of product title (in sub-category)
Cat Supply	51.14%	PETMATE 26317 Cat Crazies Cat Toy
Cat Cleaning	21.83%	LitterMaid Universal Cat Privacy Tent (LMT100)
Cat Pharmacy	15.90%	NutraMax Dasuquin For Cats
Cat Food, Treats	8.32%	Sojos Certified Organic Catnip
Cat Accessories	2.81%	SSScat - The Ultimat Cat Control System

Sub-category	Frequency (%)	Example of product title (in sub-category)
Fish Gear	78.43%	Insten Adjustable Aquarium Submersible Fish Tank Water Heater
Fish Wellness	16.26%	Nualgi Ponds Fish Health and Controls Algae
Fish Food	5.31%	Healthy Pond, 3 Lb.Bag

Sub-category	Frequency (%)	Example of product title (in sub-category)
Bird Supply	62.26%	Wesco Pet Kabob Shreddable Bird Toy
Bird Food,Treats	24.27%	Kaytee Supreme Bird Food for Parakeets
Bird Cleaning	13.46%	Poop-Off Bird Poop Remover Sprayer, 32-Ounce

Sub-category	Frequency (%)	Example of product title (in sub-category)
Reptile Decor,Accessories	66.87%	Exo Terra Reptile Cave
Reptile Cleaning	20.07%	Tetra Reptile Decorative Filter
Reptile Bedding, Substrate	8.86%	Zilla Reptile Terrarium Bedding Substrate Litter Cleaner Corner Scoop
Reptile Food, Treats	4.30%	Zoo Med Reptile Calcium Without Vitamin D3, 48 oz

[4] https://scikit-learn.org/stable/modules/generated/sklearn.feature_extraction.text.TfidfVectorizer.html; Accessed: March 1, 2023.
[5] https://docs.python.org/3/library/string.html; Accessed: March 1, 2023.
[6] https://scikit-learn.org/stable/modules/feature_extraction.html#stop-words; Accessed: March 1, 2023.

important TF-IDF features. After we vectorized each review, we normalized[7] all the vectors using l2-norm. We then trained a logistic regression model[8] (with the *max iteration* parameter set to 10000) using 5-fold cross-validation to automatically determine the optimal value for the regularization parameter. The model directly gives us a probability that a review is associated with a product sub-category, which we can apply to the test set to compute performance metrics like precision and recall (described subsequently). Other methods described below work in a similar fashion, the only difference being the manner in which the text is vectorized. We also trained text classifiers for predicting the Level 1 product category, but almost all methods were able to get near-perfect performance on this simpler problem.

2. **FastText:** A second approach that we incorporate in the study is FastText, which is a *word embedding* technique that uses a neural network (in our case, the skip-gram architecture) to learn real-valued vectors by sliding a window over the text, and optimizing a function using the neural network. FastText breaks words down into subword units (character n-grams), allowing it to capture morphological information and handle out-of-vocabulary words. We preprocessed the text by tokenizing[9], removing punctuation and stop words and setting the vector dimensionality to 100. Then, we normalize the resulting vectors and conducting stratified sampling (similar to TF-IDF), we trained a logistic regression model using similar parameter values. When embedding the text, we used a window size of 15 as it allows the model to consider more distant words as context during training. We set the training to be performed using a single worker thread and the skip-gram algorithm.
3. **FastText (pre-trained):** This method is similar to the one described above. The only difference is that we used a pre-trained model instead. We first prepossessed the text data by tokenizing, and removing punctuation and stop words. Rather than using pet review data for deriving the embeddings, we used the pre-trained FastText model[10] available in the gensim Python package. We then use these embeddings to convert each review-text into a 300-dimensional vector, similar to the previously described FastText method, but with the only difference being that the embeddings were obtained from the pre-trained model, rather than being optimized from scratch on the review text-corpus. The same logistic regression methodology was used to train the model, just like with earlier text classification methods.
4. **GLoVE (Global Vectors for Word Representation):** We employed the 50-dimensional GloVe model[11] for generating word embeddings of customer reviews related to pet products. We followed the same preprocessing steps

[7] https://scikit-learn.org/stable/modules/generated/sklearn.preprocessing.normalize.html; Accessed: March 1, 2023.
[8] https://scikit-learn.org/stable/modules/generated/sklearn.linear_model.LogisticRegressionCV.html; Accessed: March 1, 2023.
[9] https://www.nltk.org/api/nltk.tokenize.html; Accessed: March 1, 2023.
[10] Specifically, "fasttext-wiki-news-subwords-300" accessed at https://fasttext.cc/docs/en/english-vectors.html.
[11] https://nlp.stanford.edu/projects/glove/; Accessed: March 1, 2023.

as described earlier, and then converted each review into a text vector by averaging the GloVe embeddings of the constituent words in the review text. The vectorized reviews were used to train a logistic regression model, exactly as described earlier.

5. **BERT:** BERT, or Bidirectional Encoder Representations from Transformers, uses transformer neural networks and significantly improved over the previous generation of word embeddings. It was introduced in a paper by Jacob Devlin and his colleagues from Google in 2018. As our final text classification method, we used BERT. In consideration of the high computational cost of using BERT, we used 25% of our manually labeled sample for training and testing, rather than the entire dataset. For deriving the vectors, we used a publicly available pre-trained model[12]. DistilBERT is a smaller, faster, cheaper, and lighter version of BERT. It is a result of a research project by Hugging Face and is described in [18]. To maintain conformance with the previous text classification methods, we continue to divide the manually labeled sample into an 80-20 split for training and testing purposes, followed by using a regularized logistic regression for the actual classification.

Of these five methods, we found that the FastText (pre-trained) generally performed comparably to (or even outperformed) both the FastText and GLoVE methods on all performance metrics. As all three of these methods rely on a previous generation of word embeddings (prior to the development of transformers, and subsequently, large language models), we only show results for the FastText (pre-trained) out of the three due to space limitations. However, we do comment on the difference between the FastText (pre-trained) and the other two omitted word embedding methods in the main text.

Finally, we note that the classification problem considered in this paper is a *multi-class* problem, in that each review-text can be assigned exactly one label from a set of *more than two* labels, distinguishing it from binary classification. This is because, even when controlling for the Level 1 category, there are usually more than two sub-categories per Level 1 category. To account for this, we train and apply the logistic regression using the *one vs. all* methodology; namely, for each sub-category, a logistic regression model is independently trained and evaluated. All review-texts with that sub-category as label are considered to have the positive label, and all other review-texts are considered to have a single 'negative' label (although technically, they are all negative in 'different' ways, since they are split across the other sub-categories). The *one vs. all* methodology is fairly standard in multi-class classification, although an alternative methodology, such as *one vs. one* could also be considered in future work.

For evaluating each of the classifiers, we use standard performance metrics, such as precision, recall and F1-measure (or harmonic mean, of precision and recall). Because the negative samples per classifier far outstrip the positive samples, the use of ordinary accuracy is not appropriate, as random performance is

[12] Specifically, the distilbert-base-uncased model accessed at https://huggingface.co/distilbert-base-uncased.

well above 90%. Similar such issues arise in related KG problem domains like entity resolution and information extraction [19–21]. Precision is defined as the ratio of true positives to the sum of true positives and false positives, while recall is the ratio of true positives to the sum of true positives and false negatives. The true negatives do not affect either calculation. We also show the *support* when reporting the results, defined as the number of actual (or positive) instances in the specific test set being evaluated.

3.4 Experimental Methodology

Although fundamentally, semantic typing of reviews can be modeled as a text classification problem, there are important methodological issues that have to be controlled for. The main reason is that a given product can have many independent reviews. To formalize the problem, let us assume that a pet product p (e.g., *Tick Collar for Cats and Kittens*) with sub-category P (e.g., *Cat Accessories*) is associated with a *set* R of $n >= 1$ reviews[13] $\{r_1, \ldots, r_n\}$. With this basic framework in place, we consider three reasonable experimental methodologies in this paper, denoted as RRR, RRP and PPP:

1. In the RRR approach, training, testing and evaluation are all conducted at the level of individual reviews. The text classifier, using one versus all logistic regression as described earlier, is set up so that it takes as input a single review text, and outputs a label (whether it belongs to the product sub-category associated with the trained logistic regression, or not). Each review in the test set is considered as a single data point. The product p is therefore a latent variable, and plays no role in training or testing.
2. In the RRP approach, the model is trained exactly as in the RRR approach. However, when evaluating the model on the test set, the product p is no longer consider as a latent variable, but as the unit of classification (i.e., each product p is now considered a unique data point, rather than each review corresponding to the product). Because the classifier is only trained to output a label per review, we derive a label for the product by taking the majority-label of the product's reviews as the predicted label for that product. In principle, this method is similar to ensemble methods, although we are aggregating over labels output by the same classifier over multiple reviews (inputs) rather than labels output by multiple classifiers over the same input.
3. In the PPP approach, we first consolidate all the reviews of a product into a single document, so that, in effect, there is only one 'review' per product. We do this both for the training and test sets. The label is still the product sub-category. Training, testing and evaluation are now all at the level of products than at the level of individual reviews. In principle, this method can be directly compared to the previous method, as the evaluation is at the

[13] Note that, because we have partitioned products into training and test sets, reviews would not 'straddle' the two sets: either *all* n reviews for a product would be allocated to the training set partition, or to the test set partition.

product level for both methods, and they are technically using the same 'information sets' even though they are using them differently. Neither method can be directly compared to the first (RRR) method, and we only include it in the experiments by way of comparison. In practice, either the RRP or PPP method would be selected for a semantic typing implementation, and one of our key experimental goals is to determine which one empirically yields higher performance.

4 Results

Table 3 shows the experimental results for three selected text classification methods at the sub-category level, assuming the RRR methodology. The table shows that classic methods, such as bag-of-words, tend to work quite well across the different sub-categories compared to the other methods. However, as expected, not all sub-categories are equally difficult. Even in sub-categories for popular pets like dogs and cats, some sub-categories (e.g., *Dog Pharmacy*) have performance that is surprisingly lower (70% F-Measure), at least on a relative basis. Generally, we found that when one method tends to decline in performance on a given sub-category, other methods do so as well, although the drop is not as visible. Later, in Sect. 5, we further elaborate on the performance-correlation between the methods.

Considering the two methods not shown in the table (GLoVE and corpus-trained FastText), we found that the corpus-trained FastText did outperform the pre-trained FastText (shown in the table) when using the RRR methodology, but it also tended to fail when too little data available was available for training, as is the case for the *Reptile Food, Treats* sub-category. Interestingly, these methods also tend to out-compete a much larger language model, appropriately fine-tuned, such as BERT. One reason may be that the data available is not enough for BERT to fine-tune on. Evidence for this can be found in the higher performance seen for BERT in sub-categories with thousands of samples in the support, compared to those with lower numbers of samples. Even in these cases, however, the bag-of-words is still found to perform quite well, attesting to its reputation as a method of first resort on problems like these.

Tables 4 and Table 5 respectively show the experimental results for the text classification methods at the sub-category level, assuming the RRP and PPPP methodology. Recall that, for these methodologies, evaluation was done at the product level, in contrast with the previous methodology (where evaluation was done at the review level). Because there are far fewer products than reviews, the support is much smaller in both Tables 4 and 5. The performance of the bag-of-words is now found to be near-perfect in Table 4, and improvements are also noted for some of the other methods (especially BERT, where for some sub-categories, near-perfect performance is observed). Nevertheless, the bag-of-words continues to obtain the best performance compared to the other methods.

In Table 5, differences in performance are more apparent, and the bag-of-words also does not perform as well. Note that one key difference between the

PPP and RRP methodology is that training is done at the 'product' level for the former, since all reviews for the product are now concatenated into one large text document, which is then tagged with the product sub-category. In contrast, for RRP, we first train and classify at the individual review level, and only aggregate these 'ensemble' classifications into a single product sub-category classification after the fact. The performance-contrast of PPP with RRP is especially noteworthy as it provides clear evidence that training and validating on reviews, even if the ultimate evaluation is conducted on product classification, is the best way to proceed methodologically rather than a more 'symmetric' or direct training (and testing) on products.

5 Discussion

Returning to the two motivating research questions with which we had begun this paper, the results showed that, for the first research question, classic methods like bag-of-words should be considered as a good initial baseline for addressing the semantic typing problem. The evidence does not directly lend itself to trying out more advanced methods like BERT without carefully evaluating whether the bag-of-words method achieves the desired performance that an e-commerce

Table 3. Results (Precision/Recall/F-Measure/Support) across three representative text classification methods using the RRR experimental methodology.

Sub-category	Bag-of-words	FastText (pre-trained)	BERT
Dog Supply	0.97/0.96/0.97/8224	0.8/0.8/0.8/8224	0.74/0.76/0.75/2055
Dog Accessories	0.96/0.98/0.97/8099	0.82/0.86/0.84/8099	0.78/0.82/0.8/2027
Dog Food, Treats	0.95/0.95/0.95/4701	0.77/0.79/0.78/4701	0.75/0.78/0.77/1179
Dog Cleaning	0.98/0.95/0.96/2127	0.85/0.75/0.8/2127	0.8/0.69/0.74/525
Dog Pharmacy	0.69/0.72/0.7/878	0.74/0.51/0.6/878	0.69/0.34/0.45/222
Cat Supply	0.96/0.97/0.97/12511	0.86/0.93/0.89/12511	0.84/0.94/0.89/3115
Cat Cleaning	0.94/0.94/0.94/5342	0.8/0.78/0.79/5342	0.8/0.74/0.77/1344
Cat Pharmacy	0.88/0.88/0.88/3890	0.76/0.73/0.74/3890	0.74/0.7/0.72/967
Cat Food, Treats	0.8/0.83/0.82/2035	0.74/0.57/0.65/2035	0.73/0.6/0.66/513
Cat Accessories	0.66/0.54/0.6/687	0.72/0.45/0.55/687	0.91/0.35/0.51/177
Fish Gear	1/1/1/284	0.88/0.96/0.92/284	0.81/95/0.88/74
Fish Wellness	0.98/0.92/0.95/59	0.71/0.59/0.65/59	0.25/0.08/0.12/13
Fish Food	0.82/0.95/0.88/19	0.67/0.11/0.18/19	0/0/0/4
Bird Supply	0.98/1/0.99/291	0.82/0.9/0.86/291	0.8/0.89/0.84/72
Bird Food, Treats	1/0.96/0.98/114	0.74/0.68/0.71/114	0.67/0.67/0.67/10
Bird Cleaning	0.98/0.95/0.97/63	0.76/0.56/0.64/63	0.86/0.4/0.55/15
Reptile Décor, Accessories	0.98/1/0.99/199	0.87/0.96/0.91/199	0.82/0.92/0.87/51
Reptile Cleaning	1/1/1/60	0.89/0.85/0.87/60	0.73/0.57/0.64/14
Reptile Bedding, Substrate	1/0.96/0.98/26	0.94/0.67/0.74/26	0.25/0.17/0.2/6
Reptile Food, Treats	1/0.77/0.87/13	0.67/0.15/0.25/13	0.33/0.25/0.29/14

Table 4. Results (Precision/Recall/F-Measure/Support) across three representative text classification methods using the RRP experimental methodology.

Sub-category	Bag-of-words	FastText (pre-trained)	BERT
Dog Supply	0.98/0.99/0.98/94	0.95/0.77/0.85/92	0.8/0.72/0.76/74
Dog Accessories	0.99/0.99/0.99/113	0.86/0.97/0.91/117	0.78/0.9/0.84/100
Dog Food, Treats	0.98/1/0.99/65	0.83/0.97/0.9/66	0.79/0.86/0.83/58
Dog Cleaning	1/1/1/18	0.89/0.84/0.86/19	0.81/0.76/0.79/17
Dog Pharmacy	1/0.92/0.96/25	0.92/0.52/0.67/21	0.83/0.29/0.43/17
Cat Supply	0.99/1/0.99/94	0.85/1/0.92/94	0.8/0.99/0.89/90
Cat Cleaning	1/1/1/41	0.92/0.8/0.86/41	0.89/0.92/0.91/37
Cat Pharmacy	1/1/1/31	0.83/0.91/0.87/33	0.89/0.86/0.87/28
Cat Food, Treats	1/1/1/47	0.98/0.83/0.9/48	0.96/0.58/0.72/43
Cat Accessories	1/0.94/0.97/17	0.9/0.53/0.87/17	0.9/0.64/0.75/14
Fish Gear	1/1/1/13	0.71/0.94/0.81/18	0.61/1/0.76/14
Fish Wellness	1/1/1/10	0.75/0.67/0.71/9	0/0/0/6
Fish Food	1/1/1/3	0/0/0/5	0/0/0/3
Bird Supply	1/1/1/25	0.87/0.93/0.9/29	0.58/0.93/0.72/15
Bird Food, Treats	1/1/1/16	0.92/0.85/0.88/13	0.83/0.42/0.56/12
Bird Cleaning	1/1/1/5	0.6/0.5/0.55/6	1/0.4/0.57/5
Reptile Décor, Accessories	1/1/1/10	0.62/1/0.77/10	0.55/0.67/0.6/9
Reptile Cleaning	1/1/1/1	1/1/1/1	0/0/0/1
Reptile Bedding, Substrate	1/1/1/2	1/0.25/0.4/4	0/0/0/3
Reptile Food, Treats	1/1/1/5	1/0.4/0.57/5	1/0.5/0.67/4

organization might desire. These methods also tend to be more computationally intensive compared to algorithms like TF-IDF. More generally, however, the results suggest that semantic typing is feasible using review text alone. To the best of our knowledge, such feasibility has not been demonstrated in prior studies; rather, far greater focus was placed on semantic typing using text that was believed to be directly pertinent to semantic typing, such as a description of the product. An intriguing question arises as to whether it might be possible to semi-automatically 'construct' a product ontology using only a large corpus of review texts. We leave investigating such a possibility for future research, but we note that it would have significant implications for e-commerce organizations, as it would allow them to use reviews for semantic typing, which they may not have not considered before.

Concerning the second research question, we find that some techniques do encounter the 'long-tail' problem when semantically typing review texts corresponding to product sub-categories like *Reptile Cleaning*. The bag-of-words is found to be more robust, perhaps due to it not being as parameter-heavy as the newer neural networks. Finally, we find that the manner in which the training and testing is set up (the three experimental methodologies i.e., RRR, RRP

Table 5. Results (Precision/Recall/F-Measure/Support) across three representative text classification methods using the PPP experimental methodology.

Sub-category	Bag-of-words	FastText (pre-trained)	BERT
Dog Supply	0.89/0.76/0.82/21	0.94/0.76/0.84/21	0.5/0.38/0.43/21
Dog Accessories	0.9/0.96/0.93/27	0.93/0.96/0.95/27	0.7/0.85/0.77/27
Dog Food, Treats	0.79/0.79/0.79/14	0.93/0.93/0.93/14	0.71/0.84/0.77/14
Dog Cleaning	1/0.8/0.89/5	1/0.6/0.75/5	0/0/0/5
Dog Pharmacy	0.67/0.86/0.75/7	0.5/0.86/0.63/7	0.5/0.43/0.46/7
Cat Supply	0.9/0.9/0.9/20	0.91/1.0/0.95/20	0.65/0.75/0.7/20
Cat Cleaning	0.86/0.75/0.8/8	1/0.75/0.86/8	0.5/0.5/0.5/8
Cat Pharmacy	1/0.71/0.83/7	0.71/0.71/0.71/7	0.29/0.29/0.29/7
Cat Food, Treats	0.79/1/0.88/11	0.91/0.0.91/0.91/11	0.6/0.55/0.57/11
Cat Accessories	0.5/0.5/0.5/4	0.75/0.75/0.75/4	0/0/0/4
Fish Gear	0.75/1/0.86/6	0.86/1/0.92/6	0.6/1/0.75/6
Fish Wellness	1/0.33/0.5/3	0.67/0.67/0.67/3	0/0/0/3
Fish Food	1/1/1/1	0/0/0/1	0/0/0/1
Bird Supply	0.88/0.88/0.88/8	1/0.75/0.86/8	0.8/1/0.89/8
Bird Food, Treats	1/0.75/0.86/4	0.8/1/0.89/4	0.67/0.5/0.57/4
Bird Cleaning	0.5/1/0.67/1	0.5/1/0.67/1	0/0/0/1
Reptile Décor, Accessories	0.6/0.75/0.67/4	0.8/1/0.89/4	0.67/1/0.8/4
Reptile Cleaning	0/0/0/0	NA/NA/NA/0	NA/NA/NA/0
Reptile Bedding, Substrate	0/0/0/1	1/1/1/1	0/0/0/1
Reptile Food, Treats	0/0/0/1	0/0/0/1	0/0/0/1

and PPP) can significantly influence performance estimates. Since the ultimate goal is to semantically type reviews into product sub-categories, the RRP and PPP methodologies are more aligned with the goal than the RRR methodology. Between the RRP and PPP methodology, the former is clearly a better way of modelling the problem, as it allows for a higher quality output without using any extra information.

Finally, it is instructive to consider the *correlation* between the different text classifiers. Specifically, when one method performs badly on a sample, do other methods do so as well, or are different methods labelling different instances incorrectly? Table 6 provides a 'confusion matrix' of instance-counts where one method is correct (column) but another is incorrect (row). Congruent with the results in Sect. 4, we find that there are few instances where the bag-of-words is incorrect and the other methods are correct; however, these are also not trivially small (>50), and on a larger sample, would still count for significant errors. At the same time, we do not see a correlation or pattern in the confusion matrix (for the non bag-of-words methods) that would seem to suggest a bias.

We also provide some qualitative examples in Tables 7 and 8 for specific reviews where none of the methods obtained a correct result, and a case where

2 out of 4 methods obtained a correct result. We also indicate which systems classified incorrectly, and what the 'wrong' label was. These instances illustrate the difficulty of the problem, and potential avenues for future research to improve the outcome.

Table 6. A count of the number of test instances, assuming the RRR methodology, that the method indicated in the column header classified *correctly*, but that the method indicated in the row classified *incorrectly*. The BERT system is not included, as it is not tested on exactly the same sample as the other methods (only 25% of the sample, as detailed earlier).

Model	Bag-of-words	FastText	FastText (pre-trained)	GLoVE
Bag-of-words	NA	67	68	61
FastText	7019	NA	1855	1707
FastText-pre-trained	9052	3887	NA	2390
GLoVE	14907	9061	8252	NA

Table 7. Examples of a review (Cat Accessories sub-category), with product title, and other details of which system(s) wrongly classified the sub-category, with the wrong classification.

Cat Accessories	Review Body	Product Title	System Classified Correctly	Wrong Classification
All Incorrect	Great product! For nine months we had a problem with our 5-year old cat potying in our kitchen rather than the catbox. I took her to the vet several times, spent hundreds of dollars on vet bills looking for a reason in order to treat it. Nothing made a difference. I bought this product 2 months ago. IT WORKED! She is back to using the catbox and our kitchen smells like a kitchen again.	Catit Design Senses Food Maze	0	Food and treats
2/4 incorrect	Collars are great for my cats. The break-away feature is great un case they get caught on something. Takes care of ticts and fleas says mosquitoes too. I buy these every year. They also last 8 months and Adam Plus Breakaway Flea	Tick Collar for Cats and Kittens, 13"	Bag-of-words, FastText	Pharmacy

Table 8. Examples of a review (*Dog Pharmacy* sub-category), with product title, and other details of which system(s) wrongly classified the sub-category, with the wrong classification.

Dog Pharmacy	Review Body	Product Title	System Classified Correctly	Wrong Classification
All Incorrect	I own stud dogs and breed English bulldogs. It's always been important for me to have healthy quality dogs but, I've had a real problem with one of my males. He's always been thin and has looked unhealthy I've really strugled with him. I've taken him to vets, Ive given him high calorie foods I've tried everything and haven't been able to get him looking the way I'd like him to look. Luckily Im always on Amazon just looking at thing and I came across the bully supplements. The price was right so I figured I'd give it a try I had nothing to lose. It's been a little over a month and suprisingly Ive seen quiet some improvement. Walter is looking better and eating more then normal I can't wait to see him in 6 mobths. What's been a real suprise is that I also started my other male on the supplements and wow he's looking really good. I call him my little beefcake. I highly recommend this product to everyone owning a bully breed. You can't go wrong.	Vita Bully Vitamins for Bully Breeds: Pit Bulls, American Bullies, Exotic Bullies, Bulldogs, Pocket Bullies, Made in the USA	0	Food and treats
2/4 incorrect	The product is great. The seal On 1 bottle was broken. Half of the bottle leaked out. Thanks, Sharon Ames I am not returning it. Can you replace it?Thanks again and EcoEars Dog Ear Cleaner Infection Formula. For Itch, Head Shaking, Discharge	EcoEars Dog Ear Cleaner Infection Formula. For Itch, Head Shaking, Discharge and Smell. Natural Multi Symptom Ear Cleaner for Cleaning Away Most Dog Ear Problems. 100 Guaranteed	Bag-of-words, FastText	Food and treats, Supply

6 Conclusion

This paper considered the problem of semantically typing pet products using only reviews provided for them, rather than descriptions or product titles. Results were promising and showed that this is indeed feasible, and that classic methods can serve as robust solutions. Language models and word embeddings are also promising, but are computationally more intensive, and are susceptible to the long-tail phenomenon. The experimental results also suggest many possibilities for future research, including conducting semantic typing using a much larger taxonomy or ontology, deeper error analysis, and strategic use of generative large language models like ChatGPT.

References

1. Kejriwal, M.: Domain-Specific Knowledge Graph Construction. Springer, Cham (2019). https://doi.org/10.1007/978-3-030-12375-8
2. Ehrig, M.: Ontology Alignment: Bridging the Semantic Gap. Springer, New York (2006). https://doi.org/10.1007/978-0-387-36501-5
3. Kejriwal, M., Shen, K., Ni, C.-C., Torzec, N.: An evaluation and annotation methodology for product category matching in e-commerce. Comput. Ind. **131**, 103497 (2021)
4. Cho, Y.H., Kim, J.K.: Application of web usage mining and product taxonomy to collaborative recommendations in e-commerce. Expert Syst. Appl. **26**(2), 233–246 (2004)
5. Kejriwal, M., Shen, K., Ni, C.-C., Torzec, N.: Transfer-based taxonomy induction over concept labels. Eng. Appl. Artif. Intell. **108**, 104548 (2022)
6. Zhang, W., Cao, H., Lin, L.: Analysis of the future development trend of the pet industry. In: 2022 7th International Conference on Financial Innovation and Economic Development (ICFIED 2022), pp. 1682–1689. Atlantis Press (2022)
7. Bakos, Y.: The emerging landscape for retail e-commerce. J. Econ. Perspect. **15**(1), 69–80 (2001)
8. Nakashole, N., Tylenda, T., Weikum, G.: Fine-grained semantic typing of emerging entities. In: Proceedings of the 51st Annual Meeting of the Association for Computational Linguistics (Volume 1: Long Papers), pp. 1488–1497 (2013)
9. Kejriwal, M., Szekely, P.: Supervised typing of big graphs using semantic embeddings. In: Proceedings of the International Workshop on Semantic Big Data, pp. 1–6 (2017)
10. Kapoor, R., Kejriwal, M., Szekely, P.: Using contexts and constraints for improved geotagging of human trafficking webpages. In: Proceedings of the Fourth International ACM Workshop on Managing and Mining Enriched Geo-Spatial Data, pp. 1–6 (2017)
11. Kejriwal, M., Szekely, P.: Scalable generation of type embeddings using the ABox. Open J. Semant. Web (OJSW) **4**(1), 20–34 (2017)
12. Gene Ontology Consortium: Creating the gene ontology resource: design and implementation. Genome Res. **11**(8), 1425–1433 (2001)
13. Kejriwal, M., Knoblock, C.A., Szekely, P.: Knowledge Graphs: Fundamentals, Techniques, and Applications. MIT Press, Cambridge (2021)
14. Dong, X.L.: Challenges and innovations in building a product knowledge graph. In: Proceedings of the 24th ACM SIGKDD International Conference on Knowledge Discovery & Data Mining, pp. 2869–2869 (2018)
15. Kowsari, K., Jafari Meimandi, K., Heidarysafa, M., Mendu, S., Barnes, L., Brown, D.: Text classification algorithms: a survey. Information **10**(4), 150 (2019)
16. Minaee, S., Kalchbrenner, N., Cambria, E., Nikzad, N., Chenaghlu, M., Gao, J.: Deep learning-based text classification: a comprehensive review. ACM Comput. Surv. (CSUR) **54**(3), 1–40 (2021)
17. Grandini, M., Bagli, E., Visani, G.: Metrics for multi-class classification: an overview. arXiv preprint arXiv:2008.05756 (2020)
18. Sanh, V., Debut, L., Chaumond, J., Wolf, T.: Distilbert, a distilled version of bert: smaller, faster, cheaper and lighter. arXiv preprint arXiv:1910.01108 (2019)
19. Gheini, M., Kejriwal, M.: Unsupervised product entity resolution using graph representation learning. In: eCOM@ SIGIR (2019)

20. Sarawagi, S., et al.: Information extraction. Found. Trends® Databases **1**(3), 261–377 (2008)
21. Kejriwal, M.: A meta-engine for building domain-specific search engines. Softw. Impacts **7**, 100052 (2021)

A Modular Ontology for MODS – Metadata Object Description Schema

Rushrukh Rayan[1(✉)], Cogan Shimizu[2], Heidi Sieverding[3], and Pascal Hitzler[1]

[1] Kansas State University, Manhattan, KS 66502, USA
{rushrukh,hitzler}@ksu.edu
[2] Wright State University, Dayton, OH 45435, USA
cogan.shimizu@wright.edu
[3] South Dakota School of Mines & Technology, Rapid City, SD 57701, USA
heidi.sieverding@sdsmt.edu

Abstract. The Metadata Object Description Schema (MODS) was developed to describe bibliographic concepts and metadata and is maintained by the Library of Congress. Its authoritative version is given as an XML schema based on an XML mindset which means that it has significant limitations for use in a knowledge graphs context. We have therefore developed the Modular MODS Ontology (MMODS-O) which incorporates all elements and attributes of the MODS XML schema. In designing the ontology, we adopt the recent Modular Ontology Design Methodology (MOMo) with the intention to strike a balance between modularity and quality ontology design on the one hand, and conservative backward compatibility with MODS on the other.

1 Introduction

XML – a markup language – is designed to organize information [6]. The main design goal is to store and share information while maintaining human and machine readability. Also, the purpose of XML Schema is to serve as a description for an XML document, within it detailing the constraints on the structure, syntax, and content type. The schema outlines rules and constraints for elements, attributes, data types and relationships between them. It also helps ensure that the XML document conforms with the expected structure, serving as a way of validation. It is important to note that XML structures information in a hierarchical form, essentially representing a tree structure.

The Metadata Object Description Schema (MODS) [7] is an XML schema developed by the Library of Congress' Network Development in 2002 to be used to describe a set of bibliographic elements. MODS contains a wide range of elements and attributes using which a well-rounded description can be provided about bibliographic elements. For instance, it has *elements* to describe Title Information, Type of Resource, Genre of Resource, Origin Information, Target Audience, Access Restrictions of the material, etc. Furthermore, MODS also has

F. Ortiz-Rodriguez et al. (Eds.): KGSWC 2023, LNCS 14382, pp. 168–182, 2023.
https://doi.org/10.1007/978-3-031-47745-4_13

attributes to outline additional important information, to name a few: Display Label (describes how the resource under description is being displayed), Lang (points to the language that is used for the content of an element: imagine a book title that is French), Authority (specifies the organization that has established the usage of, for instance, an acronym), etc. General example use-cases of MODS lie within the realm of describing metadata of Journal Publications (one or more), Research Projects, Experiments, Books, etc.

While XML schema does a decent job in imposing structure on XML data, it lacks some desirable features. In the age of data, where cleaning, pre-processing, and managing data takes up a large chunk of resources in data operation, it is desirable to have the ability to organize data in such a way that allows semantic expressiveness of the data and conveys information on relationships between various *concepts* by means of a *graph structure* [5] as opposed to the XML tree structure, in the sense of modern knowledge graphs [3], e.g. based on RDF [2] and OWL [8]. An XML schema

- lacks semantic expressiveness to convey relationship among concepts, context of data;
- lacks native support for automated reasoning and inference;
- lacks a common framework that allows integration of data from various sources;
- possesses a hierarchical nature with a rigid structure which makes it rather less flexible with respect to incorporation of different perspectives;
- and lacks native support for querying.

Ontologies as knowledge graph schemas, on the other hand, provide a structured and graph-based way to represent knowledge in an application domain. By defining the necessary vocabulary, concepts, entities, and relationship between concepts, ontologies allow a meaningful interpretation of the data.

The reason we have developed the Modular MODS Ontology (MMODS-O) is to address some of the challenges which the MODS XML schema exhibits. Indeed MMODS-O is designed to strike a balance between conservative backward-compatibility with the MODS XML schema and quality modular ontology design principles following the MOMo methodology [10,11]. The modular structure in particular is supportive of simplified extending, modifying or removing parts of the ontology.

We have created 34 modules and patterns to capture the entire MODS XML schema. To provide semantic robustness, we have re-engineered some of the modules from their XML schema definition. The schema is expressed in the form of an OWL Ontology and extensive documentation is available on Github[1].

One of our target use-cases for MMODS-O is to provide a metadata structure to a large-scale collaborative research project, where the knowledge graph would contain information such as different research groups, experiments performed, geo-location information, associated publications, presentations, book-chapters, collaborators etc.

[1] https://github.com/rushrukh/mods_metadata_schema/tree/main/documentation.

We would like to point out that this is not the first attempt towards developing a MODS Ontology. However, our version is an improvement over the existing ontology across multiple aspects, including modular structure, adherence to MOMo quality control principles, rich axiomatization, extensive documentation. We will outline some of the key improvements over previous work in Sect. 3. In general, our contributions are:

1. Development of the modular ontology, where some of the modules differ significantly from the original MODS XML schema in order to reflect good ontology design principles.
2. Carefully considered and rich axiomatization to scope intended usage and to provide automated reasoning capabilities.
3. Complete documentation of the graph schema outlining each of the modules, associated axioms, competency questions.

The rest of the paper is organized as follows. Section 2 contains the description of key modules from our ontology. In Sect. 3, we describe related work and highlight some of the key differences of our modeling with previous efforts. We conclude in Sect. 4. The ontology is available as serialized in the Web Ontology Language OWL from https://github.com/rushrukh/mods_metadata_schema/tree/main/modules.

2 Description of the MODS Ontology

The general usage of the MMODS-O (and MODS) lies in the realm of expressing bibliographic metadata. Indeed, the details in the XML schema reflect the association with bibliographic applications. From the top level elements and their attributes in the MODS XML schema, we have identified 34 modules to be part of MMODS-O. Some of the key modules are briefly described below. The primary goal of using formal axiomatization[2] in MOMo is to limit unintended use and to disambiguate the modules, but axioms can also be used for logical inferences [4]. The axioms are expressed using the OWL 2 DL profile [8]. Note that for all the modules outlined here, the list of axioms is not complete as we only highlight some of the most important axioms for brevity. The complete list of axioms and modules can be found in the documentation pointed to earlier.

The modules that we selected for presentation in this paper include some that deviate most from the underlying MODS XML schema. We touch upon the differences throughout and will discuss them further in Sect. 3.

We make extensive use of schema diagrams when discussing modules following the suggested visual coding from the MOMo methodology [10] where further explanations can be found: orange (rectangular) boxes indicated classes; teal (dashed) boxes indicate other modules (and usually also the core class of that module); purple (dashed) boxes with *.txt* indicate controlled vocabularies (i.e., formally, classes with pre-defined individuals as members, which have

[2] A primer on description logic and the notation can be found in [1,5].

meaning that is defined outside the ontology); yellow (ovals) indicate datatype values; white-headed arrows are rdfs:subClassOf relationships, all other arrows are object or data properties, depending on type of node pointed to.

2.1 Overview of the Modules in the Ontology

Figure 1 represents a brief overview of all the modules that are part of the ontology. Each of the modules has its separate schema. MODS Item is a reference to the MODS resource under description. The ontology has 34 modules, while we highlight some of the key modules later in the paper, details about the other modules are available in the documentation. Figure 1 suggests almost a tree structure, which is actually not the case but this is not quite apparent from this high-level perspective.

2.2 Role-Dependent Names

Role-Dependent Names is an ontology design pattern [4, 10] that is useful when there is an Agent Role that is performed by an Agent. Naturally, Agent will have a Name. There are instances when an Agent assumes a Role under a particular Name, but the same Agent will assume a different role under a different Name. An example for such a scenario would be a writer writing different books under different pseudonyms. For example, Ian Banks publishes science fiction as "Iain M. Banks" and mainstream fiction as "Iain Banks". Another example use case within the application scope we are primarily interested in could be as follows: if the resource under description refers to a journal publication, there would be Agent Roles for authors, which would be assumed by Agents under some name. Note that names associated with an author may differ between different publications for a variety of reasons, including different transcriptions from other languages, inclusion or not of middle names, name changes, etc., and the MODS XML schema reflects this. While we do not discuss the ontology design pattern at length here, details can be found in [9] (Fig. 2).

Selected Axioms

$$\top \sqsubseteq \leq 1 \mathsf{providesAgentRole}^{-}.\top \quad (1)$$

$$\mathsf{AgentRole} \sqsubseteq \geq 0 \mathsf{hasRoleUnderName}.\mathsf{Name} \quad (2)$$

$$\exists \mathsf{assumesAgentRole}.\mathsf{Agent} \sqsubseteq \mathsf{AgentRole} \quad (3)$$

$$\mathsf{AgentRole} \sqsubseteq \leq 1 \mathsf{assumesAgentRole}^{-}.\mathsf{Agent} \quad (4)$$

$$\mathsf{Agent} \sqsubseteq \geq 0 \mathsf{assumesAgentRole}.\mathsf{AgentRole} \quad (5)$$

$$\mathsf{Agent} \sqsubseteq \exists \mathsf{hasName}.\mathsf{Name} \quad (6)$$

$$\mathsf{assumesAgentRole} \circ \mathsf{hasRoleUnderName} \sqsubseteq \mathsf{hasName} \quad (7)$$

$$\mathsf{hasName} \circ \mathsf{hasRoleUnderName}^{-} \sqsubseteq \mathsf{assumesAgentRole} \quad (8)$$

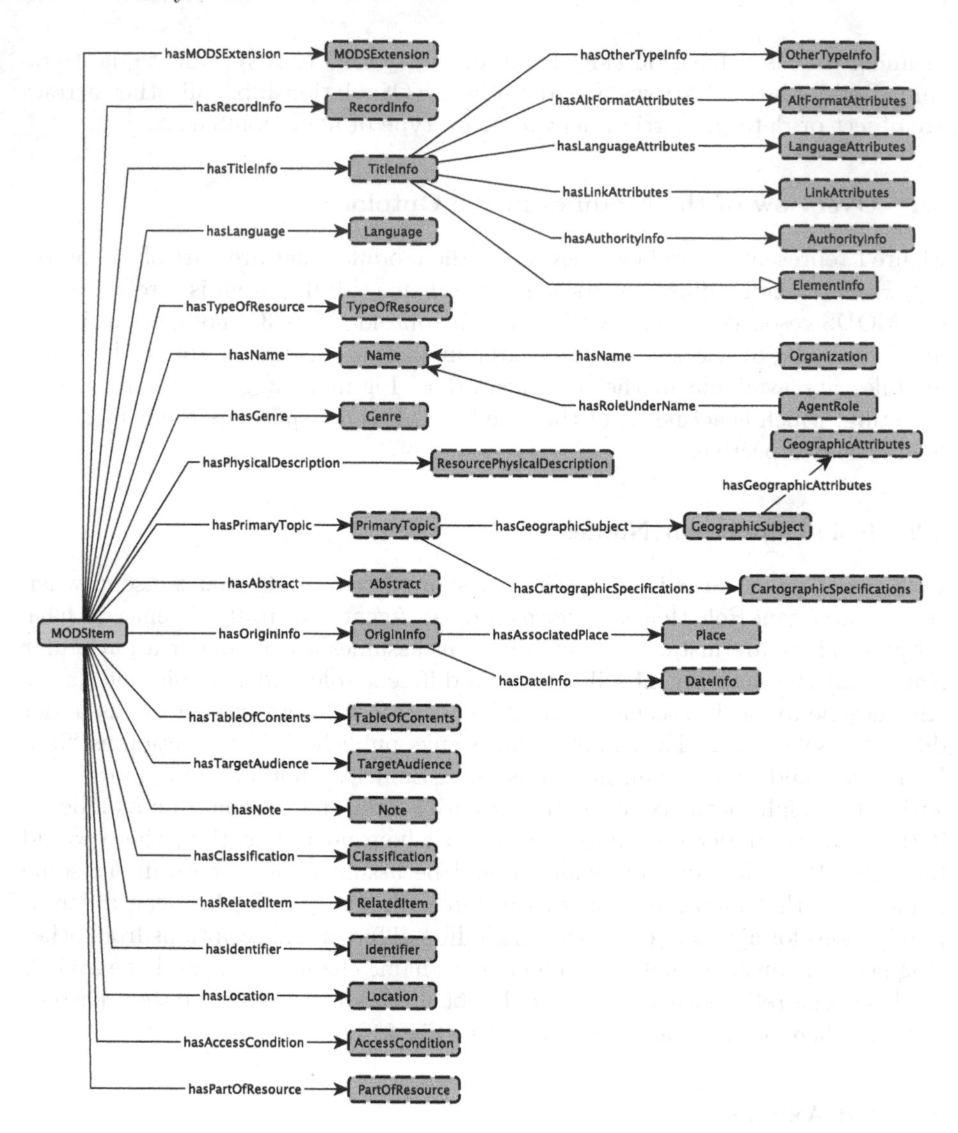

Fig. 1. An Overview of all the Modules

If an Agent Role is provided, we argue that there must be at most 1 entity that provides the role which is expressed using an *inverse functionality* in (1). Furthermore, we claim that if an Agent Role is assumed, there can be at most 1 Agent who assumes the role, expressed through an *inverse qualified scoped functionality* in (4). Axioms (1) and (4) essentially state that an AgentRole is unique to both the Agent and the entity providing the role, i.e., these axioms give guidance as to the graph structure for the underlying data graph. It is not necessary for an Agent Role to be assumed under a Name which is why we use a *structural*

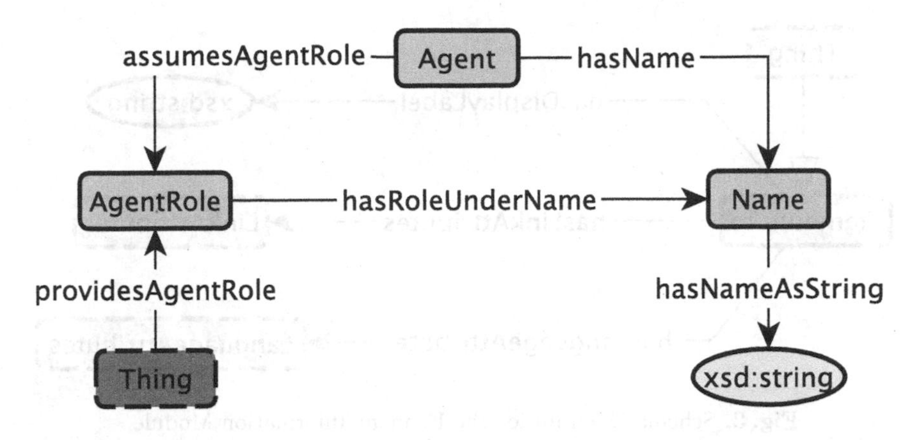

Fig. 2. Schema Diagram for the Role-Dependent Names Pattern

tautology in (2).[3] We also argue that, naturally, an Agent must have a name. Hence we use an *existential* to convey that in (6).

The Role-Dependent Names module exemplifies very well why an RDF graph structure is much more natural than an XML tree structure for expressing relevant relationships. In particular, the *triangular* relationships indicated by the role chain axioms (7) and (8) cannot be naturally captured in a tree structure, but really demand a directed graph.

2.3 Element Information

There are many elements within the MODS XML schema which may have a display label, a combination of attributes that provide external links, and a set of attributes to describe the language for the resource under description. The Element Information module is created such that the aforementioned connections can be expressed conveniently. Concretely, whenever in a module it needs to be said that the module may have a Display Label, Link Attributes, and Language Attributes, we use the module to be a sub-class of the module Element Information which is expressed using a *sub-class of* relationship in (9) (Fig. 3).

Selected Axioms

$$\top \sqsubseteq \mathsf{ElementInfo} \tag{9}$$

$$\top \sqsubseteq {\leq}1\mathsf{hasLinkAttributes}.\top \tag{10}$$

$$\mathsf{ElementInfo} \sqsubseteq {\geq}0\mathsf{hasLinkAttributes}.\mathsf{LinkAttributes} \tag{11}$$

$$\top \sqsubseteq \forall \mathsf{hasLanguageAttributes}.\mathsf{LanguageAttributes} \tag{12}$$

$$\top \sqsubseteq {\leq}1\mathsf{hasLanguageAttributes}.\top \tag{13}$$

$$\mathsf{ElementInfo} \sqsubseteq {\geq}0\mathsf{hasLanguageAttributes}.\mathsf{LanguageAttributes} \tag{14}$$

[3] Structural tautologies are logically inert, however they provide structural guidance on use for the human using an ontology; see [10].

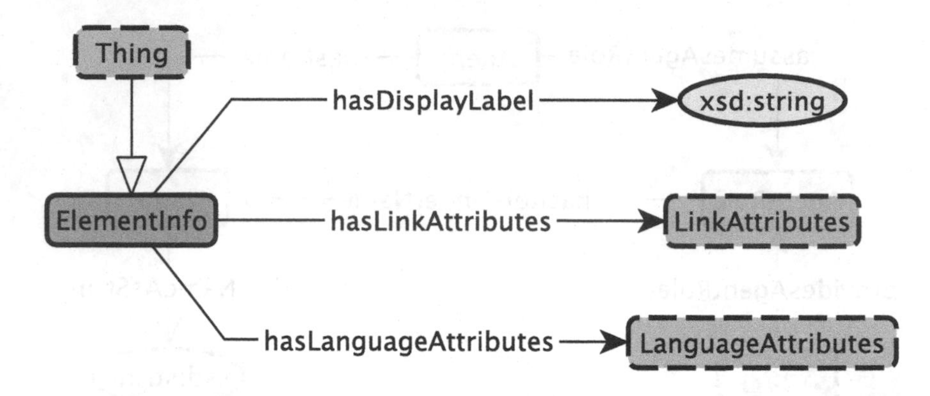

Fig. 3. Schema Diagram for the Element Information Module

A module which is a sub-class of Element Information can have at most 1 set of Link Attributes and 1 set of Language Attributes which in axioms have been conveyed using *functionalities* in (10) and (13). Additionally, it is not mandatory for a module to have a set of Link Attributes and Language Attributes, therefore we make use of *structural tautologies* in (11) and (14).

2.4 Organization

The Organization module works in conjunction with the Role-Dependent Names and Name module. It is important to note that the MODS XML schema does not have an element named Organization. In order to instill natural semantics into the ontology, we introduce the Organization module to replace the attribute "Affiliation" and element "Alternative Names". The concrete differences are outlined in Sect. 3. Organization is used as the main entity which provides an Agent Role. Naturally, it makes sense for an organization to have a Name. In the case where an organization is referred to using different names, we denote the primary name with *hasStandardizedName* and the rest of the names using *hasName* (Fig. 4).

Selected Axioms

$$\mathsf{Organization} \sqsubseteq \geq 0 \mathsf{providesAgentRole.AgentRole} \tag{15}$$

$$\mathsf{Organization} \sqsubseteq \exists \mathsf{hasName.Name} \tag{16}$$

$$\mathsf{Organization} \sqsubseteq \geq 0 \mathsf{hasStandardizedName.Name} \tag{17}$$

$$\top \sqsubseteq \leq 1 \mathsf{hasLinkAttributes}.\top \tag{18}$$

$$\mathsf{Organization} \sqsubseteq \geq 0 \mathsf{hasLinkAttributes.LinkAttributes} \tag{19}$$

It is not necessary that the Organization under description must provide an Agent Role. It can be referred in any general context, as such we say in (15) that an Organization *may* provide an Agent Role by using a *structural tautology.*

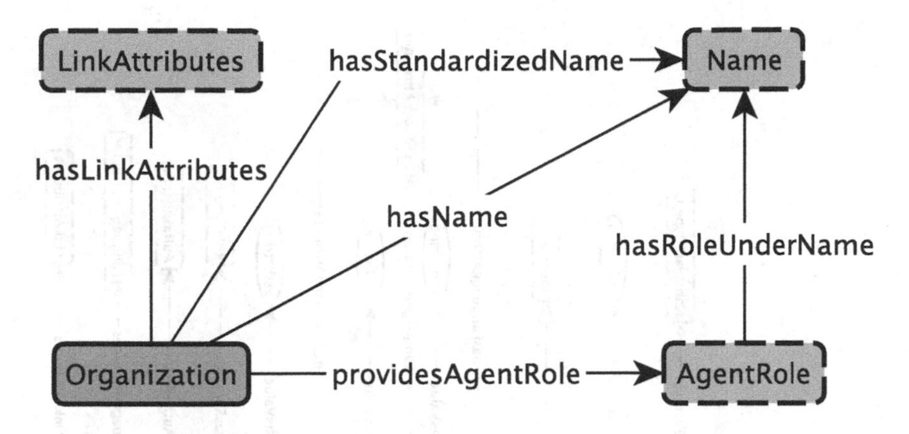

Fig. 4. Schema Diagram for the Organization Module

Furthermore, we argue that an Organization, naturally, must have a name and express that using an *existential* in (16). To distinguish between different names and the standardized name, we use (17) to say that the Organization *may* have a Standardized Name. Also an Organization *may* have a set of Link Attributes to provide additional information (19).

2.5 Name

The Name module is intended to be used for describing entities associated with the resource under description which may have one or more names. A necessary element of the Name module is Name Part. All the parts of a name (one or more) are described through Name Parts. In some cases, a name can refer to an acronym which is dictated by some Authority where the information regarding authority is expressed using Authority Information module. It is not uncommon for a name to have a specific form to display (e.g. Last name, First name), which is specified using Display Form. Furthermore, if a name has an associated identifier (e.g. ISBN, DOI), it is expressed using Name Identifier which is a sub-class of the module Identifier.

In the Name module, there are a few controlled vocabulary nodes (purple nodes in Fig. 5). To begin with, a Name can be assigned with a Name Type. MODS XML schema allows 4 name types: Personal, Corporate, Conference, Family. To let the user select a value from the available options, we make use of controlled vocabulary. Similarly, if among multiple instances of names, one particular name is to be regarded as the primary instance, the controlled vocabulary Usage is used to identify that. Another example of controlled vocabulary's usage can be seen in Name Part Type. To identify a part of name to be first name, middle name, or last name the Name Part Type controlled vocabulary can be used.

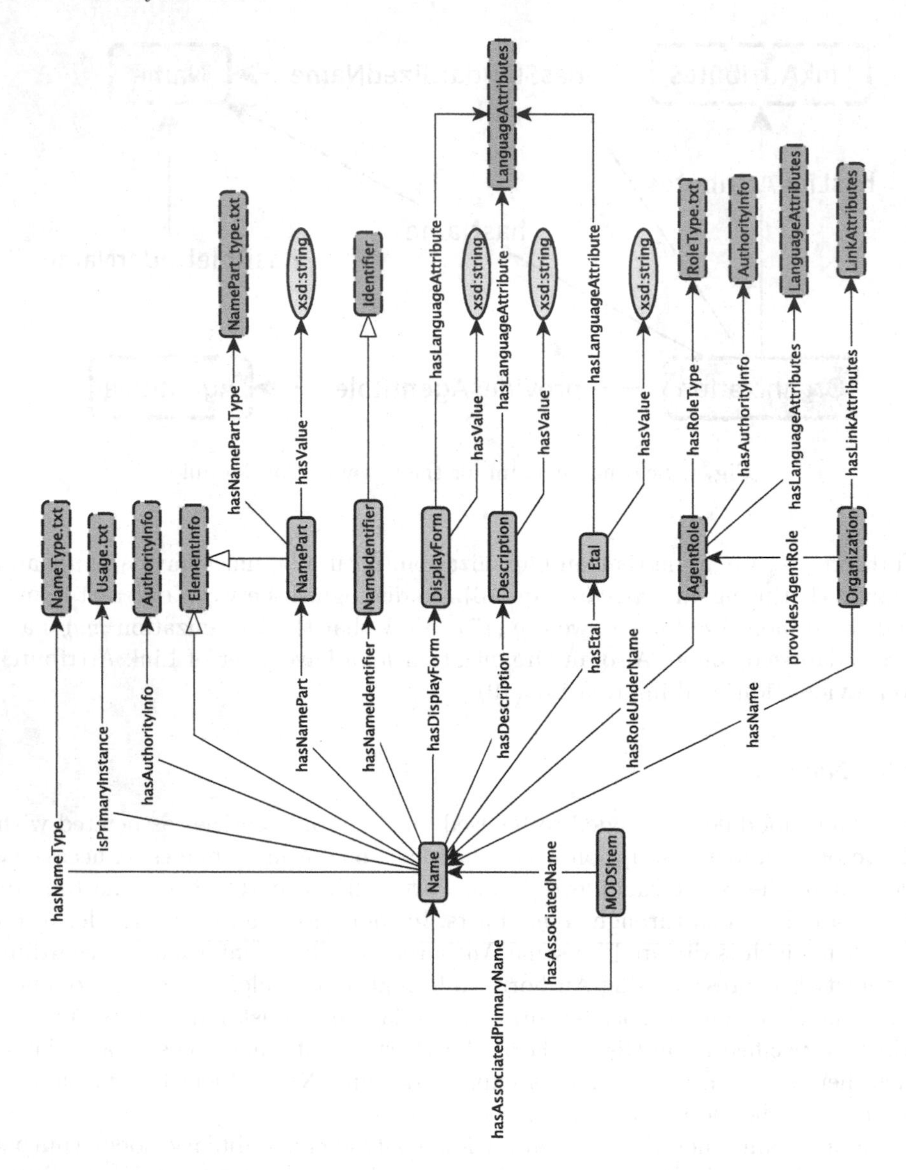

Fig. 5. Schema Diagram for the Name Module

Selected Axioms

$$\mathsf{Name} \sqsubseteq \exists \mathsf{hasNamePart.NamePart} \quad (20)$$

$$\mathsf{NamePart} \sqsubseteq \mathsf{hasNamePart}^{-}.\mathsf{Name} \quad (21)$$

$$\top \sqsubseteq \leq 1 \mathsf{hasNamePart}^{-}.\top \quad (22)$$

$$\mathsf{Name} \sqsubseteq \geq 0 \mathsf{hasNamePart.NamePart} \quad (23)$$

$$\top \sqsubseteq \forall \mathsf{hasNamePartType.NamePartType.txt} \quad (24)$$

$$\mathsf{Name} \sqsubseteq \geq 0 \mathsf{hasDescription.Description} \quad (25)$$

$$\mathsf{Name} \sqsubseteq \geq 0 \mathsf{hasNameType.NameType.txt} \quad (26)$$

$$\mathsf{Name} \sqsubseteq \geq 0 \mathsf{isPrimaryInstance.Usage.txt} \quad (27)$$

$$\top \sqsubseteq \leq 1 \mathsf{hasAuthorityInfo}.\top \quad (28)$$

$$\mathsf{Name} \sqsubseteq \geq 0 \mathsf{hasAuthorityInfo.AuthorityInfo} \quad (29)$$

$$\mathsf{NamePart} \sqsubseteq \mathsf{ElementInfo} \quad (30)$$

$$\mathsf{NamePart} \sqsubseteq \neg(\exists \mathsf{hasLinkAttributes}.\exists \mathsf{hasID}.\top) \quad (31)$$

$$\mathsf{NameIdentifier} \sqsubseteq \mathsf{Identifier} \quad (32)$$

As described in the beginning of this module, a Name must have at least one NamePart. Otherwise, having a Name which does not have any string value as part of it would not be natural. We express this using an *existential* in (20). On the other hand, to restrict the usage of NamePart outside of Name, we use an *inverse existential* to convey that if there is a hasNamePart property, its domain must be a Name. A Name can also have any number of NameParts, to allow which we use *structural tautology* in (23). Axioms (20) and (23) together mean that there can be one or more NameParts.

The Name module is a *sub-class* of Element Information (30) which says that a Name instance may have a set of Link Attributes and/or Language Attributes. One axiom to note here is (31) which essentially says that, an instance of a Name cannot have an ID which is a part of Link Attributes. The Link Attributes module has not been discussed here, we refer to the documentation for further details.

2.6 Date Information and Date Attributes

Date Information is a key module that has numerous usage within MMODS-O. A Bibliographic resource may have associated date information to express the timeline of creation, last updated, physical and/or digital origin information, etc. Throughout the MODS XML schema, all the date information under different names follow more or less a similar structure. That is why, we realized the necessity of having a Date Information module which conforms with our general intention of having a modular, reusable design. Primarily, a DateInfo instance may have a set of Language Attributes (e.g. date mentioned in multiple languages), some essential Date Attributes. We have created a Date Attributes module to further aid reusability and compact design. Another important aspect of the DateInfo module is that it must have a type of DateInfoType. Note, that there is no DateInfoType available in MODS XML schema. We outline the differences in detail in Sect. 3 (Fig. 6).

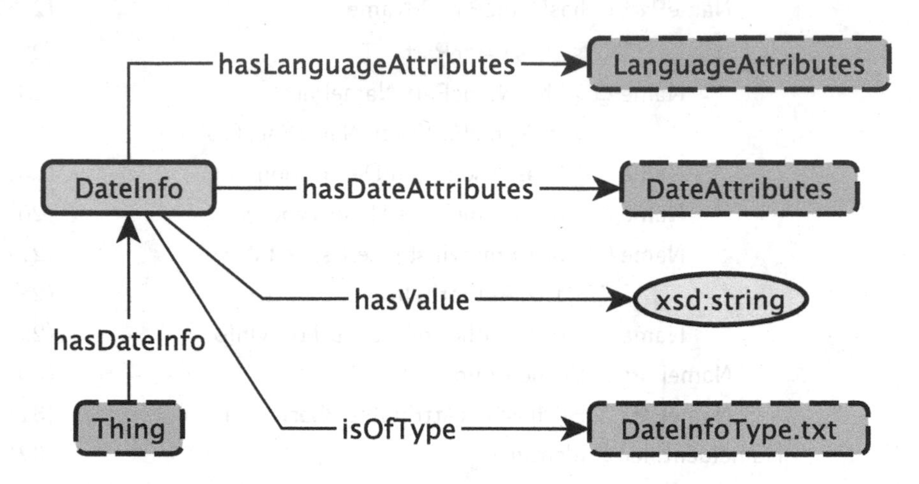

Fig. 6. Schema Diagram for the Date Info Module

Different types of dates across the MODS XML schema generally offer a similar set of attributes, as such we make use of the DateAttributes module. The Qualifier identifies the date under description to be either *approximate*, *inferred*, or *questionable* which is why this is a controlled vocabulary in Fig. 7. The DateEncoding controlled vocabulary identifies the encoding type of the date (e.g. *w3cdtf*, *iso8601*). It is also possible to identify one DateInfo instance to be the Key Date among different instances of DateInfo using the DateAttributes with the property isKeyDate which provides a boolean value.

Selected Axioms

$$\top \sqsubseteq \leq 1 \mathsf{hasDateInfo}^{-}.\top \tag{33}$$

$$\mathsf{Thing} \sqsubseteq \geq 0 \mathsf{hasDateInfo}.\mathsf{DateInfo} \tag{34}$$

$$\mathsf{DateInfo} \sqsubseteq \exists \mathsf{hasDateAttributes}.\mathsf{DateAttributes} \tag{35}$$

$$\top \sqsubseteq \leq 1 \mathsf{hasDateAttributes}.\top \tag{36}$$

$$\mathsf{DateInfo} \sqsubseteq \geq 0 \mathsf{hasDateAttributes}.\mathsf{DateAttributes} \tag{37}$$

$$\mathsf{DateInfo} \sqsubseteq \exists \mathsf{isOfType}.\mathsf{DateInfoType.txt} \tag{38}$$

$$\mathsf{DateInfo} \sqsubseteq \exists \mathsf{hasValue}.\mathsf{xsd{:}string} \tag{39}$$

$$\mathsf{DateAttributes} \sqsubseteq \geq 0 \mathsf{hasDateEncodingType}.\mathsf{DateEncoding.txt} \tag{40}$$

$$\mathsf{DateAttributes} \sqsubseteq \geq 0 \mathsf{isKeyDate}.\mathsf{xsd{:}boolean} \tag{41}$$

$$\mathsf{DateAttributes} \sqsubseteq \geq 0 \mathsf{isStartOrEndPoint}.\mathsf{Point.txt} \tag{42}$$

$$\mathsf{DateAttributes} \sqsubseteq \geq 0 \mathsf{hasAlternativeCalendar}.\mathsf{Calendar.txt} \tag{43}$$

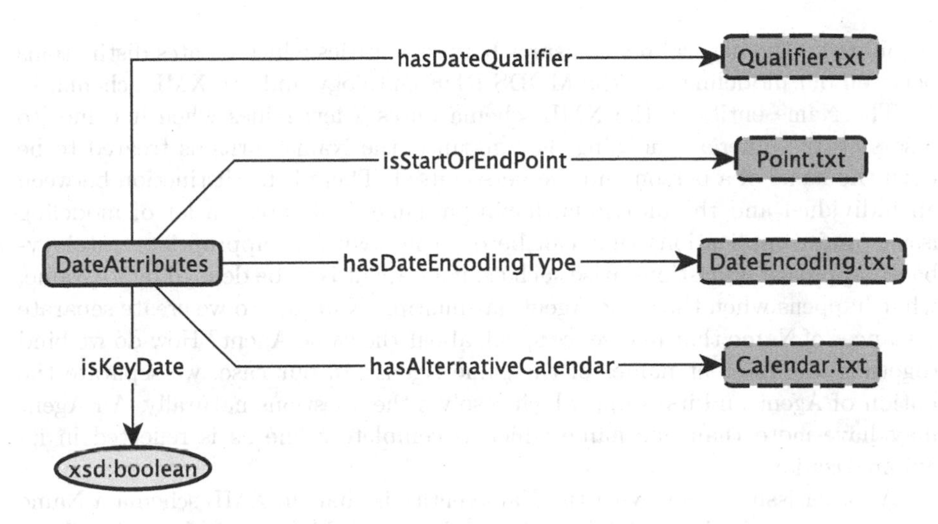

Fig. 7. Schema Diagram for the Date Attributes Module

In order to formalize the intended use, the property hasDateInfo can only be associated with at most one instance of *Thing*, expressed using an *inverse functionality* (33) wherein a *Thing* can have 0 or more instances of DateInfo, expressed using a *structural tautology* (34). An instance of DateInfo must have exactly one set of DateAttributes which is conveyed by using a combination of *existential* (35) and *functionality* (36). Furthermore, a DateInfo must have a DateInfo type (38). The DateInfo type is a controlled vocabulary that contains a list of Date elements available in MODS XML schema, for example: dateIssued, dateCreated, dateCaptured, dateModified, dateValid, etc.

We have outlined 7 out of the 34 modules we have created as part of the MMODS-O ontology. In those 7 modules, we have only discussed the formal axioms which we considered the most interesting. The documentation contains a detailed description of all the modules including a comprehensive formalization.

3 Related Work and Comparison with Previous Work

To the best of our knowledge, there is very few published work available regarding ontologies based on MODS. The closest effort appears to be the MODS RDF Ontology[4] available from Library of Congress pages. It appears to be a mostly straightfoward transcription of the XML schema without significant effort to make modifications to adjust to the ontology paradigm. We will use this for comparison; as it is very close to the MODS XML schema, we make only reference to the XML schema in the discussion.[5] Our ontology design in many cases

[4] https://www.loc.gov/standards/mods/modsrdf/primer.html.

[5] We also found http://arco.istc.cnr.it:8081/ontologies/MODS which appears to be abandoned work-in progress without meaningful documentation.

accounts for the natural relationships between entities which creates distinctions between our modeling and the MODS RDF Ontology and the XML schema.

The Name entity in the XML schema raises a few issues when it comes to assessing the inherent meaning. For instance, the Name entity is treated to be both the name of a person and the person itself. There is no distinction between an individual and the individual having a name. This poses a lot of modeling issues and complications that can be overcome with an appropriate ontology-based approach. Questions arise such as: if an Agent is to be defined by its Name, what happens when the same Agent has multiple Names? Do we create separate instances of Name that in essence speak about the same Agent? How do we bind together the different names of the same Agent? In our case, we separate the notion of Agent and its Name which resolves the questions naturally. An Agent may have more than one name which is completely fine as is reflected in its axiomatization.

Another issue we see with the Name entity is that, in XML schema a Name entity has an affiliation which is again another Name-like entity. Much like above, if we associate the name, the agent, and the affiliation all together with the name and agent, one may ask: if the agent has multiple names, do we create separate instances of names and write the same affiliations in all name instances? Perhaps more importantly, does it make more sense semantically to have an Organization entity that provides an affiliation? We argue that, an Agent, much less a Name, should not have an affiliation which is a Name, rather an Agent has an affiliation with an Organization, and that Organization will have a Name.

Furthermore, the XML schema states that the Name entity has a Role. We argue that it is more natural for an Agent to have a Name and for that same Agent to assume a particular Role. There are cases where it is possible for the same Agent to assume multiple roles under different pseudonyms. The XML schema and the existing RDF Ontology do not account for such intricate scenarios. The XML schema also allows for Names to have Alternative Names. It can be easily seen that it is not the Name which has Alternative Names, rather it is an Agent or an Organization which may have Alternative Names.

Another instance where we argue that our approach is more modularized and has reusable aspects is concerning DateInfo. Both the XML schema and the MODS RDF Ontology use separate elements of dates to convey different use-cases of dates. Namely, dateIssued, dateCreated, dateCaptured, dateModified, dateValid, etc. What we have done instead is, we have created a common module for DateInfo, where for each of the use-cases of dates can just be defined as a type of date through the use of controlled vocabularies. This module also recognizes the fact that all date-related elements within MODS share the same set of attributes, which gives rise to the DateAttributes model.

In our opinion, it is important to define and limit the applicability of modules within an ontology which we achieve through our carefully thought-out axiomatizations. It is imperative to leverage the different types of axioms available such as Scoped Domain, Scoped Range, Existential, Inverse Existential, Functionalities, Inverse Functionalities in order to formalize the scopes and boundaries.

The existing RDF Ontology only uses Domain, Range, and Subproperties as formalization of the ontology, which in our opinion does often not suffice [4].

4 Conclusion

We have presented the MMODS-O Ontology which has been developed from the MODS XML schema that has general use-cases in dealing with bibliographic metadata. We have developed the ontology in a way such that it is modularized, the distinct modules are reusable, and it paves the way for future improvement and module additions to the ontology. It incorporates modules that are concerned with Title information, Origin information, Geographic location, Target audience, Name, Subject, etc., of the resource under description. The ontology is serialized in OWL and has been formalized by extensive axiomatization.

Acknowledgement. The authors acknowledge funding under the National Science Foundation grants 2119753 "RII Track-2 FEC: BioWRAP (Bioplastics With Regenerative Agricultural Properties): Spray-on bioplastics with growth synchronous decomposition and water, nutrient, and agrochemical management" and 2033521: "A1: KnowWhereGraph: Enriching and Linking Cross-Domain Knowledge Graphs using Spatially-Explicit AI Technologies".

References

1. Baader, F., Calvanese, D., Mcguinness, D., Nardi, D., Patel-Schneider, P.: The Description Logic Handbook: Theory, Implementation, and Applications (2007)
2. Guha, R., Brickley, D.: RDF Schema 1.1. W3C Recommendation, W3C (2014). https://www.w3.org/TR/2014/REC-rdf-schema-20140225/
3. Hitzler, P.: A review of the semantic web field. Commun. ACM **64**(2), 76–83 (2021)
4. Hitzler, P., Krisnadhi, A.: On the roles of logical axiomatizations for ontologies. In: Hitzler, P., Gangemi, A., Janowicz, K., Krisnadhi, A., Presutti, V. (eds.) Ontology Engineering with Ontology Design Patterns - Foundations and Applications, Studies on the Semantic Web, vol. 25, pp. 73–80. IOS Press (2016). https://doi.org/10.3233/978-1-61499-676-7-73
5. Hitzler, P., Krötzsch, M., Rudolph, S.: Foundations of Semantic Web Technologies. Chapman and Hall/CRC Press (2010)
6. Maler, E., Bray, T., Paoli, J., Yergeau, F., Sperberg-McQueen, M.: Extensible Markup Language (XML) 1.0 (Fifth Edition). W3C Recommendation, W3C (2008). https://www.w3.org/TR/2008/REC-xml-20081126/
7. McCallum, S.H.: An introduction to the metadata object description schema (MODS). In: Proceedings of the 10th Workshop on Ontology Design and Patterns (WOP 2019) co-located with 18th International Semantic Web Conference (ISWC 2019), Auckland, New Zealand, 27 October 2019, pp. 82–88. Emerald Group Publishing Limited (2019). https://doi.org/10.1108/07378830410524521
8. Parsia, B., Krötzsch, M., Hitzler, P., Rudolph, S., Patel-Schneider, P.: OWL 2 Web Ontology Language Primer (Second Edition). W3C Recommendation, W3C (2012). https://www.w3.org/TR/2012/REC-owl2-primer-20121211/
9. Rayan, R., Shimizu, C., Hitzler, P.: An ontology design pattern for role-dependent names (2023). arXiv:2305.02077. https://arxiv.org/abs/2305.02077

10. Shimizu, C., Hammar, K., Hitzler, P.: Modular ontology modeling. Semant. Web **14**(3), 459–489 (2023)
11. Shimizu, C., Hirt, Q., Hitzler, P.: MODL: a modular ontology design library. In: Janowicz, K., Krisnadhi, A.A., Poveda-Villalón, M., Hammar, K., Shimizu, C. (eds.) Proceedings of the 10th Workshop on Ontology Design and Patterns (WOP 2019) co-located with 18th International Semantic Web Conference (ISWC 2019), Auckland, New Zealand, 27 October 2019. CEUR Workshop Proceedings, vol. 2459, pp. 47–58. CEUR-WS.org (2019). https://ceur-ws.org/Vol-2459/paper4.pdf

A Topic Model for the Data Web

Michael Röder([✉]), Denis Kuchelev, and Axel-Cyrille Ngonga Ngomo

DICE, Paderborn University, Paderborn, Germany
michael.roeder@uni-paderborn.de

Abstract. The usage of knowledge graphs in industry and at Web scale has increased steadily within recent years. However, the decentralized approach to data creation which underpins the popularity of knowledge graphs also comes with significant challenges. In particular, gaining an overview of the topics covered by existing datasets manually becomes a gargantuan if not impossible feat. Several dataset catalogs, portals and search engines offer different ways to interact with lists of available datasets. However, these interactions range from keyword searches to manually created tags and none of these solutions offers an easy access to human-interpretable categories. In addition, most of these approaches rely on metadata instead of the dataset itself. We propose to use topic modeling to fill this gap. Our implementation LODCat automatically creates human-interpretable topics and assigns them to RDF datasets. It does not need any metadata and solely relies on the provided RDF dataset. Our evaluation shows that LODCat can be used to identify the topics of hundreds of thousands of RDF datasets. Also, our experiment results suggest that humans agree with the topics that LODCat assigns to RDF datasets. Our code and data are available online.

Keywords: RDF datasets · topic modeling · Data Web

1 Introduction

With the growth of the size and the increase in the number of knowledge graphs available on the Web comes the need to process this data in a scalable way [17]. The large number of datasets that are available online and their sheer size make it costly or even infeasible to handle each of these datasets manually without the support of proper tools. A particularly important issue is that the mere identification of relevant datasets for a particular task (e.g., data integration [24], question answering [32], machine learning [14], etc.) may become challenging. Indeed, domain experts who plan to use knowledge graphs for a task may be able to read Resource Description Framework (RDF) data but will not have the time to read through hundreds of thousands of datasets to determine whether they are relevant. Hence, *we need to be able to characterize RDF datasets so that users can easily find datasets of interest.*

A similar problem is already known from the processing of large amounts of human-readable documents. Most users might be able to read all books within

F. Ortiz-Rodriguez et al. (Eds.): KGSWC 2023, LNCS 14382, pp. 183–198, 2023.
https://doi.org/10.1007/978-3-031-47745-4_14

a library. However, they may not have the time to do so just to identify the books that they are interested in. Although there are search engines that allow the indexing of documents, users would have to know the right keywords to find documents that they are interested in [15]. Hence, "search engines are not the perfect tool to explore the unknown in document collections" [15]. However, today's dataset search engines mainly rely on keyword searches on the dataset's metadata and user-created tags although both suffer from the aforementioned drawback. At the same time, RDF datasets may not have rich metadata that could be used for such a search to improve their findability [23,37]. For example, the Vocabulary of Interlinked Datasets (VoID) vocabulary offers ways to add metadata in form of descriptions that can be indexed by a search engine. However, Paulheim et al. [27] show that a best practice proposed in the VoID specification [2]—i.e., to use the `/.well-known/void` path on a Web server to provide an RDF file with VoID information about datasets hosted on the server—is not adopted on a large scale. Similarly, Schmachtenberg et al. [33] report that only 14.69% of 1014 datasets that they crawled provide VoID metadata.

We tackle this gap with our topic-modeling-based approach LODCat. Topic modeling algorithms can be used to infer latent topics in a given document collection. These topics can be used to structure the document collection and enable users to focus on subsets of the collection, which belong to their area of interest. Our main contribution in this publication is the application of topic modeling to a large set of RDF datasets to support the exploration of the Data Web based on human-interpretable topics. To this end, we tackle the challenge of transforming the RDF datasets into a form that allows the application of a topic modeling algorithm. Our evaluation shows that this approach can be applied to hundreds of thousands of RDF datasets. The results of a questionnaire suggest that humans generally agree with the topics that our approach assigns to a sample of example datasets.

The following section describes related work before Sect. 3 describes the single steps of our approach. Section 4 describes the setup and results of our evaluation before we conclude with Sect. 5.

2 Related Work

In their survey of data search engines, Chapman et al. [10] divide these engines into four categories. The first category are database search engines. They are used with structured queries that are executed against a database back end. The second set of search engines are information retrieval engines. These are integrated into data portals like CKAN[1] and offer a keyword-based search on the metadata of datasets. The third category are entity-centric search engines. The query of such an engine comprises entities of interest and the search engine derives additional information about these entities. The last category is named tabular search. A user of such a search engine tries to extend or manipulate one or more existing tables by executing search queries.

[1] https://ckan.org/.

The second category represents the most common approach to tackle the search for datasets on the web. Several open data portals exist that offer a list of datasets and a search on the dataset's metadata. Examples are the aforementioned CKAN, kaggle[2] or open government portals like the european data portal[3]. The Google dataset search presented by Brickley et al. [8] works in a similar way but uses the Google crawler to collect the data from different sources. Singhal et al. [34] present DataGopher—a dataset search that is optimized for research datasets. Devaraju et al. [12] propose a personalized recommendation of datasets based on user behavior. Our approach differs to these approaches as we focus on RDF datasets and rely on the dataset itself instead of only using metadata. In addition, we do not rely on a keyword search or user created tags but automatically generated topics that are assigned to the datasets.

Kunze et al. [19] propose an explorative search engine for a set of RDF datasets. This engine is mainly based on filters that work similar to a faceted search. For example, one of these filters is based on the RDF vocabularies that the datasets use. Vandenbussche et al. [39] present a web search for RDF vocabularies.[4] Kopsachilis et al. [18] propose GeoLOD—a dataset catalog that focuses on geographical RDF datasets. LODAtlas [28] combines several features of the previously mentioned systems into a single user interface. These approaches have similar limitations as the generic open data portals described above. While some of them offer additional features, non of them offers human-interpretable categories that go beyond manually created tags.

Topical profiling of RDF datasets [36] is very closely related to our work. The task is defined as a single- or multi-label classification problem. Blerina et al. [36] propose a benchmark that is based on the manually created classes of the Linked Open Data cloud project [22]. In a recent work, Asprino et al. [3] tackle the multi-classification task and extend the benchmark dataset. Some of the features that are used for the classification, like the virtual documents that are generated by the approach of Asprino et al. are similar to the documents our approach creates. However, our approach is unsupervised while the proposed approaches are supervised and rely on training data, that has been created manually.

Several approaches exist to explore RDF datasets. Tzitzikas et al. [38] define a theoretical framework for these explorative search engines and compare several approaches. However, all these approaches focus on exploring a single RDF dataset while our goal is to enable users to derive topically interesting RDF datasets from a set of datasets.

Röder et al. [30] propose the application of topic models to identify RDF datasets as candidates for link prediction. However, their work relies solely on the ontologies of datasets and the topics are used as features for a similarity calculation while we need high-quality topics that can be shown to users.[5] Sleeman et al. [35] use topic modeling to assign topics to single entities of an RDF

[2] https://www.kaggle.com/datasets.

[3] https://data.europa.eu/en.

[4] https://lov.linkeddata.es/dataset/lov.

[5] Chang et al. [9] show that there is a difference between these two usages of topics.

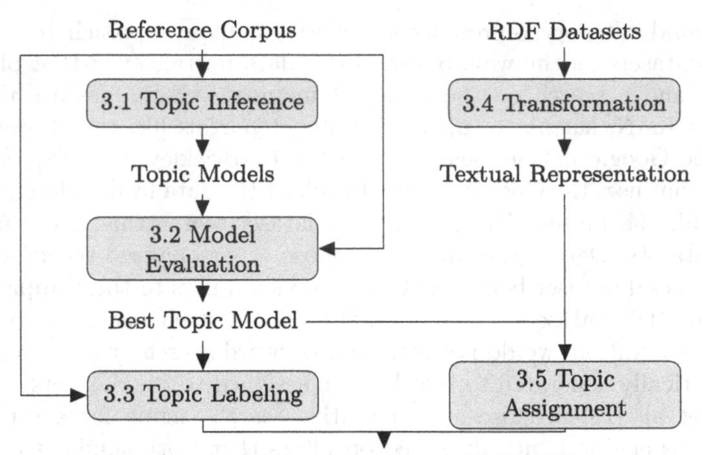

Fig. 1. Overview of the workflow of LODCat.

dataset. Thus, both of the aforementioned approaches use topic modeling with different aims and are not comparable to our approach.

3 LODCat

Figure 1 shows an overview of our proposed approach LODCat. It relies on a reference text corpus (e.g., Wikipedia) as a source of general knowledge and uses topic modeling to assign human-interpretable topics to the single RDF datasets. First, we use the reference corpus to generate several topic models. Thereafter, the single models are evaluated and the best model is chosen for further processing. For each topic of this model, a label is generated to make the complex probability distributions human-readable. In parallel, the RDF datasets are transformed into a textual representation (i.e., documents). Based on the chosen topic model, a topic distribution is assigned to each of the generated documents. At the end, each RDF dataset has a set of topics that are dominant for that dataset and that are described by their labels. This data is used to provide a faceted search, which helps the user to find datasets related to their field of interest. The single steps of our approach are described in more detail in the following.[6]

3.1 Topic Inference

Our current version of LODCat relies on the topic modeling approach Latent Dirichlet Allocation (LDA) [6,7]. This approach assumes that there are latent topics that are defined as distributions over words ϕ, i.e., each word type w has

[6] LODCat is open source at https://github.com/dice-group/lodcat.

a probability representing the likeliness to encounter this word while reading about the topic. The topics are derived from a given corpus D, i.e., a set of documents. Each document d_i has a topic distribution θ_i, i.e., each topic has a probability how likely it is to occur within the document. LDA connects these distributions by defining that each word token w has been created by a single topic and assigning the ID of this topic to the token's z variable. Let $w_{i,j}$ be the j-th word token in the i-th document, let $\mathrm{w}_{i,j}$ be its word type, and let $z_{i,j}$ denote the id of the topic from which the word type of this token has been sampled. Let ϱ be the number of topics and let $Z = \left\{z_{1,1}, \ldots, z_{|D|,|d_{|D|}|}\right\}$ be the set of the topic indices of all word tokens in the corpus D. Let $\Phi = \{\phi_1, \ldots, \phi_\varrho\}$ be the set of word distributions and $\Theta = \{\theta_1, \ldots, \theta_{|D|}\}$ be the set of topic distributions. LDA is based on the following joint distribution [6]:

$$\mathbb{P}(\Phi, \Theta, Z, D) = \left(\prod_{i=1}^{\varrho} \mathbb{P}(\phi_i)\right)\left(\prod_{i=1}^{|D|} \mathbb{P}(\theta_i)\left(\prod_{j=1}^{|d_i|} \mathbb{P}(z_{i,j}|\,\theta_i)\,\mathbb{P}\left(\mathrm{w}_{i,j}\,|\phi_{z_{i,j}}\right)\right)\right). \quad (1)$$

We use the LDA inference algorithm proposed by Hoffman et al. [16] that takes a corpus and the number of topics ϱ as input. The output is a topic model comprising the topics' distributions Φ.[7] Since the best number of topics ϱ is unknown we generate several models with different ϱ values.

3.2 Model Evaluation

In this step, we choose the best model from the set of generated topic models. To this end, we use topic coherence measures to evaluate the human-readability and interpretability of the model's topics. We represent each topic by its 10 top words $\bar{W}$, i.e., the 10 words that have the highest probability in the topic's word distribution Φ. We use these top words as input for two coherence measures proposed by Röder et al. [31], namely C_P and a variant of the C_V measure that we call C_{V2}.[8] Hence, we get two coherence values for each topic. For each measure, we sort all models based on the average coherence value of their topics. The model that achieves the best rank on average for both coherence measures is the model that will be used for further processing.

In addition, we use the coherence values to identify low-quality topics. These are topics that should not be shown to the user. We define a topic to be of low-quality if its C_{V2} or C_P value is below 0.125 or 0.25, respectively.

3.3 Topic Labeling

For each topic of the chosen model, we assign label that can be used to present the topic to users. For our current implementation, we use the Neural Embedding

[7] Due to space limitations, we refer the interested reader to Blei et al. [6] and Hoffman et al. [16] for further details about LDA and the used inference algorithm.

[8] The C_{V2} measure has the same definition as the C_V measure but uses the S_{all}^{one} segmentation [31]. The variant showed a better performance in our experiments.

Topic Labelling (NETL) approach of Bathia et al. [5] since 1) their evaluation shows that NETL outperforms the approach of Lau et al. [20] and 2) the approach is available as open-source project.[9] NETL generates label candidates for a given topic from a reference corpus and ranks them according to a trained model. Following Bathia et al. [5], we use the English Wikipedia as reference corpus and use their pre-trained support vector regression model to rank the label candidates. We also use the topics top words as additional topic descriptions.

3.4 RDF Dataset Transformation

Our goal in this step is to transform given RDF datasets into a textual representation that can be used in combination with the generated topic model. This step relies on the Internationalized Resource Identifiers (IRIs) that occur in the datasets. We determine the frequency f of each IRI in the dataset (either as subject, predicate or object of a triple). IRIs of well-known namespaces that do not have any topical value like `rdf`, `rdfs` and `owl` are filtered out. After that, the labels of each IRI are retrieved. This label retrieval is based on the list of IRIs that have been identified as label-defining properties by Ell et al. [13]. Additionally, we treat values of `rdfs:comment` as additional labels. If there are no labels available, the namespace of the IRI is removed and the remaining part is used as label. If this generated label is written in camel case or contains symbols like underscores, it is split into multiple words. The derived labels are further preprocessed using a tokenizer and a lemmatizer [21]. The derived words inherit the counts f of their IRI. If IRIs share the same word their counts are summed up to derive the count of this word.

However, we do not use the counts directly for generating a document since some IRIs may occur hundreds of thousand times within a dataset. Their words would dominate the generated document and marginalize the influence of other words. In addition, large count values could lead to very long documents that may create further problems with respect to the resource consumption in later steps. To reduce the influence of words with very high f values we determine the frequency ψ of word type w for the document d'_i of the i-th dataset as follows:

$$\psi_{i,\mathrm{w}} = r(\log_2(f_{\mathrm{w}}) + 1)\,, \tag{2}$$

where r is the rounding function which returns the closest integer value preferring the higher value in case of a tie [1].[10] The result of this step is a bag of words representation of one document for each RDF dataset.

3.5 Topic Assignment

The last step is the assignment of topics to the documents that represent the RDF datasets. For each created synthetic document d'_i, we use the chosen topic

[9] https://github.com/sb1992/NETL-Automatic-Topic-Labelling-.

[10] The transformation of counts into occurrences in a synthetic document is similar to the logarithmic variant of the approach described by Röder et al. [30].

model to infer a topic distribution θ_i'. This distribution is used to derive the dataset's top topics, i.e., the topics with the highest probabilities in the distribution. The labels and top words of these topics are used as a human-readable representation of the RDF dataset.

4 Evaluation of LODCat

We evaluate LODCat in a setup close to a real-world scenario.[11] We start with the English Wikipedia as corpus and process more than 600 thousand RDF datasets using LODCat following the steps described in Sect. 3. The evaluation can be separated into the following three consecutive experiments:

1. The generation and selection of the topic model,
2. The RDF dataset transformation and the topic assignment, and
3. The evaluation of the assigned topics based on a user study.

4.1 Datasets

For our evaluation, we use two types of data—a reference corpus to generate the topic model and the set of RDF datasets, which should be represented in a human-interpretable way. We use the English Wikipedia as reference corpus.[12] We preprocess the dump file by removing Wikimedia markup, removing redirect articles and handling each remaining article as an own document. Each document is preprocessed as described in Sect. 3.1. From the created set of documents, we derive all word types and count their occurrence. Then, we filter the word types by removing 1) common English terms based on a stop word list, and 2) all word types that occur in more than 50% of the documents or 3) in less than 20 documents.[13] From the remaining word types, we select the 100 000 word types with the highest occurrence counts and remove all other from the documents. After that, we remove empty documents and randomly sample 10% of the remaining documents. Finally, we get a corpus with 619 475 documents and 190 million word tokens.

We gather 623 927 real-world RDF datasets from the LOD Laundromat project [4].[14] These datasets should be represented in a human-interpretable way. Note that these datasets have been stored without any metadata that could be used. Figure 2 shows the size of the RDF datasets. The largest dataset has 43 million triples while the majority has between 100 and 10 000 triples. In total, the datasets comprise 3.7 billion triples.[15]

[11] The RDF datasets we use are available at https://figshare.com/s/af7f18a7f3307cc86bdd while the results as well as the Wikipedia-based corpus can be found at https://figshare.com/s/9c7670579c969cfeac05.
[12] We use the dump of the English Wikipedia from September 1st 2021.
[13] The stop word list can be found online. We will add the link after the review phase.
[14] We downloaded the datasets in January 2018.
[15] Note that we do not deduplicate the triples across the datasets.

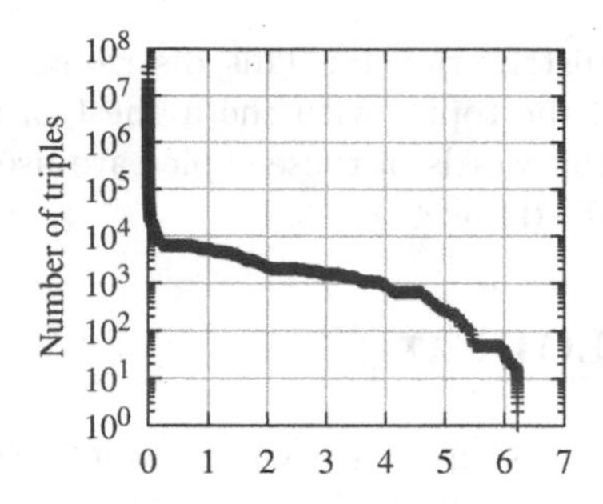

Fig. 2. The sizes of the RDF datasets (the x-axis is the dataset ID in 10^5).

4.2 Setup

Experiment I. In the first experiment, we infer the topic models based on the English Wikipedia corpus. We infer several models with different numbers of topics.[16] For this evaluation, we use $\varrho = \{80, 90, 100, 105, 110, 115, 120, 125, 135\}$ and generate three models for each number of topics. We choose the best topic model as described in Sect. 3.2, analyze this model with respect to the model's coherence values and show example topics.

Experiment II. Based on the best topic model created in the first experiment, we process each of the 623 927 RDF datasets by LODCat. During this step, we remove datasets that lead to an empty document. The result comprises a topic distribution for each dataset based on the used topic model. We analyze these distributions by looking at their top topics.

Experiment III. Finally, we evaluate the assignment of the topics to the datasets. Chang et al. [9] propose the topic intruder experiment to evaluate the assignment of topics to documents. They determine the top topics of a document and insert a randomly chosen topic from the same topic model that is not one of the document's top topics. This randomly chosen topic is called intruder topic. After that, volunteers are given the created list of topics and the document, and are asked to identify the intruder. The more often the intruder is successfully identified, the better is the topic assignment of the topic model. We use the same approach to evaluate whether a topic model can assign meaningful topics to an RDF dataset. We sample 60 datasets that have more than 100 and less than 10 000 triples. For each of the sampled datasets, we derive the three topics with the highest probability. Based on the dataset content and the quality of their top topics, we choose 10 datasets that 1) have at least two high-quality topics among the top three topics, 2) have a high-quality topic as highest ranked topic, 3) have a content that can be understood without accessing further sources, and 4) have not exactly the same top topics as the already chosen datasets. For each

[16] We use the Gensim library [29] with hyper parameter optimization. https://radimrehurek.com/gensim/index.html.

chosen dataset, we sample an intruder topic from the set of high-quality topics that are not within the top three topics of the dataset.

We create a questionnaire with 10 questions. Each question gives the link to one of the chosen datasets and a list of topics comprising the top topics of the dataset and the intruder topic in a random order. 5 chosen datasets have three high-quality topics while the other 5 datasets have one top topic with a low coherence value. We remove the topics with the low values. Hence, 5 question comprise 4 topics and the other 5 questions have 3 topics from which a user should choose the intruder topic. For the questionnaire, the topics are represented in the human-readable way described in Sect. 3.3, i.e., with their label and their top words. The participants of the questionnaire are encouraged to look into the RDF dataset. However, they should not include further material. We sent this questionnaire to several mailing lists to encourage experts and experienced users of the Semantic Web to participate.

Following Chang et al. [9], we calculate the topic log odds to measure the agreement between the topic model and the human judgments that we gather with our questionnaire. Let θ'_i be the topic distribution of the i-th document d'_i. Let $\theta'_{i,k}$ be the probability of the k-th topic for document d'_i. Let $Y_i = \{y_{i,1}, \ldots\}$ be the bag of all user answers for document d'_i, i.e., the j-th element is the id of the topic that the j-th user has chosen as intruder topic for this document. Let x_i be the id of the real intruder topic for document d'_i. Chang et al. [9] define the topic log odds $\mathfrak{o}$ for the i-th document as the average difference between the probabilities of the chosen intruder topics compared to the real intruder topic:

$$\mathfrak{o}(\theta'_i, Y_i, x_i) = \frac{1}{|Y_i|} \sum_{j=1}^{|Y_i|} \left(\log(\theta'_{i,x_i}) - \log(\theta'_{i,y_{i,j}}) \right) . \quad (3)$$

A perfect agreement between the human participants and the topic model would lead to $\mathfrak{o} = 0$. In practice, this is only reached if all participating volunteers find the correct intruder topic.

4.3 Results

Experiment I. From the 27 generated topic models, the model that received the best average ranks according to both topic coherence measures is a model with 115 topics. Figure 3 shows the coherence values of this model's topics for both coherence measures. The dashed line shows the threshold used to distinguish between high and low-quality topics. Based on the two thresholds, 74 topics are marked as high-quality topics while the remaining 41 topics are treated as low-quality topics. Table 1 shows the model's topics with the 5 highest and the 5 lowest C_{V2} coherence values. While the first 5 topics seem to focus on a single topic the topics with the low coherence scores comprise words that seem to have no strong relation to each other.

Experiment II. LODCAT is able to assign topics to 561 944 of the given 623 927 RDF datasets, which are > 90%. 61 983 datasets lead to the creation of empty

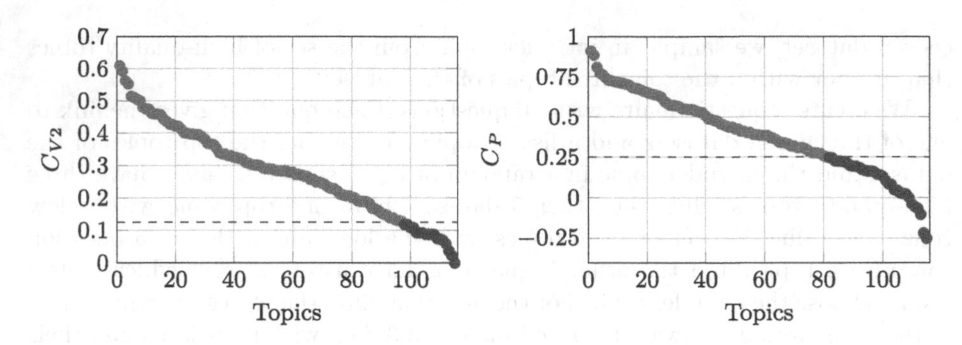

Fig. 3. Topics of the best performing model sorted by their C_{V2} and C_P coherence values, respectively. The dashed line shows the threshold used to separate high-quality topics (green) from low-quality topics (red). (Color figure online)

Table 1. The top words of the 5 topics of the chosen topic model with the highest and lowest C_{V2} values, respectively.

C_{V2}	$\bar{W}$
0.60942	canadian, canada, quebec, ontario, montreal, toronto, ottawa, nova, scotia, alberta
0.59273	album, song, release, band, music, chart, record, single, track, records
0.56269	age, population, household, female, city, male, family, census, average, year
0.55282	chinese, china, singapore, li, wang, shanghai, chen, beijing, hong, zhang
0.51424	league, club, player, football, season, cup, play, goal, team, first
0.06969	rank, time, men, advance, event, final, result, athlete, heat, emperor
0.05900	use, language, word, name, form, one, english, see, greek, two
0.03767	use, system, one, number, two, function, set, space, model, time
0.02269	use, health, may, child, include, provide, would, act, make, public
0.00000	j., a., m., c., r., s., l., e., p., d

documents and, hence, cannot get any topics assigned. These are mainly small datasets with IRIs that cannot be transformed into meaningful words, i.e., words that are not removed by our stop word removal step.

After generating the topic distributions for the documents created from the RDF datasets, we analyze these distributions. For each dataset, we determine its main topic, i.e., the high-quality topic with the highest probability for this dataset. Figure 4 shows the number of datasets for which each topic is the main topic. The figure shows that a single topic covers more than 508 thousand datasets. Table 2 shows the 5 topics that have the highest values in Fig. 4. We can see that a weather-related topic covers roughly 90% of the datasets to which LODCAT could assign topics. The next biggest topics are transportation- and car-related topics and each of them covers nearly 10 thousand datasets. They are followed by a computer- and a travel-related topic.

We further analyze the RDF datasets with respect to the claim that the majority of them is related to weather. We analyze the namespaces that are used

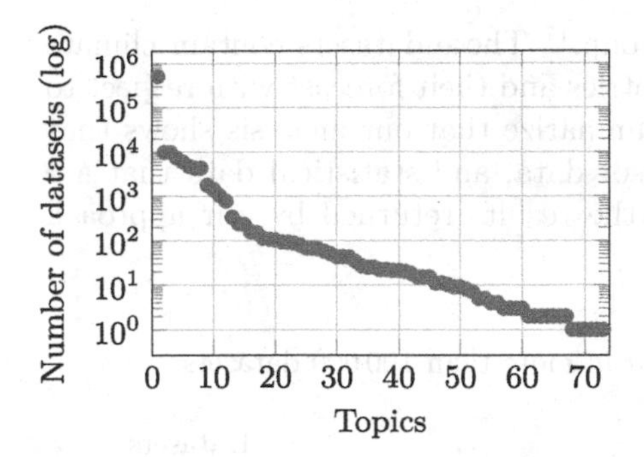

Fig. 4. Number of datasets per topic for which this topic has the highest probability sorted in descending order. Topics with no datasets have been left out.

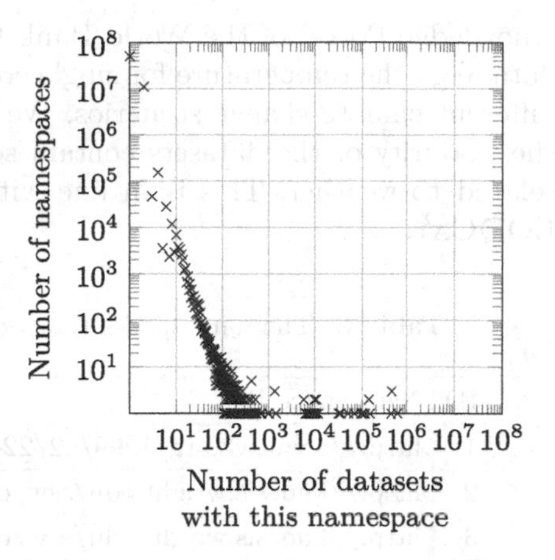

Fig. 5. The number of namespaces that occur in a number of datasets.

Table 2. Top topics with the highest number of datasets.

Datasets	$\bar{W}$
508 095	water, storm, wind, tropical, nuclear, temperature, hurricane, damage, cause, system
9 828	station, road, route, line, street, bridge, railway, city, highway, east
9 794	car, engine, model, vehicle, first, use, point, motor, design, safe
7 328	use, system, software, user, datum, computer, include, information, support, service
5 946	airport, international, brazil, portuguese, são, romanian, portugal, brazilian, language, romania

within the RDF datasets and count the number of datasets in which they occur. Figure 5 shows the result of this analysis for all 623 thousand RDF datasets. In the lower right corner of the figure, we can see that there is only a small number of namespaces that are used in many datasets. Table 3 shows the 12 namespaces that occur in more than 100 thousand datasets. The most often used namespace is the `rdf`, which is expected. The namespaces on position 2–4 occur in more than 450 thousand RDF datasets and belong to datasets with sensor data described by Patni et al. [26]. A further search revealed that the data comprises hurricane and blizzard observations from weather stations [25]. These datasets also make use of the fifth namespace from Table 3. The sixth namespace is the Data Cube namespace, which is used to described statistical data in RDF [11]. This namespace occurs often together with the remaining namespaces (7–12). They occur in datasets that origin from the Climate Change

Knowledge Portal of the World Bank Group.[17] These datasets contain climate data, e.g., the temperature for single countries and their forecast with respect to different climate change scenarios. We summarize that our analysis shows that the majority of the datasets contain sensor data, and statistical data that are related to weather. This is in line with the results returned by our approach LODCAT.

Table 3. The namespaces that occur in more than 100 000 datasets.

ID	Namespace IRI	Datasets
1	http://www.w3.org/1999/02/22-rdf-syntax-ns#	620 653
2	http://knoesis.wright.edu/ssw/ont/weather.owl#	452 453
3	http://knoesis.wright.edu/ssw/ont/sensor-observation.owl#	452 453
4	http://knoesis.wright.edu/ssw/	452 453
5	http://www.w3.org/2006/time#	442 719
6	http://purl.org/linked-data/cube#	147 731
7	http://worldbank.270a.info/property/	147 348
8	http://purl.org/linked-data/sdmx/2009/dimension#	147 305
9	http://worldbank.270a.info/dataset/world-bank-climates/	139 865
10	http://worldbank.270a.info/classification/variable/	139 865
11	http://worldbank.270a.info/classification/scenario/	114 064
12	http://worldbank.270a.info/classification/percentile/	103 202

Experiment III. Our questionnaire received 225 answers from 65 participants.[18] Fig. 6 shows the results. The left side of the figure summarizes the results for the two groups of questions—those with 3 and 4 topics, respectively. The center of the figure shows the detailed results for each of the questions. The plot shows that in the majority of cases, the intruder was successfully identified by the participants. The results look slightly different for datasets 4 and 5. In both cases, the third topic is not strongly related to the dataset and has been chosen quite often as intruder. However, since the first and second topic have been chosen much less often for these datasets, the result shows that the ranking of the topics make sense, i.e., the participants were able to identify the first two topics as related to the given dataset.

On the right of Fig. 6, there is a box plot for the topic log odd values that have been measured for the single documents. The average value across the 10 datasets is -1.23 with dataset 4 getting the worst value. This value is visible

[17] https://climateknowledgeportal.worldbank.org/.

[18] We used LimeSurvey for the questionnaire (https://www.limesurvey.org/). The questionnaire allowed users to skip questions. These skipped questions are not taken into account for the number of answers.

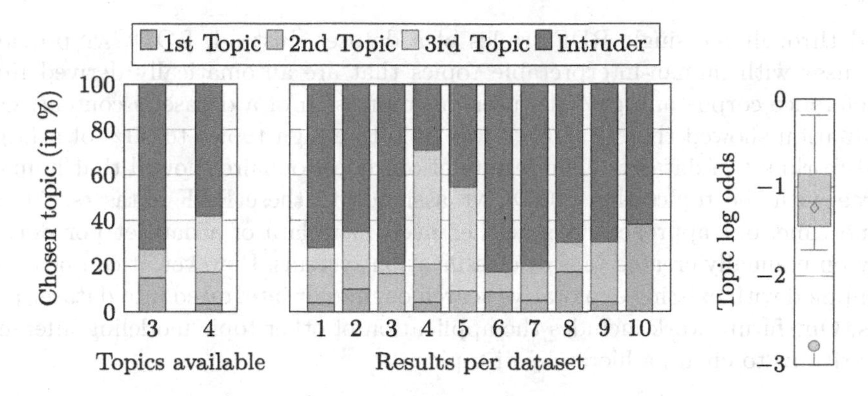

Fig. 6. Questionnaire results. Left: Average amount of topics chosen as intruder. Center: Amount of topics chosen as intruder for the single datasets. Right: The topic log odds o per dataset. The diamond marks the arithmetic mean.

as an outlier in the lower part of the plot. This result is comparable to the results Chang et al. [9] present for various topic modeling models on two different corpora. This confirms our finding that the human-readable topics fit to the RDF datasets to which they have been assigned. However, the experiment setup comes with two restrictions. First, we manually chose the RDF datasets for this experiment with the requirement that the participants of the questionnaire have to be able to easily understand the content of the chosen datasets. This may have introduced a bias. However, it can be assumed that the results would be less reliable if the datasets would have been selected randomly since the experiment setup suggested by Chang et al. [9] relies on the assumption that the participants understand the target object to which the topics have been assigned (in our case, the RDF dataset). Second, we made use of topic coherence measures to filter low-quality topics and we chose datasets that have at least two high-quality topics within their top-3 topics. It can be assumed that the topic log odd values would be lower if we would have included low-quality topics, since they are less likely interpretable by humans. However, a dataset that has mainly low-quality topics assigned could cause problems in a user application since no human-interpretable description of the dataset could be provided. We find that out of the 561 944 RDF datasets, to which LODCat could assign topics, only 220 datasets have not a single high-quality topic within their top-3 topics. Hence, the filtering of low-quality topics seems to have a minor impact on the number of RDF datasets for which LODCat is applicable.

5 Conclusion

Within this paper, we presented LODCat—an approach to support the exploration of the Data Web based on human-interpretable topics. With this approach, we ease the identification of RDF datasets that might be interesting to a user since they neither have to go through all available datasets nor do they need to read through the single RDF triples of a dataset. Instead, LODCat provides the user with human-interpretable topics that are automatically derived from a reference corpus and give the user an impression of a dataset's content. Our evaluation showed that LODCat was able to assign topics to 90% of a large, real-world set of datasets. The results of our questionnaire showed that humans agree with the topics that LODCat assigned to these RDF datasets. At the same time, our approach does neither need metadata of a dataset nor does it rely on manually created tags or classification systems. However, it can be easily combined with existing explorative search engines or integrated into dataset portals. Our future work includes the application of other topic modeling inference algorithms to create a hierarchy of topics.

Acknowledgements. This work has been supported by the Ministry of Culture and Science of North Rhine-Westphalia (MKW NRW) within the project SAIL under the grant no NW21-059D.

References

1. Java Platform Standard Ed. 8: Class Math. Website (2014). https://docs.oracle.com/javase/8/docs/api/java/lang/Math.html. Accessed 18 May 2022
2. Alexander, K., Cyganiak, R., Hausenblas, M., Zhao, J.: Describing linked datasets with the void vocabulary. W3C Note, W3C, March 2011. http://www.w3.org/TR/2011/NOTE-void-20110303/
3. Asprino, L., Presutti, V.: Observing IoD: its knowledge domains and the varying behavior of ontologies across them. IEEE Access. **11**, 21127–21143 (2023)
4. Beek, W., Rietveld, L., Bazoobandi, H.R., Wielemaker, J., Schlobach, S.: LOD Laundromat: a uniform way of publishing other people's dirty data. In: Mika, P., et al. (eds.) ISWC 2014. LNCS, vol. 8796, pp. 213–228. Springer, Cham (2014). https://doi.org/10.1007/978-3-319-11964-9_14
5. Bhatia, S., Lau, J.H., Baldwin, T.: Automatic labelling of topics with neural embeddings. In: Proceedings of COLING 2016, the 26th International Conference on Computational Linguistics: Technical Papers, pp. 953–963. The COLING 2016 Organizing Committee, Osaka, Japan, December 2016
6. Blei, D.M.: Probabilistic topic models. Commun. ACM **55**(4), 77–84 (2012)
7. Blei, D.M., Ng, A.Y., Jordan, M.I.: Latent Dirichlet allocation. J. Mach. Learn. Res. **3**, 993–1022 (2003)
8. Brickley, D., Burgess, M., Noy, N.: Google dataset search: Building a search engine for datasets in an open web ecosystem. In: The World Wide Web Conference, pp. 1365–1375. WWW 2019, Association for Computing Machinery (2019)
9. Chang, J., Gerrish, S., Wang, C., Boyd-graber, J.L., Blei, D.M.: Reading tea leaves: How humans interpret topic models. In: Advances in Neural Information Processing Systems, vol. 22, pp. 288–296. Curran Associates, Inc. (2009)
10. Chapman, A., et al.: Dataset search: a survey. Int. J. Very Large Data Bases **29**, 251–272 (2020)
11. Cyganiak, R., Reynolds, D.: The RDF data cube vocabulary. W3c Recommendation, January 2014. http://www.w3.org/TR/2014/REC-vocab-data-cube-20140116/

12. Devaraju, A., Berkovsky, S.: A hybrid recommendation approach for open research datasets. In: Proceedings of the 26th Conference on User Modeling, Adaptation and Personalization, pp. 207–211. ACM, UMAP 2018 (2018)
13. Ell, B., Vrandečić, D., Simperl, E.: Labels in the web of data. In: Aroyo, L., et al. (eds.) ISWC 2011. LNCS, vol. 7031, pp. 162–176. Springer, Heidelberg (2011). https://doi.org/10.1007/978-3-642-25073-6_11
14. Heindorf, S., et al.: EvoLearner: learning description logics with evolutionary algorithms. In: Proceedings of the ACM Web Conference 2022, pp. 818–828 (2022)
15. Hinneburg, A., Preiss, R., Schröder, R.: TopicExplorer: exploring document collections with topic models. In: Flach, P.A., De Bie, T., Cristianini, N. (eds.) ECML PKDD 2012. LNCS (LNAI), vol. 7524, pp. 838–841. Springer, Heidelberg (2012). https://doi.org/10.1007/978-3-642-33486-3_59
16. Hoffman, M., Bach, F., Blei, D.: Online Learning for Latent Dirichlet Allocation. In: Advances in Neural Information Processing Systems. Curran Associates (2010)
17. Ji, S., Pan, S., Cambria, E., Marttinen, P., Philip, S.Y.: A survey on knowledge graphs: representation, acquisition, and applications. IEEE Trans. Neural Netw. Learn. Syst. **43**, 494–512 (2021)
18. Kopsachilis, V., Vaitis, M.: GeoLOD: a spatial linked data catalog and recommender. Big Data Cogn. Comput. **5**(2), 17 (2021)
19. Kunze, S., Auer, S.: Dataset retrieval. In: 2013 IEEE Seventh International Conference on Semantic Computing (ICSC), pp. 1–8, September 2013
20. Lau, J.H., Grieser, K., Newman, D., Baldwin, T.: Automatic labelling of topic models. In: Proceedings of the 49th Annual Meeting of the Association for Computational Linguistics: Human Language Technologies - Volume 1, pp. 1536–1545. HLT 2011, Association for Computational Linguistics, USA (2011)
21. Manning, C.D., Surdeanu, M., Bauer, J., Finkel, J.R., Bethard, S., McClosky, D.: The stanford corenlp natural language processing toolkit. In: Proceedings of the 52nd Annual Meeting of the Association for Computational Linguistics: System Demonstrations, pp. 55–60. Association for Computational Linguistics (2014)
22. McCrae, J.P.: The Linked Open Data Cloud. Website, May 2021. https://www.lod-cloud.net/. Accessed 24 Aug 2021
23. Mohammadi, M.: (semi-) automatic construction of knowledge graph metadata. In: The Semantic Web: ESWC 2022 Satellite Events, pp. 171–178 (2022)
24. Ngomo, A.C.N., et al.: LIMES-a framework for link discovery on the semantic web. J. Web Semant. **35**, 413–423 (2018)
25. Patni, H.: Linkedsensordata. Website in the web archive, September 2010. https://web.archive.org/web/20190816202119/http://wiki.knoesis.org/index.php/SSW_Datasets. Accessed 11 May 2022
26. Patni, H., Henson, C., Sheth, A.: Linked sensor data. In: 2010 International Symposium on Collaborative Technologies and Systems, pp. 362–370 (2010)
27. Paulheim, H., Hertling, S.: Discoverability of SPARQL endpoints in linked open data. In: Proceedings of the ISWC 2013 Posters & Demonstrations Track, vol. 1035, pp. 245–248. CEUR-WS.org, Aachen, Germany, Germany (2013)
28. Pietriga, E., et al.: Browsing linked data catalogs with LODAtlas. In: Vrandečić, D., et al. (eds.) ISWC 2018. LNCS, vol. 11137, pp. 137–153. Springer, Cham (2018). https://doi.org/10.1007/978-3-030-00668-6_9
29. Řehůřek, R., Sojka, P.: Software framework for topic modelling with large corpora. In: Proceedings of the LREC 2010 Workshop on New Challenges for NLP Frameworks, pp. 45–50. ELRA, May 2010
30. Röder, M., Ngonga Ngomo, A.C., Ermilov, I., Both, A.: Detecting similar linked datasets using topic modelling. In: ESWC (2016)

31. Röder, M., Both, A., Hinneburg, A.: Exploring the space of topic coherence measures. In: Proceedings of the WSDM (2015)
32. Saxena, A., Tripathi, A., Talukdar, P.: Improving multi-hop question answering over knowledge graphs using knowledge base embeddings. In: Proceedings of the 58th Annual Meeting of the Association for Computational Linguistics (2020)
33. Schmachtenberg, M., Bizer, C., Paulheim, H.: Adoption of the linked data best practices in different topical domains. In: The Semantic Web - ISWC 2014 (2014)
34. Singhal, A., Kasturi, R., Srivastava, J.: DataGopher: context-based search for research datasets. In: Proceedings of the 2014 IEEE 15th International Conference on Information Reuse and Integration, pp. 749–756. IEEE IRI 2014 (2014)
35. Sleeman, J., Finin, T., Joshi, A.: Topic modeling for RDF graphs. In: ISWC (2015)
36. Spahiu, B., Maurino, A., Meusel, R.: Topic profiling benchmarks in the linked open data cloud: issues and lessons learned. Semant. Web **10**(2), 329–348 (2019)
37. Spahiu, B., Porrini, R., Palmonari, M., Rula, A., Maurino, A.: ABSTAT: ontology-driven linked data summaries with pattern minimalization. In: ESWC (2016)
38. Tzitzikas, Y., Manolis, N., Papadakos, P.: Faceted exploration of RDF/S datasets: a survey. J. Intell. Inf. Syst. **48**(2), 329–364 (2017)
39. Vandenbussche, P.Y., Atemezing, G.A., Poveda-Villalón, M., Vatant, B.: Linked open vocabularies (LOV): a gateway to reusable semantic vocabularies on the Web. Semant. Web **8**(3), 437–452 (2017)

KnowWhereGraph-Lite: A Perspective of the KnowWhereGraph

Cogan Shimizu[1(✉)], Shirly Stephen[2], Antrea Christou[1], Kitty Currier[2], Mohammad Saeid Mahdavinejad[3], Sanaz Saki Norouzi[3], Abhilekha Dalal[3], Adrita Barua[3], Colby K. Fisher[4], Anthony D'Onofrio[5], Thomas Thelen[6], Krzysztof Janowicz[2,7], Dean Rehberger[5], Mark Schildhauer[8], and Pascal Hitzler[3]

[1] Wright State University, Dayton, USA
cogan.shimizu@wright.edu
[2] University of California, Santa Barbara, Santa Barbara, USA
[3] Kansas State University, Manhattan, USA
[4] Hydronos Labs, Princeton, USA
[5] Michigan State University, East Lansing, USA
[6] Independent Researcher, Carmel, USA
[7] University of Vienna, Vienna, Austria
[8] National Center for Ecological Analysis & Synthesis, Santa Barbara, USA

Abstract. KnowWhereGraph (KWG) is a massive, geo-enabled knowledge graph with a rich and expressive schema. KWG comes with many benefits including helping to capture detailed context of the data. However, the full KWG can be commensurately difficult to navigate and visualize for certain use cases, and its size can impact query performance and complexity. In this paper, we introduce a simplified framework for discussing and constructing *perspectives* of knowledge graphs or ontologies to, in turn, construct simpler versions; describe our exemplar KnowWhereGraph-Lite (KWG-Lite), which is a perspective of the KnowWhereGraph; and introduce an interface for navigating and visualizing entities within KWG-Lite called KnowWherePanel.

1 Introduction

KnowWhereGraph[1] (KWG) is one of the largest, publicly available geospatial knowledge graphs in the world [7,17]. KWG generally supports applications in the food, agriculture, humanitarian relief, and energy sectors and their attendant supply chains; and more specifically supports environmental policy issues relative to interactions among agricultural sustainability, soil conservation practices, and farm labor; and delivery of emergency humanitarian aid, within the US and internationally. To do so, KWG brings together over 30 datasets related to observations of natural hazards (e.g., hurricanes, wildfires, and smoke plumes), spatial characteristics related to climate (e.g., temperature, precipitation, and

[1] https://knowwheregraph.org/.

F. Ortiz-Rodriguez et al. (Eds.): KGSWC 2023, LNCS 14382, pp. 199–212, 2023.
https://doi.org/10.1007/978-3-031-47745-4_15

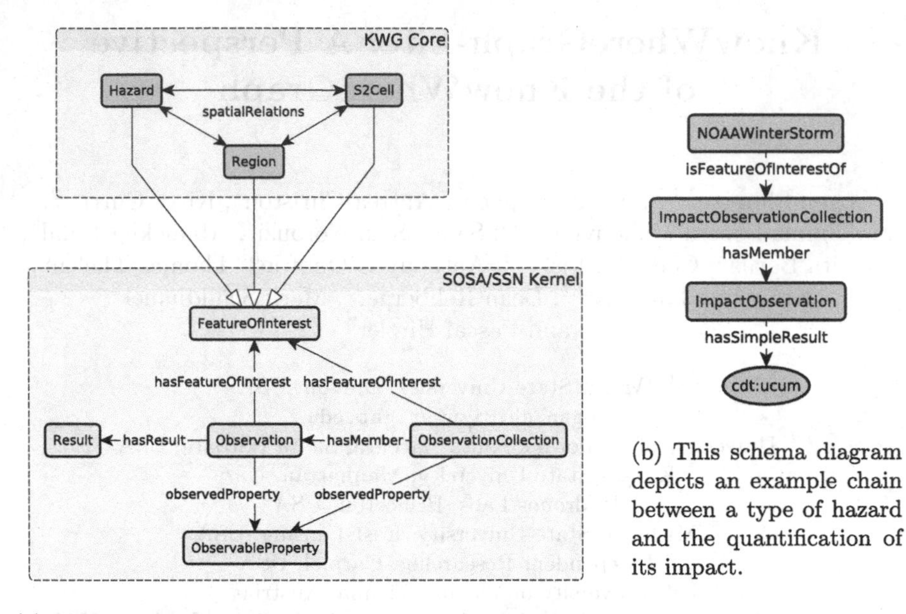

(a) This schema diagram depicts the two central modules of the KnowWhereGraph ontology.

(b) This schema diagram depicts an example chain between a type of hazard and the quantification of its impact.

Fig. 1. The KnowWhereGraph Ontology thoroughly encodes context, which, while effective, can result in a complex ontological structure.

air quality), soil properties, crop and land-cover types, demographics, human health, and spatial representations of human-meaningful places, resulting in a knowledge graph with over 16 billion triples. To integrate these data, we have added an additional layer: KWG uses a schema that provides a thorough and rich[2] ontological representation formally describing the relationships between the types of data. The geospatial integration is performed by a consistent alignment to a Discrete Global Grid (DGG) [2], where we partition the surface of the Earth into small squares. These squares form an approximation of the spatial extent of physical and regional phenomena within the graph and act as a geospatial backbone.

However, due to the size and complexity of the graph and its schema, this can result in a steep learning curve and usability obstacles for those who are not well versed in ontologies, knowledge graphs, or SPARQL [4]. Likewise, for certain use cases a deep or rich ontological representation is not necessary; for some users visualizing or navigating graph data values can be unintuitive; and finally, the use of the DGG can be a barrier itself, as it can result in long and expensive queries. For example, a central piece of our schema is depicted in Fig. 1a. Learning the specific value of an observation pertaining to a particular feature of interest is already a complex, multi-hop query. Figure 1b shows how a particular instance

[2] The OWL ontology has over 300 classes and about 3,000 axioms.

of a Hazard finally relates to an observation on its impact, which is encoded as a data type. This rather complex way of representing the data is necessary because some of the target applications of our graph are aimed at specialists who require a high level of detail. However, for less involved application use cases, we aim to simplify this process and reduce the conceptual barrier to entry, as well as improve performance for certain use cases. To do so, we have created a simplified version of KWG, which we call KnowWhereGraph Lite (KWG-Lite) and is a *perspective* of the base graph. The "lite" version of the graph is constructed from a series of SPARQL CONSTRUCT queries, which, in turn, are formulated from the sets of shortcuts and views that define the perspective.

To quickly access knowledge and data from KWG-Lite, we have developed a visualization interface, which we call KnowWherePanel (KWP), inspired by Google's knowledge panels[3] and Wikipedia's infoboxes.[4] Entities within KWG-Lite can be viewed through KWP, which shows the simplified views in an easily consumable tabular format. Furthermore, the tool provides a way to export embeddable, styled snapshots of each panel, similar to how one can embed Tweets from Twitter, with a styled, living view of the data.

In summary, our contributions are as follows.

1. Perspective Development – A method for defining, creating, and utilizing perspectives over a graph;
2. Perspective Deployment – a method for creating an effective, efficient, and powerful UI for defined user groups from large knowledge graphs;
3. KnowWhereGraph-Lite – a subgraph of the KnowWhereGraph that is intended for simple queries, easy visualization, and quick consumption;
4. KnowWherePanel – a visual interface for generating "panels," which contain a view of an entity; these panels can then be embedded in other media.

The rest of this paper is organized as follows. Section 2 presents the underlying technique used to generate and describe this perspective of the KnowWhereGraph, which is, in turn, used to create KnowWhereGraph-Lite – which we introduce in Sect. 4. Section 3 briefly discusses related work. Entities in KWG-Lite can be viewed in a number of different ways. In Sect. 5, we present the KnowWherePanel interface for transferable and embeddable views of the entity. In Sect. 6 we discuss resource availability. Finally, in Sect. 7, we conclude with future work and next steps.

2 Technical Foundation

The technical basis for constructing the KWG-Lite graph is relatively straightforward. The process is to reduce an expressive structure (i.e., one that encodes context, provenance, and so on in a rich manner) into something more easily queryable, approachable by humans, and easily visualized. We often call such

[3] https://support.google.com/knowledgepanel/answer/9163198?hl=en.
[4] https://en.wikipedia.org/wiki/Infobox.

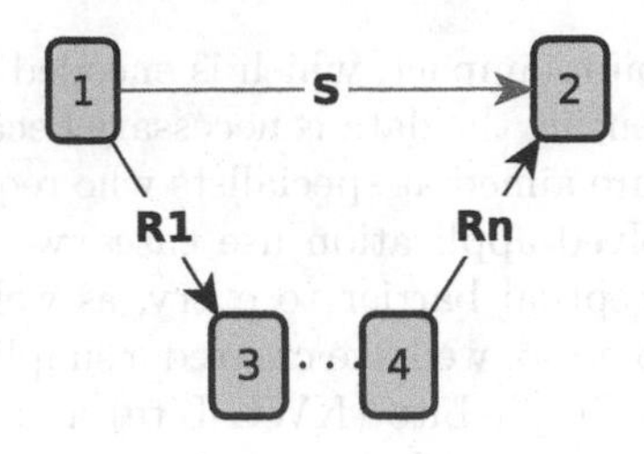

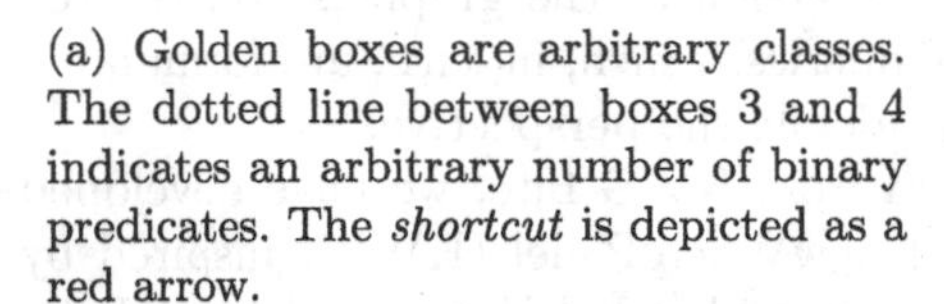

(a) Golden boxes are arbitrary classes. The dotted line between boxes 3 and 4 indicates an arbitrary number of binary predicates. The *shortcut* is depicted as a red arrow.

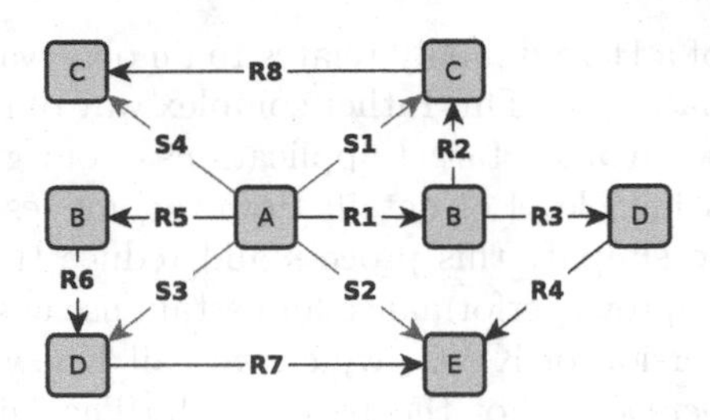

(b) This figure displays a *view* for the class A. Essentially, it constructs a star pattern out of the more expressive structure.

Fig. 2. Depictions of a shortcut and a view.

simplified structure a *star pattern*, due to its shape. That is, important entities are directly related to each other (or explicit data values) with minimal context (or the context is encoded into the name of the relationship, reducing machine interpretability).

Our approach to this process is to define a *perspective* of an expressive ontology. A *perspective* is defined as a set of *shortcuts* that link classes and datatypes by removing or reducing contextual information (i.e., the "skipped-over" nodes in the graph).

A *shortcut* is effectively a bidirectional role chain that is intended to reduce the complexity of an expressive ontological structure, either by facilitating querying or improving human understanding of the encoded data. However, due to the limitations on tractability, we cannot actually express within OWL-DL [5] that a shortcut also implies the expressive structure [10]. We are limited to expressing only that the shortcut is the super-property of the role chain. This role chain,

$$\mathsf{R}_1 \circ \cdots \circ \mathsf{R}_n \sqsubseteq \mathsf{S},$$

is depicted graphically in Fig. 2a. As a rule, it can also be written as

$$\mathsf{R}_1(x_0, x_1) \wedge \cdots \wedge \mathsf{R}_n(x_{n-1}, x_n) \rightarrow \mathsf{S}(x_0, x_n).$$

The body (left-hand side) of the rule, as well as its head (right-hand side), can also take more complex forms, e.g., by adding type information to the variables, such as in

$$\begin{aligned} \mathsf{C}_0(x_0) \wedge \mathsf{R}_1(x_0, x_1) \wedge \mathsf{C}_1(x_1) \wedge \cdots \wedge \mathsf{C}_{n-1}(x_{n-1}) \wedge \mathsf{R}_n(x_{n-1}, x_n) \wedge \mathsf{C}_n(x_n) \\ \rightarrow \mathsf{D}(x_0) \wedge \mathsf{S}(x_0, x_n) \wedge \mathsf{E}(x_n), \end{aligned} \quad (1)$$

or even more complex rule bodies (or heads). In the case of KWG-Lite that we present in this paper, we cast such rules into SPARQL CONSTRUCT queries (discussed further below), but we would like to also remark that many (but not all) of these rules can be expressed in OWL-DL using a technique called *rolification*, which was introduced in [14] (or see [9] for a tutorial). Rules that cannot be converted that way to OWL can be approximated using so-called *nominal schemas* [11].

A *view* is a set of such shortcuts for a particular class in the knowledge graph. This is shown in Fig. 2b. A *perspective* of an ontology (or knowledge graph) is defined as a set of views.[5] Essentially, for some (sub)set of the classes in the ontology, we construct simplified views of the data or knowledge. Intuitively, one now has *a* perspective of the graph. Indeed, we provide an example of one such perspective in Sect. 4 as a core contribution of this paper: KnowWhereGraph-Lite.

In some cases, it is undesirable to materialize directly, or even include, these shortcuts in the base ontology or schema. Nor is it always desirable to even attach formal semantics (in the form of ontology axioms) to the shortcut. Frequently, a shortcut will encode context in the name of the connecting predicate. For example, in our KWG-Lite, we have averageHeatingDegreeDaysPerMonthAug2021. As such, we can also leave annotations within the base ontology, indicating where convenient (and human-meaningful) shortcuts exist, and thus also leave tooling to enable the retrieval and rendering of a view.

In doing so, these annotations can be consumed and mapped to SPARQL CONSTRUCT queries to construct a materialized perspective. We provide examples of such in Sect. 4.

3 Related Work

The concept of leveraging short paths through a graph for aiding understanding, navigation, visualization, or publishing is not altogether new.

The concept of shortcuts, as it applies to linked data, was first explored in [13]. There, shortcuts are intrinsically tied to the notion of pattern-flattening and pattern-expansion, which are methods for publishing and ingesting data at different levels of conceptual granularity. A similar but less principled approach was taken in creating the GeoLink Base Ontology from the GeoLink Modular Ontology [19]. In our case, we have built on these concepts to go beyond a simple pattern-based method to produce a navigable "lite" version of KnowWhereGraph. In [10] the formal logical underpinnings of shortcuts, in the context of OWL, were discussed.

From a tooling perspective, WebVOWL [12] offers a customizable view of an instance graph by automatically flattening or condensing paths of certain lengths. This can be exceptionally useful when examining the graph in its force-directed layout. However, in our case, not all shortcuts in the perspective are of the same length. Equally as important, it would be effectively impossible to use a WebVOWL view of an instance graph to render KnowWhereGraph meaningfully.

[5] The general identification of *which* classes should feature prominently in the perspective is a human-centric process, which is outside of the scope of this paper.

WiSP, or "Weighted Shortest Paths" [18], is a strategy for finding "interesting" connections across the graph. In their discussion, these tend to be meaningful to humans and sometimes even insightful. However, due to how they are constructed, they are not guaranteed to return the same sort of information for entities of the same type. WiSP would be an interesting component to add from a visualization and discovery perspective, but for the formulation of a consistent "lite" graph, it will not work.

4 KnowWhereGraph-Lite

For KnowWhereGraph-Lite (KWG-Lite), two classes were selected to serve as the central concepts of the perspective: HazardEvent and Place, which correspond to the classes Hazard and Region from KnowWhereGraph. We opted to modify the names of the classes to make it human-meaningful and to reduce confusion between to the two graphs.

The schema diagram for KWG-Lite is shown in Fig. 3. Note the distinctive (mostly) star shape of the diagram. As we have stated above, we have removed much of the contextual information, for example, by collapsing the SOSA/SSN kernel into a single relation (with a complex name), such as averageHeatingDegreeDaysPerMonthAug2021. How this name is constructed is shown in Fig. 4. The ObservableProperty (i.e., averageHeatingDays) is combined with the result time (i.e., a month and year), which points directly to the value (i.e., the simple result) of the observation pertaining to that observed property. This results in a large number of unique predicates. However, since these predicates are manually curated, and from a panel perspective *all* relations are meaningful, querying to populate a panel for a particular entity is quite simple.

To construct KWG-Lite (i.e., to materialize the graph), we developed a set of SPARQL CONSTRUCT queries based on the schema diagram in Fig. 3. One example of such a query is shown in Fig. 5. This query is meant to be executed against the base graph (i.e., KnowWhereGraph) and will produce a set of triples that shall constitute a portion of the lite graph. The queries pull double duty in also rebinding the entity from KWG to the KWG-Lite namespace.

In total, we utilized over 20 queries to construct the full KWG-Lite, which are documented in our repository.[6] In all, this results in a graph that contains approximately 200,000 triples (which is about four orders of magnitude smaller than the base graph). Figure 6 shows the respective triple structures for KWG and KWG-Lite.

In addition to providing a SPARQL endpoint specifically including KWG-Lite, we provide the KnowWherePanel interface, which we describe in the next section.

[6] https://github.com/KnowWhereGraph/knowwheregraph-lite/blob/main/construct-queries.md.

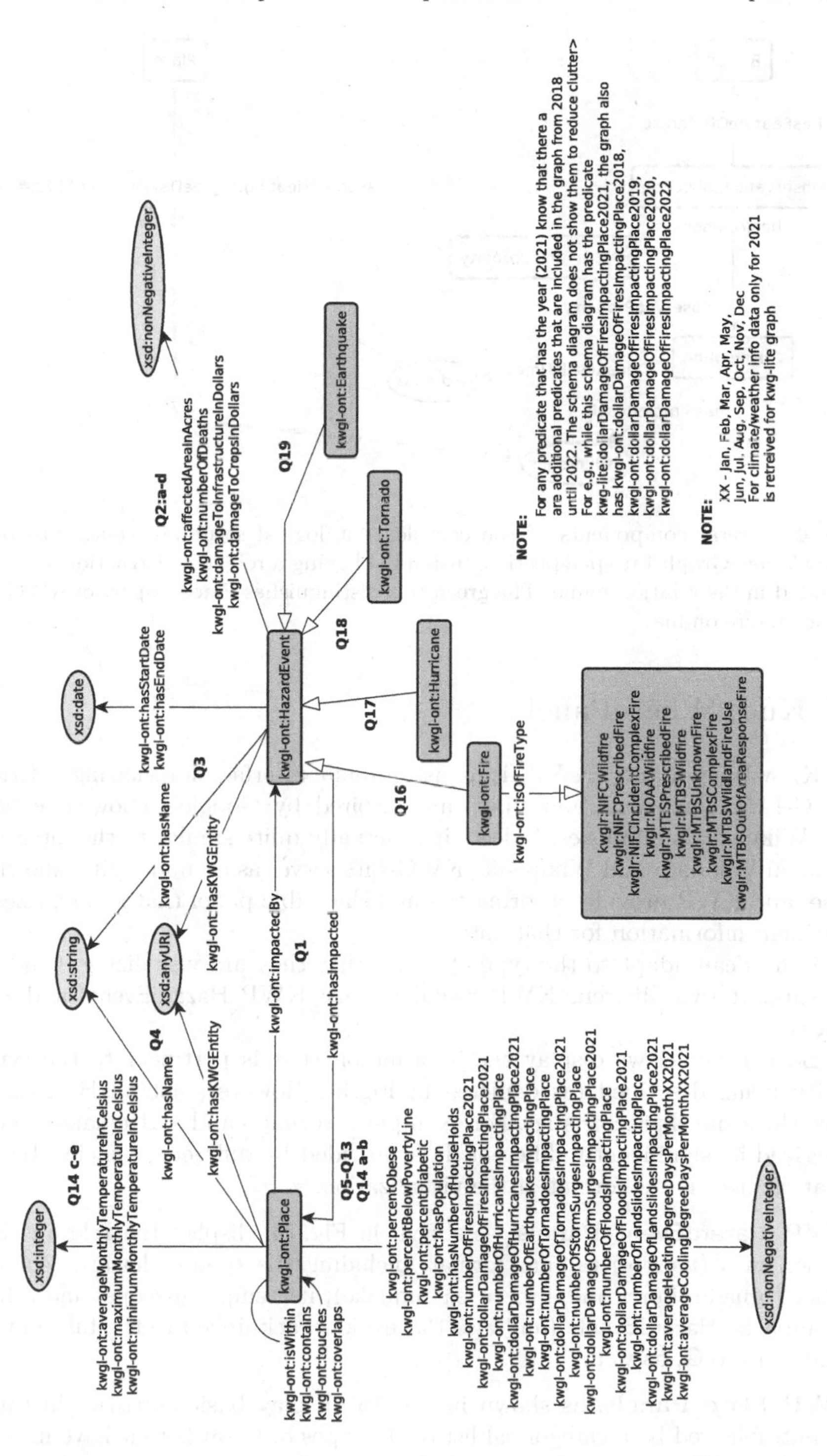

Fig. 3. This schema diagram depicts the entities and relations within KnowWhereGraph-Lite.

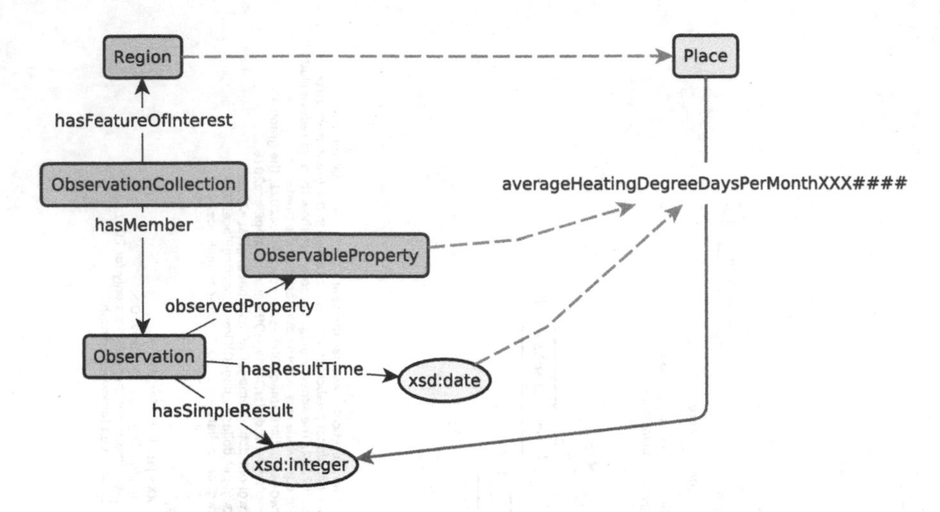

Fig. 4. Several components of the complex ontological structure contribute to the KnowWhereGraph-Lite perspective. Instead of having a reified construction, context is encoded in the relation name. The green box distinguishes Place as part of KWG-Lite. (Color figure online)

5 KnowWherePanel

KnowWherePanel (KWP) is a custom-built interface for viewing entities in KWG-Lite. The name and shape are inspired by Google's knowledge panels and Wikipedia's infoboxes. Indeed, it is actually quite similar to the interaction between Wikidata and Wikipedia. KWG-Lite serves as an underlying knowledge base, and KWP provides a formatted and shareable panel that summarizes the pertinent information for that entity.

Panels can adapt to the type of entity that they are visualizing. Currently, we support two different KWP visualizations: KWP HazardEvents and KWP Places.

Both panel views display the information that is pertinent to the type of entity being described (as depicted in Fig. 3). However, not all HazardEvents have the same characteristics. For example, a tornado and earthquake are characterized by significantly different data provided by different agencies. Instead, what changes is the format of the visualization.

KWP HazardEvent Panels, as shown in Fig. 7a, display the relevant information in a (mostly) tabular format, including the type of hazard (currently either tornado, hurricane, fire, or earthquake), its temporal scope, and a list of its impacts. HazardEvents (and also Places) are back-linked to the full representation in KWG, as well.

KWP Place Panels, as shown in Fig. 7b, display basic statistics in tabular format, followed by a categorical list of the types of hazards that have impacted that place. The resultant statistics are summed per entity type and displayed

```
CONSTRUCT {
    ?p kwgl-ont:impactedBy ?h.
    ?h kwgl-ont:hasImpacted ?p.
}
WHERE {
    ?place a kwg-ont:Region.
    ?place kwg-ont:spatialRelation ?hazard.
    ?hazard a kwg-ont:Hazard.
    BIND(
      STRAFTER(STR(?place),
               "http://stko-kwg.geog.ucsb.edu/lod/resource/")
               as ?placeName)
    BIND(
      CONCAT("http://stko-kwg.geog.ucsb.edu/lod/lite-resource/",
             ?placeName)
             as ?litePlace)
    BIND(IRI(?litePlace) AS ?p ).

    BIND(
      STRAFTER(STR(?hazard),
               "http://stko-kwg.geog.ucsb.edu/lod/resource/")
               as ?hazardName)
    BIND(
      CONCAT("http://stko-kwg.geog.ucsb.edu/lod/lite-resource/",
             ?hazardName)
             as ?liteHazard)

    BIND( IRI(?liteHazard) AS ?h ).
}
```

Fig. 5. This is an example SPARQL CONSTRUCT query that identifies Places that have been impacted by a HazardEvent.

as time-series data, as in Fig. 7c. In most cases, KWG (and thus KWG-Lite) have data up to 2022. However, the query is, itself, dynamic and will update its temporal scope as new data become available.

KWP is, itself, implemented as a static website that creates browsable panels for Places and Hazards in KWG-Lite. The site operates fully in JavaScript: using jQuery/AJAX to query the public graph, and Fuse.js, which provides a fuzzy search functionality.

KWG triple structure:

```
kwgr:noaaClimateDiv.403 sosa:isFeatureOfInterestOf
    kwgr:noaaClimateDivObservationCollection.403 .
kwgr:noaaClimateDivObservationCollection.403 sosa:hasMember
    kwgr:noaaClimateDivPropObservationCollection.403.hdd .
kwgr:noaaClimateDivPropObservationCollection.403.hdd sosa:hasMember
    kwgr:noaaClimateDivObservation.403.HDD.202108 .
kwgr:noaaClimateDivObservation.403.HDD.202108 sosa:hasResult
    kwgr:noaaClimateDivObservationResult.403.hdd.202108 .
kwgr:noaaClimateDivObservationResult.403.hdd.202108 kwg-ont:value
    kwgr:noaaClimateDivObservationQuantityValue.403.hdd.202108 .
noaaClimateDivObservationQuantityValue.403.hdd.202108 qudt:numericValue
    "59.0"^^xsd:float .
```

KWG-Lite triple structure:

```
kwglr:noaaClimateDiv.403
  kwgl-ont:averageHeatingDegreeDaysPerMonthAug2021 "59"^^xsd:float .
```

Fig. 6. This block shows the respective triple structure for KWG and KWG-Lite, as indicated by Figs. 3 and 4.

6 Resource Availability

We have attempted to follow as many applicable best practices as possible for the deployment and provisioning of KWG-Lite and KWP. It should be noted that KWG and, thus, KWG-Lite are both RDF graphs, leveraging W3C standards (e.g., OWL:Time [1], SOSA/SSN [6], and PROV-O [15]), and designed using practices already anchored in the literature [13,16,17].

From an availability standpoint, we have deployed KWG-Lite in a GraphDB [3] repository, which can be directly queried from a SPARQL endpoint.[7] Documentation for KWG-Lite (e.g., its schema diagram), as well as our set of SPARQL queries used to construct KWG-Lite, are housed in our public repository.[8] The KWP interface has open source code[9] and is publicly available for use.[10] KWG-Lite and KWP are released under the CC-BY-4.0 license.[11]

KWG-Lite and KWP are managed under the auspices of the KnowWhere-Graph project. As such, both resources described herein will be maintained under the same sustainability plan.

[7] https://stko-kwg.geog.ucsb.edu/workbench/ and choosing KWG-Lite as the repository (top-right).

[8] https://github.com/KnowWhereGraph/knowwheregraph-lite.

[9] https://github.com/KnowWhereGraph/kw-panels.

[10] https://knowwheregraph.org/kw-panels/.

[11] https://creativecommons.org/licenses/by/4.0/.

NOAATornado Occurred in EARLY from 2005-01-13-1758 to 2005-01-13-1802, EST

Tornado

Start Date: Thu Jan 13 2005 07:58:00 GMT-0500
End Date: Thu Jan 13 2005 08:02:00 GMT-0500
Death toll: 2
Loss of Infrastructure: $250,000.00

Full Entity URI Embed Entity

(a) This figure depicts a KnowWherePanel for a HazardEvent from KWG-Lite.

Caldwell

Type: Administrative Region Level 3 - County/District/Equivalent-level

Population Below Poverty Line: 16.80%
Obesity: 38.10%
Diabetes: 13.80%

Full Entity URI Embed Entity

Hazard Data (2018-2022)	'18 #	'18 $	'19 #	'19 $	'20 #	'20 $	'21 #	'21 $	'22 #	'22 $
Floods	9	$67,000.00	3	$41,000.00	5	$156,000.00	1	$500.00		

(b) This figure depicts a KnowWherePanel for a Place from KWG-Lite.

Fresno

Administrative Region Level 3 - County/District/Equivalent-level

Population Below Poverty Line: 21.10%
Obesity: 33.00%
Diabetes: 9.60%

Full Entity URI Embed Entity

Hazard Data (2018-2022)	'18 #	'18 $	'19 #	'19 $	'20 #	'20 $	'21 #	'21 $	'22 #	'22 $
Fires	38		107		143		460		398	
Tornadoes	1		3	$20,000.00	1					
Floods	15		16		4		24			
Debris Flow Events	2		4				6	$200.00		

(c) This figure depicts a KnowWherePanel for a Place from KWG-Lite that has had multiple types of HazardEvents impact it. Currently, missing data are left blank, as it is not currently known if data are missing or if there were simply no impacts, at all.

Fig. 7. Examples of different KnowWherePanels and the information they are capable of displaying.

- Resource repositories and documentation will remain available in perpetuity.
- Services (e.g., the endpoints and websites) are hosted on institutional, archival resources. The URIs are expected to be indefinitely available.
- Maintenance and updates to the resources are guaranteed through 2025 or longer, pending Foundation establishment.

Finally, we intend for this paper to serve as the canonical citation.

7 Conclusion

KnowWhereGraph is a massive knowledge graph [7] with a rich and expressive schema [8,17]. This comes with many benefits, insofar as it helps to capture provenance, lineage, and spatiotemporal context of the data, and other aspects relevant for expert-level applications. However, it can be commensurately difficult to navigate and visualize for certain (generally simpler) use cases, and its size can impact query performance and complexity.

In this paper, we have introduced a simplified framework for discussing and constructing *perspectives* of knowledge graphs or ontologies which allow us to construct simpler versions of the graph or ontology; introduced our exemplar KnowWhereGraph-Lite, which is a perspective of the KnowWhereGraph; and introduced, as well, our interface for navigating and visualizing entities within KWG-Lite: KnowWherePanel.

Future Work

We have identified the following items for potential next steps regarding KWG-Lite and KnowWherePanel:

1. include additional places and types of hazards (e.g., drought zones) as part of KWG-Lite,
2. include additional top-level entities for KWG-Lite (e.g., create a panel view for the units – Cells or squares – composing the DGG), and
3. develop additional alternative adaptive views for KnowWherePanel, for either the new top-level entities or alternative visualizations for existing entity types.

Finally, we note that KnowWhereGraph-Lite and KnowWherePanel are both living entities; the schema and materialization of KWG and thus the queries are expected to evolve. We have tried to keep a snapshot of KWG, KWG-Lite, and KWP versioned, but these *are* expected to change in the future.

Acknowledgement. This work was funded by the National Science Foundation under Grant 2033521 A1: KnowWhereGraph: Enriching and Linking Cross-Domain Knowledge Graphs using Spatially-Explicit AI Technologies. Any opinions, findings, conclusions, or recommendations expressed in this material are those of the authors and do not necessarily reflect the views of the National Science Foundation.

References

1. Cox, S., Little, C.: Time ontology in OWL. W3C recommendation, W3C, October 2017. https://www.w3.org/TR/2017/REC-owl-time-20171019/
2. Goodchild, M.F.: Discrete global grids for digital earth. In: International Conference on Discrete Global Grids, pp. 26–28. Citeseer (2000)
3. GraphDB. http://graphdb.ontotext.com/
4. Harris, S., Seaborne, A.: SPARQL 1.1 query language. W3C recommendation, W3C, March 2013. https://www.w3.org/TR/2013/REC-sparql11-query-20130321/
5. Hitzler, P., Parsia, B., Rudolph, S., Patel-Schneider, P., Krötzsch, M.: OWL 2 web ontology language primer (second edition). W3C recommendation, W3C, December 2012. https://www.w3.org/TR/2012/REC-owl2-primer-20121211/
6. Janowicz, K., Haller, A., Cox, S., Lefrançois, M., Phuoc, D.L., Taylor, K.: Semantic sensor network ontology. W3C recommendation, W3C, October 2017. https://www.w3.org/TR/2017/REC-vocab-ssn-20171019/
7. Janowicz, K., et al.: Know, know where, Knowwheregraph: a densely connected, cross-domain knowledge graph and geo-enrichment service stack for applications in environmental intelligence. AI Mag. **43**(1), 30–39 (2022). https://doi.org/10.1609/aimag.v43i1.19120
8. Janowicz, K., et al.: Diverse data! Diverse schemata? Semant. Web **13**(1), 1–3 (2022). https://doi.org/10.3233/SW-210453
9. Krisnadhi, A., Maier, F., Hitzler, P.: OWL and rules. In: Polleres, A., et al. (eds.) Reasoning Web 2011. LNCS, vol. 6848, pp. 382–415. Springer, Heidelberg (2011). https://doi.org/10.1007/978-3-642-23032-5_7
10. Krisnadhi, A.A., Hitzler, P., Janowicz, K.: On the capabilities and limitations of OWL regarding typecasting and ontology design pattern views. In: Tamma, V., Dragoni, M., Gonçalves, R., Ławrynowicz, A. (eds.) OWLED 2015. LNCS, vol. 9557, pp. 105–116. Springer, Cham (2016). https://doi.org/10.1007/978-3-319-33245-1_11
11. Krötzsch, M., Maier, F., Krisnadhi, A., Hitzler, P.: A better uncle for OWL: nominal schemas for integrating rules and ontologies. In: Srinivasan, S., Ramamritham, K., Kumar, A., Ravindra, M.P., Bertino, E., Kumar, R. (eds.) Proceedings of the 20th International Conference on World Wide Web, WWW 2011, Hyderabad, India, March 28 - April 1, 2011. pp. 645–654. ACM (2011). https://doi.org/10.1145/1963405.1963496
12. Lohmann, S., Negru, S., Haag, F., Ertl, T.: Visualizing ontologies with VOWL. Semant. Web **7**(4), 399–419 (2016). https://doi.org/10.3233/SW-150200
13. Rodríguez-Doncel, V., Krisnadhi, A.A., Hitzler, P., Cheatham, M., Karima, N., Amini, R.: Pattern-based linked data publication: the linked chess dataset case. In: Hartig, O., Sequeda, J.F., Hogan, A. (eds.) Proceedings of the 6th International Workshop on Consuming Linked Data co-located with 14th International Semantic Web Conference (ISWC 2105), Bethlehem, Pennsylvania, USA, October 12th, 2015. CEUR Workshop Proceedings, vol. 1426. CEUR-WS.org (2015). https://ceur-ws.org/Vol-1426/paper-05.pdf
14. Rudolph, S., Krötzsch, M., Hitzler, P.: All elephants are bigger than all mice. In: Baader, F., Lutz, C., Motik, B. (eds.) Proceedings of the 21st International Workshop on Description Logics (DL2008), Dresden, Germany, May 13–16, 2008. CEUR Workshop Proceedings, vol. 353. CEUR-WS.org (2008). https://ceur-ws.org/Vol-353/RudolphKraetzschHitzler.pdf

15. Sahoo, S., McGuinness, D., Lebo, T.: PROV-O: the PROV ontology. W3C recommendation, W3C, April 2013. http://www.w3.org/TR/2013/REC-prov-o-20130430/
16. Shimizu, C., Hammar, K., Hitzler, P.: Modular ontology modeling. Semant. Web **14**(3), 459–489 (2023). https://doi.org/10.3233/SW-222886
17. Shimizu, C., et al.: The knowwheregraph ontology. Pre-print, May 2023. https://daselab.cs.ksu.edu/publications/knowwheregraph-ontology
18. Tartari, G., Hogan, A.: WISP: weighted shortest paths for RDF graphs. In: Ivanova, V., Lambrix, P., Lohmann, S., Pesquita, C. (eds.) Proceedings of the Fourth International Workshop on Visualization and Interaction for Ontologies and Linked Data co-located with the 17th International Semantic Web Conference, VOILA@ISWC 2018, Monterey, CA, USA, October 8, 2018. CEUR Workshop Proceedings, vol. 2187, pp. 37–52. CEUR-WS.org (2018). https://ceur-ws.org/Vol-2187/paper4.pdf
19. Zhou, L., Cheatham, M., Krisnadhi, A., Hitzler, P.: GeoLink data set: a complex alignment benchmark from real-world ontology. Data Intell. **2**(3), 353–378 (2020). https://doi.org/10.1162/dint_a_00054

Automating the Generation of Competency Questions for Ontologies with AgOCQs

Mary-Jane Antia(✉) and C. Maria Keet

University of Cape Town, Cape Town, South Africa
{mantia,mkeet}@cs.uct.ac.za

Abstract. Competency Questions (CQs) are natural language questions drawn from a chosen subject domain and are intended for use in ontology engineering processes. Authoring good quality and answerable CQs has been shown to be difficult and time-consuming, due to, among others, manual authoring, relevance, answerability, and re-usability. As a result, few ontologies are accompanied by few CQs and their uptake among ontology developers remains low. We aim to address the challenges with manual CQ authoring through automating CQ generation. This novel process, called AgOCQs, leverages a combination of Natural Language Processing (NLP) techniques, corpus and transfer learning methods, and an existing controlled natural language for CQs. AgOCQs was applied to CQ generation from a corpus of Covid-19 research articles, and a selection of the generated questions was evaluated in a survey. 70% of the CQs were judged as being grammatically correct by at least 70% of the participants. For 12 of the 20 evaluated CQs, the ontology expert participants deemed the CQs to be answerable by an ontology at a range of 50%-85% across the CQs, with the remainder uncertain. This same group of ontology experts found the CQs to be relevant between 70%-93% across the 12 CQs. Finally, 73% of the users group and 69% of the ontology experts judged all the CQs to provide clear domain coverage. These findings are promising for the automation of CQs authoring, which should reduce development time for ontology developers.

1 Introduction

Ontologies have been shown to be useful in a wide range of subject domains and applications. Regarding their content, they are expected to be well-delineated, explicit, and representative of the selected subject domain. To obtain those characteristics, Competency Questions (CQs) have been proposed as important means, to aid in the development, verification, and evaluation processes [5,13,22,25]. CQs are natural language questions drawn from a given (sub)domain for use in the ontology development cycle [22]. The adoption of CQs by ontology engineers has been reported as low due to difficulties including the authoring of good quality CQs [12]. In solving problems related to CQs

F. Ortiz-Rodriguez et al. (Eds.): KGSWC 2023, LNCS 14382, pp. 213–227, 2023.
https://doi.org/10.1007/978-3-031-47745-4_16

for the ontology development cycle, the focus of several CQ-related studies has been on artefacts and processes that can enhance CQ quality after they have been manually authored [3,12,19,22] Corpus-based methods have been used in several areas such as expert systems and data mining [16,24]. They are known to provide insights into knowledge domains, yet they have not been used as for CQ development. We aim to reduce hurdles with manual CQ authoring and quality by proposing an approach that uses corpus-based methods to automate CQ authoring. We aim to answer the following research questions:

Q1: Can a corpus-based method support the automation of CQ authoring, and if so, how?
Q2: How do automated CQs fare on the key quality criteria answerability, grammaticality, scope, and relevance in relation to a given (sub)domain?

The approach taken to answer these questions is as follows. We design a novel pipeline and accompanying algorithm to automate CQ generation. It combines a text corpus with CQ templates, a CQ abstraction method, transfer learning models, and NLP techniques. This procedure was evaluated with a survey among the target groups composed of ontology experts, domain experts, and users. The results showed that 14/20 CQs had 70%-100% of participants judged it as grammatically correct. Perceived answerability by a hypothetical ontology gave mixed results. Also, the vast majority of participants found 12/20 CQs to be relevant, and 73% of users and 69% of the experienced ontology experts found the CQs to provide clear domain coverage.

The rest of the paper discusses related work (Sect. 2), the novel methodology of AgOCQs (Sect. 3) and its evaluation (Sect. 4). We close with conclusions and future work (Sect. 5).

2 Related Work

We consider both the state-of-the-art CQs in ontology engineering and the promising NLP techniques for automated question generation.

2.1 CQs in Ontology Engineering

CQs have been proposed as part of the requirement specification in ontology development [23]. CQs have been shown to play other roles, such as functional requirements for ontology development, for verification (with a focus on completeness and correctness), and providing insights into the contents of a specific ontology, especially to non-expert users [5,23]. CQs have also been used for ontology reuse and relevance [4], test-driven ontology development [13], and enhancing the agile development process for ontologies [1,18].

A number of concerns have been noted in the literature around CQs in ontology development and use. In particular, manual authoring of CQs continues to be a hindrance to their successful participation in the ontology engineering process [3,12,19,22,26]. Many ontologies tend to have accompanying CQs defined at a high level to provide an overview of what the ontology is on [8]. There is a dearth of authoring support tools and the manual process of authoring CQs is tedious and time-consuming. In addition, manually authored CQs are sometimes not answerable, not relevant, not grammatical, and not sufficiently indicative of the scope of a given ontology [3]. In a bid to address the lack of authoring support, several solutions have been proposed. Current solutions and tools have centred on evaluating manually CQs from specific ontologies [5,6], creating artefacts to support manually developed CQs for ontology use [3,12,19,22,26]; or on how to develop ontologies from CQs [1]. One such solution is exemplified by a set of core and variant CQ archetypes created for the Pizza ontology [22]. These archetypes are ontology elements of OWL class and object property with a 1:1 mapping attribute, limiting their use to only OWL ontologies with certain limited formalisation patterns.

An approach is to check manually authored CQs with a few CQ patterns, focusing on OWL ontology variables [6] however, omit "Who" and "Where" question types. Another proposal separated the CQ linguistic analysis from OWL, using linguistic pattern extraction from CQs by [19,26]. They identified nouns and noun phrases from entities along with verbs and verb phrases from predicates representing them in abstract forms for a set of 234 manual CQs that were developed and corrected for 5 different ontologies (Dem@care, Stuff, African wildlife, OntoDT and Software ontologies).

They also created 106 patterns [19,26], from which the Competency question Language for specifying Requirements for an Ontology (CLaRO) controlled natural language was developed [12].

These templates restrict the CQs types but allow for a 1:m mapping and they can be used also with other ontology languages. CLaRO templates have shown good coverage with an accuracy of over 90% to unseen CQs across several domains [3,12]. Each CLaRO template can correspond to several questions and are about 150 templates. However, for ontology developers to make use of them, CQs would still have to first be manually developed and then checked for compliance using the templates. The persistent manual-only approach is the common theme, and shortcoming, in the different solutions thus far.

2.2 Transfer Learning

With the advent of machine learning technologies, the process of question generation has had successes and failures. Transfer Learning (TL), a method which leverages knowledge learned in one domain task to perform a similar task in a different domain, has been at the centre of it [2,10]. Two main methods in use include inductive and transductive TL.

With Inductive TL, tasks from the source domain differ from the tasks of the target domain. This method works with the presence or absence of labelled data. Multi-task learning is performed with labelled data while self-taught learning is performed without labelled data. A sub-category of inductive learning methods that focuses on applying the inductive approach to unsupervised learning tasks such as clustering, dimensionality reduction, and density estimation is referred to as unsupervised TL.

With Transductive TL, the tasks from both source and target domains are the same but the domains are different. This method is widely used in cross-lingual and domain adaptation TL studies.

Instance transfer, feature representation transfer, parameter transfer and relational knowledge transfer are approaches and all applied to inductive TL, while only instance transfer and feature representation transfer can be applied to transductive TL [2,17,27,28].

Much of the success of TL is attributed to the use of Large Language Models (LLM) transformer models in general and failures to training time when using LLM that make the process computationally expensive [27]. The effectiveness of LLMs require that when training, the LLM is able to assimilate various learning points ranging from small spelling errors to the contextual meanings present in the training corpus. The TL method brings performance improvements when using LLM models because the models no longer have to be trained from scratch but can be used in a pre-trained state. Large crowd-sourced text corpora, such as the Stanford Question Answering Dataset (SQuAD) and Colossal Clean Crawled Corpus, are used to train LLMs such as BERT, RoBERTa, sBERT and Text-To-Text-Transfer-Transformers(T5) to enable them to perform several tasks [27]. The T5 model was trained using the Text-To-Text framework [15]. The T5 combines all NLP techniques such as translation, question answering, text summarisation, and document classification together in one model, thereby reducing the need to perform these tasks individually [20,28].

3 Auto-generated Ontological Competency Questions (AgOCQs)

We describe the methods used to develop AgOCQs, and use as illustration its application to a small corpus of scientific articles on COVID-19.

As part of the development process, we leveraged the abstract representation from the linguistic pattern extraction method developed by [19] and the CLaRO, a controlled language set of templates for CQs by [3,12]. These methods along with a combination that infuses NLP techniques and Transformer models make up the development processes for AgOCQs.

Figure 1 and the algorithm in Fig. 2 summarise the design of the automated process of developing CQs, called AgOCQs. It begins by extracting domain text corpus which is then pre-processed using NLP techniques such as entity and sentence extraction, stop words removal [11] and regular expressions to produce cleaned data used as the input text data. We then apply the transductive TL where the source task for the model is the same as the target task. Using the pre-trained Text-to-Text-Transfer-Transformer (T5) base model [20], which is pre-trained with the SQUAD dataset to output a context, question and answer as our source task. We pass in our cleaned input data to undertake the same task as the source task of the base model; however, we only output the context

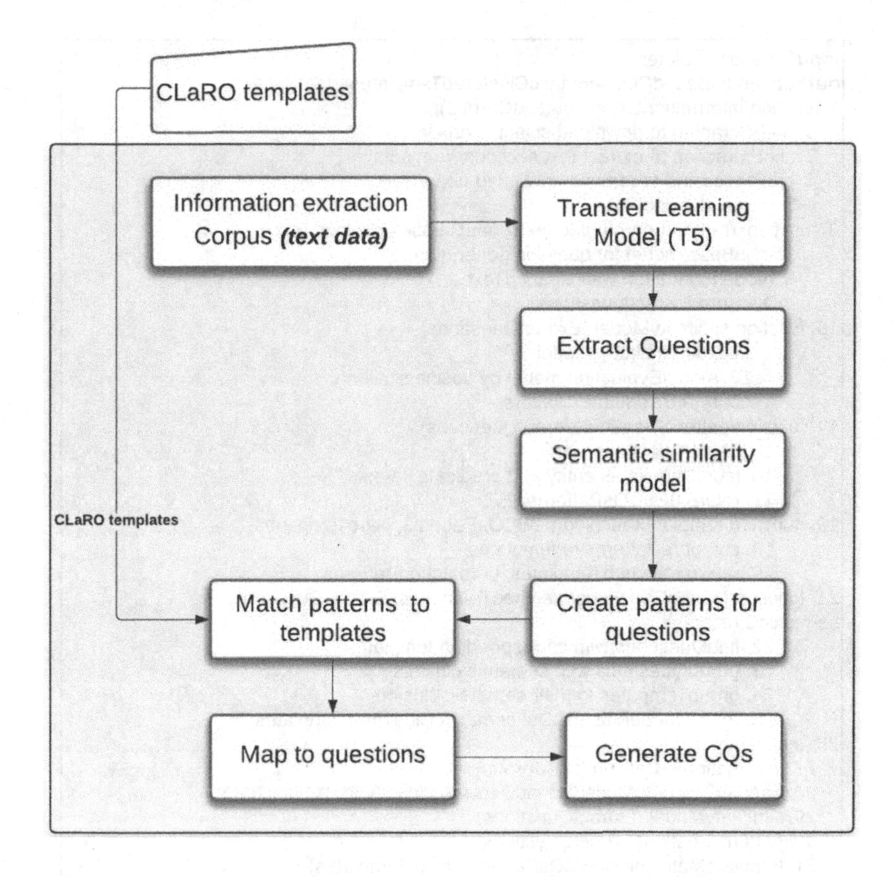

Fig. 1. Design of the pipeline architecture for automatic CQs authoring.

texts and questions in this case. The corpus of questions generated from the input data is subsequently de-cluttered to remove duplicates and meaningless questions through a semantic grouping using the paraphrased algorithm of the Sentence Transformer model [21].

Next, we represent the questions into their linguistic abstract forms using the method from [19,26]. To do this, each question is broken down into chunks and represented in the abstract form as *entity chunks(EC)* i.e., nouns/noun phrases, and *predicate chunks(PC)*, i.e., verbs/verb phrases, as illustrated in Table 1 for the use case chosen to test the approach with.

Generated questions in their abstract form are then compared to CQ templates from CLaRO that were also developed using the same abstract format [3,12]. CLaRO templates, having been assessed for grammatical correctness and answerability by an ontology, serve as a gold standard for CQs in determining if a set of questions qualifies as CQs. If the abstract form of a question matches a template, the question is then referred to as a "Competency Question". The template rules are an integral part of the development process and help ensure

```
input: claroTemplates
output: modelBasedCQ, semanticClusteredTemplates
1. function informationExtraction(textCorpus):
   2. webScraping to download scientifc articles
   3. textExtraction to extract text sections from pdfs
   4. preProcessing to remove unwanted characters
      5. return cleanedText
6. function TransferLearningModel(SquadDataset, cleanedText)
      7. trainBase model for question generation
      8. modelEvaluation with cleanedText
         9. return ExtractQuestions
10. function similarityModel (ExtractQuestions)
            11. trainSimilarity model
            12. modelEvaluation model by cosine similarity
            13. return semanticClusters
14. function patternCreation(ExtractQuestions)
         15. chunkQuestions
         16. labelChunks as entity and predicate chunks
         17. return newCQsPatterns
18.  function templateMatching(newCQsPatterns, claroTemplates)
         19. comparePatterns to templates
         20. return matchedTemplates, unmatchedPatterns
21. function MapToQuestions(matchedTemplates, ExtractQuestions,
semanticClusters)
         22. findQuestions with corresponding templates
         23. groupQuestions with semantic clusters
         24. groupTempates to their semantic clusters
         25. return modelBasedCQs, semanticClusteredTemplates
26. main()
   27. informationExtraction (textCorpus)
   28. transferLearningModel(SquadDataset, cleanedText)
   29. similarityModel (ExtractQuestions)
   30. patternCreation(ExtractQuestions)
   31. templateMatching(newCQsPatterns, claroTemplates)
   32. MapToQuestions(matchedTemplates, ExtractQuestions, semanticClusters)
   33. return modelBasedCQ, semanticClusteredTemplates
```

Fig. 2. Outline of the main algorithm for automatic CQs authoring.

that the abstract representation of the CQs produced correspond to the sentence chunk. The abstract forms can also be reviewed for any error as it is saved to a file that can be corrected and fed back into the process if needed.

Two groups emerge from this step in the pipeline: complete matches and variants (i.e., a very close match of a template). For the use case with the Covid-29 corpus (see below), only complete matches to templates were considered. With the CLaRO templates property of 1:m mapping, several questions can have abstract forms that correspond to just one of the templates. The abstract forms are then mapped back to the questions to give a set of CQs that are deemed ready for use by the ontology developer. A sample of the generated CQs for the use case and their corresponding sentence patterns mapping to templates are displayed in Table 2. All CQs can be found on Github: https://github.com/pymj/AgOCQs.

Table 1. Sampling of questions and their respective abstract representation, where the method as used on a small corpus of scientific articles on COVID-19.

Question	Abstract form
How many people have been infected with COVID-19?	How many EC1 PC1 been PC1 EC2?
What severity of the case may progress to respiratory distress or respiratory failure?	What EC1 of EC2 PC1 PC1 EC3 or EC4?
How severe is the disease related to age?	How PC1 is EC1 PC2 EC2?

Table 2. CQs and corresponding sentence patterns, generated by the AgOCQs procedure when applied to the small Covid-19 text corpus

CQs	Templates	ID
How can Coronaviruses induce psychopathological sequelae?	How PC1 EC1 PC1 EC2?	17
What is the prevalence of emergent psychiatric conditions?	What is EC1 of EC2?	60
What is the mean age range for COVID19 survivors?	What is EC1 for EC2?	38
What is the role of SARSCOV2 immuneescape mechanisms?	What is EC1 of EC2?	60
What are the mainstay of clinical treatment?	What are EC1 of EC2?	60a
What is lymphopenia?	What is EC1?	90
What is the name for the cytokine storm14?	'What is EC1 for EC2?	38
What is a potential target for IL1 IL17?	What is EC1 for EC2?	38
What is another approach to alleviate COVID19 related immunopathology?	What is EC1 PC1 EC2?	66
What is the role of standardized treatment protocols for severe cases?	What is EC1 of EC2 for EC3?	61
What percentage of the subjects reported fatigue?	What EC1 of EC2 PC1 EC3?	68
What is the spread rate of COVID19?	What is EC1 of EC2?	60
What is the duration of symptoms for mild cases?	What is EC1 of EC2 for EC3?	61
What is a blood test for COVID19?	What is EC1 for EC2?	38
What role could corticosteroids play in severe cases?	What EC1 PC1 EC2 PC1 EC3?	58
What did the severe acute respiratory syndrome SARS attack reflect?	What PC1 EC1 PC1?	41

4 Evaluation of CQs Generated with AgOCQs

The procedure described in Sect. 3 can be applied to any text corpus. The principal interest is obviously specialised subject domains, for which it is difficult to obtain extensive domain expert input. To this end, we created a small corpus of scientific articles on Covid-19, generated the questions with the proposed method, and subsequently evaluated them in a human evaluation. The evaluation approach, results, and discussion are described in the remainder of this section.

4.1 Approach

We conducted a survey to assess the CQs developed by AgOCQs. The aim of the survey was to use these target groups' responses on the answerability, relevance, scope, and grammaticality to assess the quality of these automatically generated CQs to determine how they fare on issues corresponding to some of the concerns associated with manually created CQs. The three groups are composed as follows: ontology experts, domain experts and ontology users. The participants had preliminary questions which were used to place them in the groups. The ontology users group is made up of 1) ontology professionals who do not consider themselves as experts. 2) ontology experts who are not the target domain experts. Thus, some participants in the users' group also appear in the ontology expert group.

The test corpus was created from freely available published research articles in the Covid-19 domain on the Web, in the time frame between 2020–2021. No specific criteria was applied other than the articles being a research paper from the COVID-19 domain. These articles were scraped using the PyPaperBot python tool. Seven articles only were used due to issues around compute capacity and lengthy processing time. The text corpus was created from the scientific articles' Introduction, Related work, and Methodology sections.

This corpus was then used in AgOCQs to generate candidate CQs for evaluation. Answerability, relevance, scope, and grammaticality are considered criteria for the assessment of the CQs. The target groups for the survey were Covid-19 domain experts, ontology experts, and ontology users. Users, in this case, were considered as research students working in areas directly or related to ontologies, and ontology experts working in the Covid-19 domain.

For CQs to be answerable, we presume that it should also be grammatically correct, therefore in assessing answerability we include the assessment of its grammaticality from participants with advanced English grammar competence. In terms of relevance, though the corpus used to develop the CQs is domain-specific, we still evaluate relevance as seen by domain and ontology experts as well as by ontology user groups. For the scope, all participants are asked to assess the overall CQs presented in the survey on their clarity of purpose from their standpoint. The objectives for the survey are, for each of domain experts, ontology experts, and ontology users as separate groups respectively:

1. To understand the respective judgments of the different target groups on the grammaticality and answerability of the automatically generated CQs.
2. To determine how the different groups rate the relevance of the automatically generated CQs.
3. To understand how each of the groups judge the automatic CQs as an indicator of the scope for their respective objectives.

For the analysis of the survey, we focus on the target groups' responses individually. We will analyze the overlaps of interests (if any) of the three target groups (ontology experts, Covid-19 subject experts, ontology users) and how that affects their judgments of CQs. We will also analyze how the experience of participants within each group could lead to different judgments. The survey contained question for classifying participants into different target groups as well as 20 CQs to be judged by the participants. A question was marked as a CQ as attention check, being CQ.17: *What is post-COVID similar to post-SARS syndrome?*) which serve to assist to assess some of the responses of the participants to CQs, since it is incoherent and should be judged accordingly. The last question was directed to how the participants judge the overall CQs in terms of coverage of the domain in question. A sampling of the CQs used the survey are shown in Table 3.

Table 3. A selection of the CQs used in the survey

CQ1: How can Coronaviruses induce psychopathological sequelae?
CQ2: What is the mean age range for COVID-19 survivors?
CQ4: What is the duration of symptoms for mild cases?
CQ5: What is the prevalence of emergent psychiatric disorders?
CQ7: What is lymphopenia?
CQ8: What is the current status of herd immunity?

To check for reliability, we apply Fleiss's Kappa's coefficient for inter-rater testing [7,9] which measure to underscore the results from the target groups and remove any notion of occurrence based on chance. For interpretation, we use the scale from [14], where a moderate agreement to a near-perfect agreement would mean that the participants engaged with CQs and made an informed decision in their judgment. A widely used scale for measuring the agreement between raters is as follows: Zero (0) as No agreement, 0.1 - 0.20 as Slight, 0.21 - 0.40 as Fair, 0.41 - 0.60 as Moderate, 0.61 - 0.80 as Substantial, and 0.81 - 1.00 as Near perfect agreement.

Data and results are be available at https://github.com/pymj/AgOCQs.

4.2 Results

The results of the survey is presented on the basis of the assessment criteria of answerability, grammaticality, relevance, and scope in relation to the target groups. There were 20 CQs in the survey to be judged, and a total of 17 respondents completed the survey. The domain expert group only had 1 respondent, the ontology expert group had 16 respondents, and the ontology users group (i.e., a combination of non-domain experts and non-ontology experts) had 15.

On grammatical competence and answerability, participants were asked to rate their English grammar competence (either average or very good). 75% rated their competence as *very good* while 25% rated themselves as *average*. Looking at the results by CQ from the participants that considered themselves to have very good English grammar competence, 70%-100% judged the majority of the CQs (14 of 20) as being grammatically correct Fig 3.

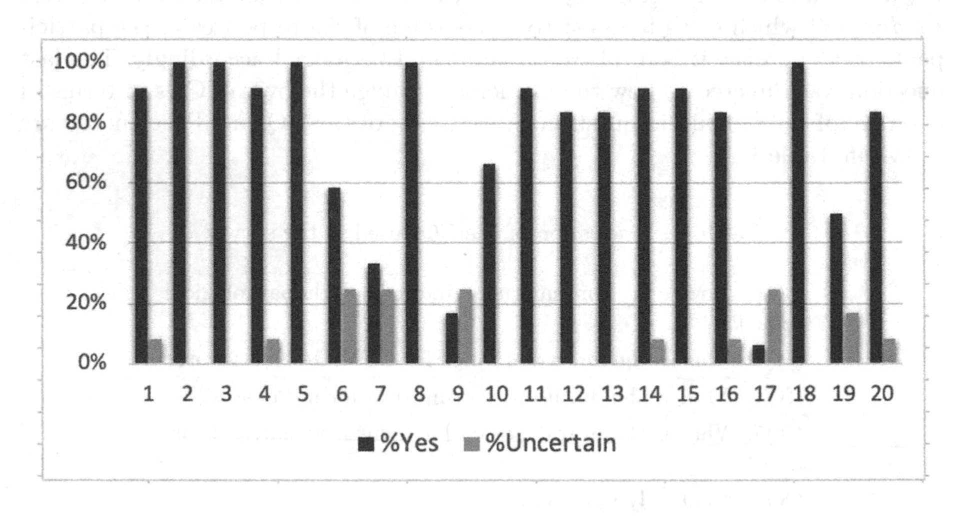

Fig. 3. Grammatical correctness by CQ

Looking at answerability on individual CQs from the same group on English grammar competence, the results showed that 50%-85% of participants deemed answerability to be positive in 12 CQs (see Fig 4).

Participants were also classified in terms of their competency in ontologies and CQs knowledge, where 81% identified as experienced and 19% as not experienced. Taking an overall view by CQs by experienced category, we observe answerability to be at an average of 50% and participants' uncertainty of answerability to be equally the same at 50% (see Fig 5).

In terms of relevance to the domain, 94% of participants classified themselves as not being experienced in the COVID-19 domain. As a result, the relevance of CQs was analyzed from the user's target group alone, which is composed of research students and ontology experts inexperienced in the COVID-19 domain.

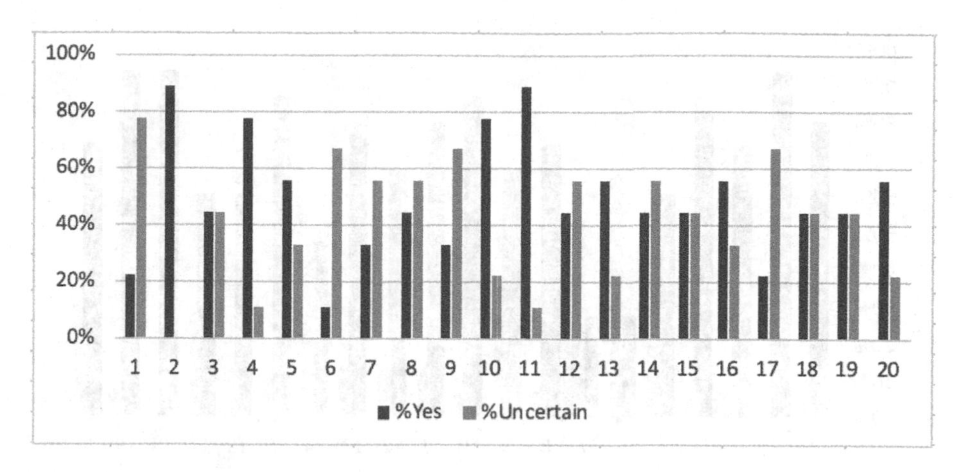

Fig. 4. Answerability by English grammar, ontology and CQs competence

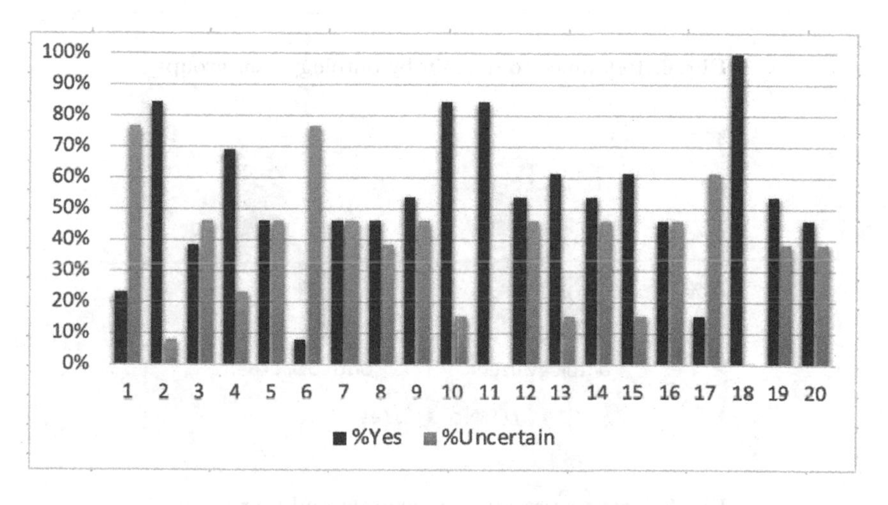

Fig. 5. Answerability by ontology and CQs experts

70% of this group judged the CQs to be relevant to the domain Fig 6. Our results also showed that 69% of ontology and CQs experts who considered themselves experienced and 73% of ontology users believed the CQs gave them a clear scope of the Covid-19 domain (see Fig 7).

To ensure that the results from our survey represent the authentic view of our participants on key criteria used and did not occur by chance, we placed *CQ.17* as an attention check. Our results showed the participants passed the attention check overwhelmingly, with 95% of them detecting the CQ to be unanswerable with poor grammar. Also, we conducted a reliability test using Fleiss's Kappa coefficient test [14]. We interpret our results based on the scale from its Wikipedia page. Our Fleiss's Kappa scores for grammatical correctness, answerability and

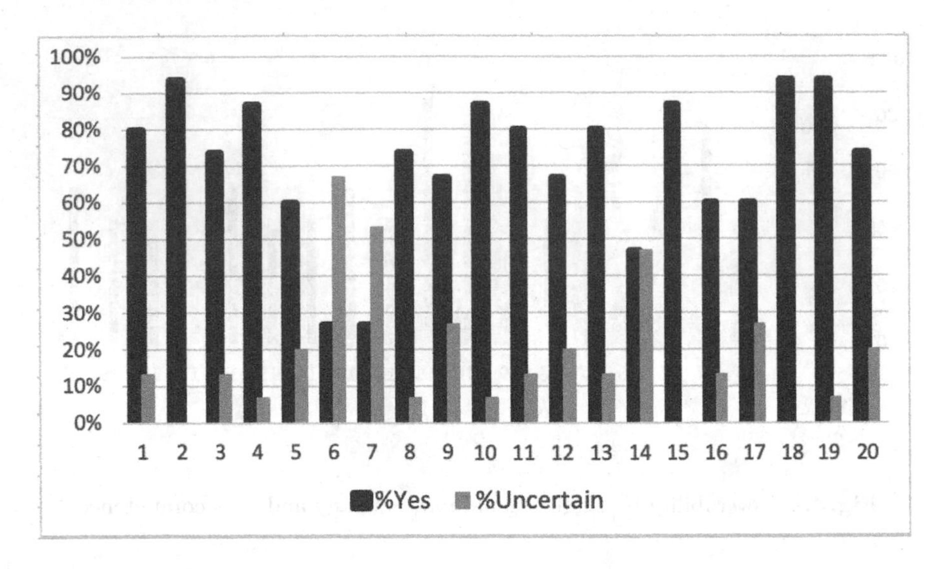

Fig. 6. Relevance to domain by ontology user group

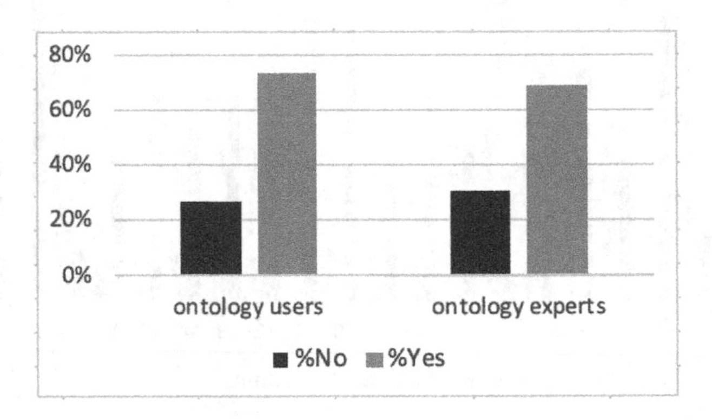

Fig. 7. Scope assessment by experts and user groups

relevance were 0.55, 0.55 and 0.77, respectively. These scores show that the responses for the survey had a moderate to a substantial degree of agreement on the judgments made by participants, thus removing any notions that the responses occurred by chance.

4.3 Discussion

As mentioned in the motivation for this research, ontology developers have indicated that the process of manual CQ development is tedious and time-consuming and produces relatively few CQs of varying quality [3,12,26] which adds to the workload of ontology development and is a major reason why many skip the use of CQs. The proposed automated process AgOCQs resolves this problem by reducing the efforts of manual CQ authoring. This may potentially increase the adoption of CQs in ontology engineering processes. Compared to their manually curated counterparts, CQs from AgOCQs are granular, providing a large pool of CQs that are more specific compared to high-level scoping CQs. In addition, since most of our user target group also judged the CQs to cover relevant information on the domain Fig 7, suggests that AgOCQs may facilitate CQ reuse across ontologies within the same domain or subdomain. The use of a text corpus in CQs development adds some of the benefits observed elsewhere [16,24] to ontologies via CQs, thereby answering our first research question.

The results of the survey also provide a window to assess human judgment on CQs from ontology experts and engaged users. AgOCQs uses a set of templates considered as ground truth because the CQ templates have been verified to be answerable by an ontology where the contents are available. Yet, there is a divide in the results from experts on the certainty of the CQs being answerable compared to novices. Also, which is not uncommon for human judgements with a limited pool of participants with varied backgrounds, it shows that the opinions of ontology experts on grammaticality and answerability do not even out. For instance, although most participants indicated having very good English grammar competence (see Fig 3), there are CQs where the judgment by these participants is questionable; e.g., CQ.9: *What severity of the case may progress to respiratory distress?* is arguably grammatically correct when taken in context, but only 13% of the participants judged it as correct.

The automated CQs also face similar issues encountered by CQs from the manual process, especially on the criteria of answerability and grammatical correctness, as can be interpreted in agreement levels from the Fleiss Kappa scores, albeit to a lesser extent.

With CLaRO templates having a property of 1:n mapping to SPARQL queries and possible axiom patterns [3,12], their use as part of the development of AgOCQs brings the possibility of realizing CQs reusability and by extension, ontology reusability within reach as This methodology can be applied to other domains with little or no reservations. However, new domain-specific templates which may currently not exist in CLaRO may cause the omission of certain questions.

A limitations of the study include the use of a small dataset to demonstrate the functionality of our method, which was due to the compute capacity available. Thus, the data used in this study is not representative of the data within the domain. We are working towards optimising the approach of the text processing as well as the use of an ontology to demonstrate its effectiveness on the issue of completeness in the ontology engineering process.

5 Conclusion and Future Work

The paper proposed AgOCQs to automate the process of creating competency questions for ontology development and selection. The results showed that CQs from AgOCQs using a domain text corpus are highly granular and provide a larger number compared to those manually developed. The evaluation indicated that it is possible to have a set of CQs that may serve a number of ontologies in the same domain. The proposed automated CQ creation process may foster CQ uptake for ontology selection, design, and evaluation. In future work, we plan to explore the effect of corpus size and genre on CQ generation.

Acknowledgements. This work was financially supported by Hasso Plattner Institute for Digital Engineering through the HPI Research School at UCT. The authors also thank the survey participants for their participation.

References

1. Abdelghany, A.S., Darwish, N.R., Hefni, H.A.: An agile methodology for ontology development. Int. J. Intell. Eng. Syst. **12**(2), 170–181 (2019)
2. Agarwal, N., Sondhi, A., Chopra, K., Singh, G.: Transfer learning: survey and classification. Smart Innov. Commun. Comput. Sci. Proc. ICSICCS **2020**, 145–155 (2021)
3. Antia, M.J., Keet, C.M.: Assessing and enhancing bottom-up CNL design for competency questions for ontologies. In: Proceedings of the Seventh International Workshop on Controlled Natural Language (CNL 2020/21). ACL, Amsterdam, Netherlands (2021). https://aclanthology.org/2021.cnl-1.11
4. Azzi, S., Iglewski, M., Nabelsi, V.: Competency questions for biomedical ontology reuse. Proc. Comput. Sci. **160**, 362–368 (2019)
5. Bezerra, C., Freitas, F.: Verifying description logic ontologies based on competency questions and unit testing. In: ONTOBRAS, pp. 159–164 (2017)
6. Bezerra, C., Santana, F., Freitas, F.: CQchecker: a tool to check ontologies in owl-dl using competency questions written in controlled natural language. Learn. Nonlin. Models **12**(2), 4 (2014)
7. Cyr, L., Francis, K.: Measures of clinical agreement for nominal and categorical data: the kappa coefficient. Comput. Biol. Med. **22**(4), 239–246 (1992)
8. Dutta, B., DeBellis, M.: CODO: an ontology for collection and analysis of COVID-19 data. arXiv preprint arXiv:2009.01210 (2020)
9. Fleiss, J.L.: Measuring nominal scale agreement among many raters. Psychol. Bull. **76**(5), 378 (1971)
10. Haller, S.: Automatic Short Answer Grading using Text-to-Text Transfer Transformer Model. Master's thesis, University of Twente, the Netherlands (2020)
11. Honnibal, M., Montani, I.: spaCY 2: natural language understanding with bloom embeddings, convolutional neural networks and incremental parsing. To Appear. **7**(1), 411–420 (2017)
12. Keet, C.M., Mahlaza, Z., Antia, M.-J.: CLaRO: a controlled language for authoring competency questions. In: Garoufallou, E., Fallucchi, F., William De Luca, E. (eds.) MTSR 2019. CCIS, vol. 1057, pp. 3–15. Springer, Cham (2019). https://doi.org/10.1007/978-3-030-36599-8_1
13. Keet, C.M., Ławrynowicz, A.: Test-driven development of ontologies. In: Sack, H., Blomqvist, E., d'Aquin, M., Ghidini, C., Ponzetto, S.P., Lange, C. (eds.) ESWC 2016. LNCS, vol. 9678, pp. 642–657. Springer, Cham (2016). https://doi.org/10.1007/978-3-319-34129-3_39
14. Landis, J.R., Koch, G.G.: An application of hierarchical kappa-type statistics in the assessment of majority agreement among multiple observers. Biometrics, pp. 363–374 (1977)

15. Li, C., Su, Y., Liu, W.: Text-to-text generative adversarial networks. In: 2018 International Joint Conference on Neural Networks (IJCNN), pp. 1–7. IEEE (2018)
16. Li, Q., Li, S., Zhang, S., Hu, J., Hu, J.: A review of text corpus-based tourism big data mining. Appl. Sci. **9**(16), 3300 (2019)
17. Pan, S.J., Yang, Q.: A survey on transfer learning. IEEE Trans. Knowl. Data Eng. **22**(10), 1345–1359 (2009)
18. Peroni, S.: A simplified agile methodology for ontology development. In: Dragoni, M., Poveda-Villalón, M., Jimenez-Ruiz, E. (eds.) OWLED/ORE -2016. LNCS, vol. 10161, pp. 55–69. Springer, Cham (2017). https://doi.org/10.1007/978-3-319-54627-8_5
19. Potoniec, J., Wiśniewski, D., Ławrynowicz, A., Keet, C.M.: Dataset of ontology competency questions to SPARQL-OWL queries translations. Data Brief **29**, 105098 (2020)
20. Raffel, C., et al.: Exploring the limits of transfer learning with a unified text-to-text transformer. J. Mach. Learn. Res. **21**(1), 5485–5551 (2020)
21. Reimers, N., Gurevych, I.: Sentence-BERT: Sentence embeddings using Siamese BERT-networks. arXiv preprint arXiv:1908.10084 (2019)
22. Ren, Y., Parvizi, A., Mellish, C., Pan, J.Z., van Deemter, K., Stevens, R.: Towards competency question-driven ontology authoring. In: Presutti, V., d'Amato, C., Gandon, F., d'Aquin, M., Staab, S., Tordai, A. (eds.) ESWC 2014. LNCS, vol. 8465, pp. 752–767. Springer, Cham (2014). https://doi.org/10.1007/978-3-319-07443-6_50
23. Suárez-Figueroa, M.C., Gómez-Pérez, A., Fernández-López, M.: The NeOn methodology for ontology engineering. In: Suárez-Figueroa, M.C., Gómez-Pérez, A., Motta, E., Gangemi, A. (eds.) Ontology Engineering in a Networked World, pp. 9–34. Springer, Heidelberg (2012). https://doi.org/10.1007/978-3-642-24794-1_2
24. Tseng, Y.H., Ho, Z.P., Yang, K.S., Chen, C.C.: Mining term networks from text collections for crime investigation. Expert Syst. Appl. **39**(11), 10082–10090 (2012)
25. Uschold, M., Gruninger, M.: Ontologies: principles, methods and applications. Knowl. Eng. Rev. **11**(2), 93–136 (1996)
26. Wiśniewski, D., Potoniec, J., Ławrynowicz, A., Keet, C.M.: Analysis of ontology competency questions and their formalizations in SPARQL-owl. J. Web Semant. **59**, 100534 (2019)
27. Zhuang, F., et al.: A comprehensive survey on transfer learning. Proc. IEEE **109**(1), 43–76 (2020)
28. Zolotareva, E., Tashu, T.M., Horváth, T.: Abstractive text summarization using transfer learning. In: ITAT, pp. 75–80 (2020)

Ontology-Based Models of Chatbots for Populating Knowledge Graphs

Petko Rutesic[1](✉), Dennis Pfisterer[2], Stefan Fischer[2], and Heiko Paulheim[3]

[1] Baden-Wuerttemberg Cooperative State University, 68163 Mannheim, Germany
petko.rutesic@dhbw-mannheim.de

[2] Institute of Telematics, University of Luebeck, Luebeck, Germany

[3] University of Mannheim, Mannheim, Germany

Abstract. Knowledge graphs and graph databases are nowadays extensively used in various domains. However, manually creating knowledge graphs using existing ontology concepts presents significant challenges. On the other hand, chatbots are one of the most prominent technologies in the recent past. In this paper, we explore the idea of utilizing chatbots to facilitate the manual population of knowledge graphs. To implement these chatbots, we generate them based on other special knowledge graphs that serve as models of chatbots. These chatbot models are created using our modelling ontology (specially designed for this purpose) and ontologies from a specific domain. The proposed approach enables the manual population of knowledge graphs in a more convenient manner through the use of automatically generated conversational agents based on our chatbot models.

Keywords: modelling · ontology · chatbots · knowledge graphs

1 Introduction

Creating user-friendly interfaces to facilitate populating knowledge graphs manually is a very demanding task. In order to create a knowledge graph, it is necessary to have expertise not only in the domain of interest but also in the field of ontology engineering. We can illustrate this using an example of flight registration, where the end-users enter flight information manually in the knowledge graph and the ultimate output of the process would be a comprehensive knowledge graph representing all existing flights. To define a specific flight, it is necessary to first choose the appropriate ontology for flight description and then create an individual of the class representing flights. Following that, the user has to know how to define departure and arrival airports, which requires the knowledge of object and data properties that can be used to describe airports. What makes this task even more complex is the need to choose whether these airports can be described as blank nodes or not, or to choose specific ontology design patterns. Creating these ontologies (editing RDF graphs) using only text

F. Ortiz-Rodriguez et al. (Eds.): KGSWC 2023, LNCS 14382, pp. 228–242, 2023.
https://doi.org/10.1007/978-3-031-47745-4_17

editors in any syntax for representing RDF graphs would be intimidating for the majority of users.

Tools for ontology engineering like Protégé, WebProtégé, TopBraid Composer and similar tools are frequently used for this purpose. Additionally, there are various approaches to modelling user interfaces for populating knowledge graphs. These user interfaces are of different kinds, from desktop and web applications to conversational agents (chatbots). Particularly, chatbots have seen great growth in popularity, especially in the last couple of years with the appearance of large language models and tools like ChatGPT that use deep learning models to generate correct humanlike responses. Chatbots are now integrated in various domains and have numerous applications, such as e-customer care services, e-commerce systems, the medical field, etc.

Our approach aims to empower end users to create knowledge graphs like the aforementioned flight knowledge graph in a simple manner. The chatbot should primarily ask simple questions like: "What is the flight number?" or "What is the flight destination?". To this end, we propose to divide population of knowledge graphs in two processes. The first process would be modelling and designing of conversational agents, and the second process is using those conversational agents by many end users to populate desired knowledge graphs. One potential group of end users who could benefit from our chatbots includes airline personnel (or flight operators) responsible for registering new flights and services related to those flights. The novelty of our idea lies in the automatic generation of chatbots from models that themselves are knowledge graphs. By adopting this approach, the design of our conversational agent models becomes the responsibility of ontology experts (chatbot model designers), while the end users of the chatbots do not necessarily need expert knowledge in ontologies. This way, the end user would be relieved from the burden of precisely knowing domain ontologies and the intricacies of ontology engineering.

The approach can be simply expressed through two functions. The first function is named the modelling function, which encompasses the process of modelling chatbot dialogues (conversations) while simultaneously defining the structure of the output knowledge graphs. The modelling function takes a set of domain ontologies and our *OBOP* ontology (Ontology for Ontology-based Ontology Population) as input parameters. The *OBOP* ontology is specifically designed for modelling purposes and can be accessed at http://purl.org/net/obop or in the GitHub repository[1].

$$f_{modelling}(DomainOntologies, OBOP) = ChatbotModel$$

The modelling process is the task for chatbot model designers (ontology experts), and the output of this process is a *ChatbotModel*, which is a knowledge graph defined using elements from *DomainOntologies* and our *OBOP* ontology. Currently, the modelling is done manually, but we have plans to design special GUI tools to support and automate this process in the future.

[1] https://github.com/ontosoft/logic-interface/blob/main/ontology/obop.owl.

The second function represents the process of data acquisition, i.e., knowledge graph population. This process is actually the use of chatbot to enter data.

$$f_{acquisition}(ChatbotModel, UserInteraction) = OutputKnowledgeGraph$$

The acquisition function takes a chatbot model created in the modeling process and user interaction (during which the data is entered) as input parameters. The output of the acquisition function is a desired knowledge graph, referred to as *OutputKnowledgeGraph*, which is defined only using elements from the domain ontologies. In Fig. 1, part of our approach that corresponds to the acquisition function is outlined. The *Chatbot generator* takes the *chatbot model* defined by elements of the OBOP ontology (colored in red) and domain ontologies (colored in black). The *output knowledge graph* is populated through the interaction of the end-user with the chatbot and contains only elements from domain ontologies.

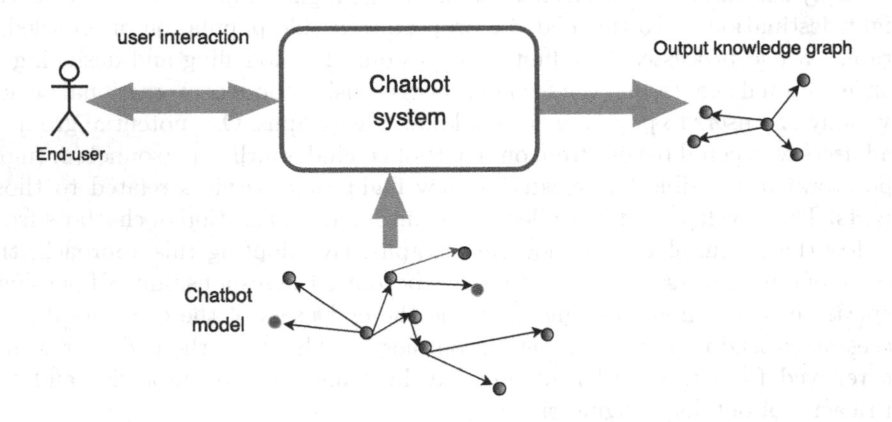

Fig. 1. A simplified representation of the acquisition function (Color figure online)

Since the main goal is to simplify the population of knowledge graphs using conversational systems (chatbots), we have explored the idea of modelling these chatbots also within knowledge graphs. To leverage the reasoning capabilities of OWL ontologies, it is reasonable for the chatbot models to be represented in the OWL DL ontology class. Therefore, the goal of our approach can be boiled down to the following set of requirements:

1. The chatbot conversation and the structure of the output knowledge graph are both specified within the same knowledge graph (e.g. RDF file), representing a model.
2. Models are defined using a dedicated ontology designed for this purpose, serving as a meta-model for generating our models.
3. The meta-model comprises elements capable of modelling main program control structures, i.e., sequential, selection (branching) and iteration control structures.

In the rest of the paper, we elaborate on the implementation of the proposed requirements. The rest of the paper is organized as follows: The next section presents a use case with flight registrations, used to illustrate our approach in subsequent sections. Then we describe the main elements of the OBOP ontology that model various aspects of chatbot conversations and showcase how the modeling function works in our use case. Each subsection focuses on specific workflows of the chatbot and explains how these workflows are modeled using entities from the OBOP ontology. Finally, the paper concludes with a section discussing our contributions.

2 Use Case

To demonstrate the applicability of our chatbot models, we introduce an example involving a chatbot designed to help create knowledge graphs for flights. Essentially, we look at process from the reverse perspective, beginning with an existing knowledge graph and assuming it to be the output knowledge graph of a chatbot. Then we demonstrate a model of this chatbot that generates this particular knowledge graph.

The knowledge graph that our chatbot has to construct is already presented in the examples of using the Ticket ontology[2]. The authors of the knowledge graph used GoodRelations ontology [6], Ticket ontology (compliant with GoodRelations) and DBpedia to describe the flight. A simplified graphical representation of the flight knowledge graph is shown in Fig. 2.

The flight knowledge graph contains data about a specific flight with the flight number LH1234, which is operated by Lufthansa airline. The flight information includes details about the departure and destination airports, as well as the types of tickets that can be booked for that flight. A flight ticket is defined as an instance of the class tio:Ticket of the Ticket ontology. In the graphical representation, individuals and blank nodes are depicted using circles, with blank nodes represented by empty circles. Ontology classes are illustrated by ovals, while literals are shown as rectangles.

During the interaction with the chatbot, the user must provide the flight number, departure and destination airports, as well as the departure and arrival times. The knowledge graphs (instances of the Ticket ontology) generated using our chatbot offer significant usefulness as they can be easily searched by customers looking to book flight tickets. The use of ontologies in the system allows for defining search operations using SPARQL queries. Moreover, customers have the option to employ conversational agents like KBot [2] to find suitable flight tickets using natural language understanding over linked data. In that case, semantic web technologies can be used directly to find suitable flights without resorting to web scraping methods as described in [12].

The primary objective of our chatbot use case is to simplify the process for end-users by posing straightforward questions, while the knowledge graph is

[2] http://purl.org/tio/ns.

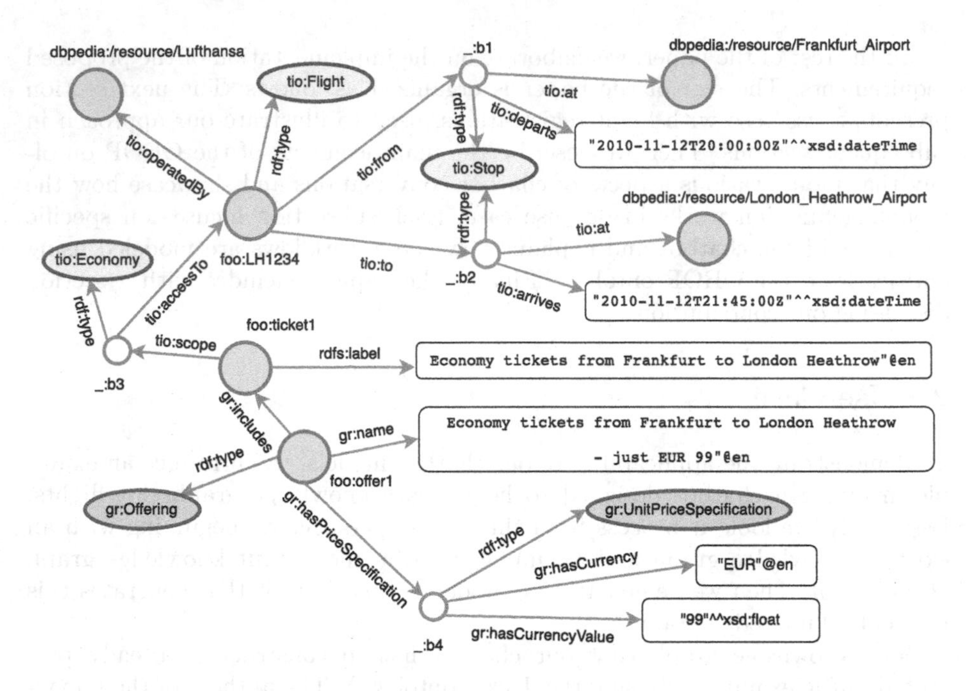

Fig. 2. A knowledge graph representing a simple flight description.

generated seamlessly in the background using answers to those questions. The chatbot must have the capability to generate this knowledge graph and also to accommodate its extensions, such as adding the definition of additional ticket types if needed. By doing so, the chatbot determines the structure of the output ontology without explicitly asking the user to specify blank nodes, instances of specific ontology classes and other details, thereby alleviating the burden of detailed ontology engineering. To achieve this level of sophistication, the chatbot model incorporates all the necessary details, and the key aspects of the model are outlined in the section dedicated to modelling architecture.

3 Related Work

One of the first known conversational agents is ALICE [13] which uses AIML (Artificial Intelligence Markup Language) which is basically XML to design conversations. The system uses a botmaster which monitors conversations to make them more appropriate. Unlike writing rules in an XML-based language, the rules in our approach are encoded in a knowledge graph based on our modeling ontology and can be stored in an RDF file. Historically, there have been various approaches to using semantic nets to generate chatbot. One of the first examples is OntBot [3] that transforms ontologies and knowledge into relational database and then uses relational database to generate chats. Another example of usage of model-driven conversational systems is presented in [9], wherein a specialized domain-specific language with components like intents, entities, actions and flows is employed to design the dialogue structure of task-oriented chatbots. The dialogue management used in our paper is also similar to the dialogue management used by the Rasa chatbot [5]. However, our approach differentiates itself from those approaches by aiming to represent all components, including intents, flows and actions through the use of knowledge graphs.

There have been many approaches to model user interfaces and web application logic using knowledge graphs. An approach described in [11] proposes a method of modelling HTML application structure and its logic using RDF graphs. In this paper, we deal with a similar question of modelling and generating chatbots used to engage in dialogs with users to acquire data. Furthermore, the acquired data is used to directly populate desired output knowledge graphs based on domain ontologies. The expressiveness of formal ontologies proves valuable in representing both the models of chatbots and the complex knowledge within the business logic of target ontologies. Chatbots generated using this approach could prove advantageous wherever the description of complex products and services using ontologies is required, as is the case in emerging business models like *Distributed Market Spaces* [10]. This approach complements the method described in [7], where the system utilizes SPARQL queries to identify appropriate complex products and services.

As stated in [1], conversational clients or chatbots are designed to be used either as task-oriented or open-ended dialog generators. In our approach, we develop a task-oriented chatbot responsible for collecting data (knowledge graph) based on particular ontologies. The actual task is described based on rules specified in the model, which is represented again as a knowledge graph defined using our OBOP ontology. Thus, our chatbot can also be classified as a rule-based chatbot. To enable human-like conversation, our task-based chatbot module is incorporated into an open-ended chatbot which is further described in the implementation section. The intention of our system is not to design a chatbot capable of convincing a human that (s)he is chatting with a human instead of a computer program. That intelligent behaviour depends on having good knowledge sets. Instead, we focus on the creation of chatbots that gather information in the form of knowledge graphs with individuals (instances) of the respective domain ontologies.

4 Model Architecture

Our meta-model (OBOP ontology) contains various structures that describe chatbot functionalities. The fundamental structure is designed to specify a simple data request. In response to this request, the user inputs a value, which is then validated and can be stored in the system as a data property, IRI, or for a similar purpose. This functionality corresponds to the process of entering data into form fields in GUI interfaces. To represent the insertion of these data values, the OBOP ontology, uses an instance of the class obop:SingleValueRequest. In cases where multiple data properties need to be entered as a group of questions, this is modeled using a conversation block. The generation of chatbot models, using constructions explained in the next sections, is the responsibility of chatbot model designers.

4.1 Conversation Block

A conversation block represents a segment of a conversation used to collect data values that are related to a specific entity. This part of the dialog is analogous to an HTML form in web applications. The chatbot poses a series of questions, and by providing answers to those questions, the system stores corresponding values. A simple model for entering the flight IRI is illustrated in Fig. 3. In the following figures, blue circles represent instances of classes of the OBOP ontology, while larger peach-colored circles (e.g., individual *flight_model_1* in Fig. 3) depict instances from domain (target) ontologies. Yellow ovals represent classes from the OBOP ontology, and blue ovals represent classes from target ontologies. Object and data properties are denoted by directed lines with labels that specify the names of those properties.

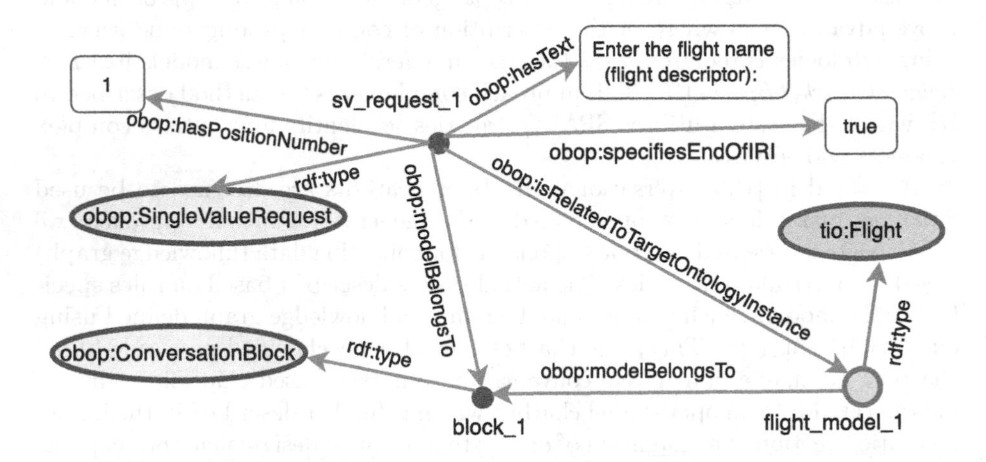

Fig. 3. Modelling a block of conversation

A segment of the model represented in Fig. 3 should initiate the following part of conversation:

```
chatbot: Please enter the flight name (flight descriptor):
user:    LH1234
```

The outcome of the previous conversation is adding of the following triple to the output knowledge graph:

```
foo:LH1234 a tio:Flight.
```

In Fig. 3 can be seen a conversation block named *block_1*, which has one instance of the class obop:SingleValueRequest called *sv_request_1*. This represents a question for entering a flight name (a unique flight descriptor). *sv_request_1* has the object property obop:specifiesEndOfIRI, which has a boolean value of true. This indicates that the provided answer to the question serves as the ending part of the IRI representing the flight instance. The model of the instance of tio:Flight class that will be generated in this process is represented by the *flight_model_1* instance. During the conversation with the chatbot from this instance will be generated the instance with the name foo:LH1234. Should the inserted string need to be the value of a data property instead of being part of the IRI, this would be indicated by the *obop:containsDatatype* object property of the *sv_request* instance which would specify the name of the wanted data property. The object property *obop:isRelatedToTargetOntologyInstance* indicates what instance will be transformed with the entered data. The data property *obop:hasPositionNumber* specifies that this question is the first one in this conversational block.

4.2 Branching Control Structure

To enable chatbot users to make decisions and select different paths of execution, the OBOP ontology has mechanisms for modeling branching control structures. The class *obop:Branching* represents a conditional structure. A simple branching structure in our example is explained in the case of selecting an airline that operates the given flight. The chatbot user is therefore prompted to choose only one airline company from the presented options. One possible conversation is presented in the following listing:

```
chatbot: Enter the airline that operates your flight.
         Choose one of the following options:
         1. Lufthansa
         2. Ryanair
user:    Lufthansa
```

The chatbot asks a question and specifies a list of possible options. The user responds by writing one among those listed names or by writing the ordinary number in front of the corresponding name. For the sake of brevity, we decided to present only two possible options (two airlines) to choose from. This statement is similar to the "switch" statement in programming languages. As the result of the execution of this part of the chatbot the output graph can be extended with the following triple:

```
foo:LH1234
  tio:operatedBy <http://dbpedia.org/resource/Lufthansa>.
```

The segment of the model that generates the previous chatbot question and adds the specified triple, as the consequence of the chosen option, is represented in Fig. 4. It can be observed that the previous conversational block *block_1* is used, as it is regarded a part of the same conversation section. The question to choose one out of several possible values is denoted by an instance of the obop:Question class, called *question_1*. *question_1* has the data property *obop:hasText* with the text of the question. Additionally, the question has the value of 2 for the *obop:hasPositionNumber* data property, which serves to denote that this question has second place in the conversation block. The question *question_1* has an instance of the class *obop:hasBranching* called *branching_1* and this branching is specified by the instance of the *obop:Connection* class named *conn_1*.

The obop:Connection class specifies how the two instances of the ontology classes should be related. An instance of this class has functional properties that specify the source and destination of object properties. In our case, the source is an instance called *flight_model_1* and the destination is *airline_model_1*. The latter serves as a model for an instance of the *dbpedia:Agent* class, which will be created during the interaction with the chatbot. Additionally, the object property that has to be inserted in the output graph between these two instances is specified by the *obop:containsDatatype* object property and it is, in our context, *tio:operatedBy* property. This particular object property is denoted by a dashed line in Fig. 4, although this edge is not part of the actual model graph.

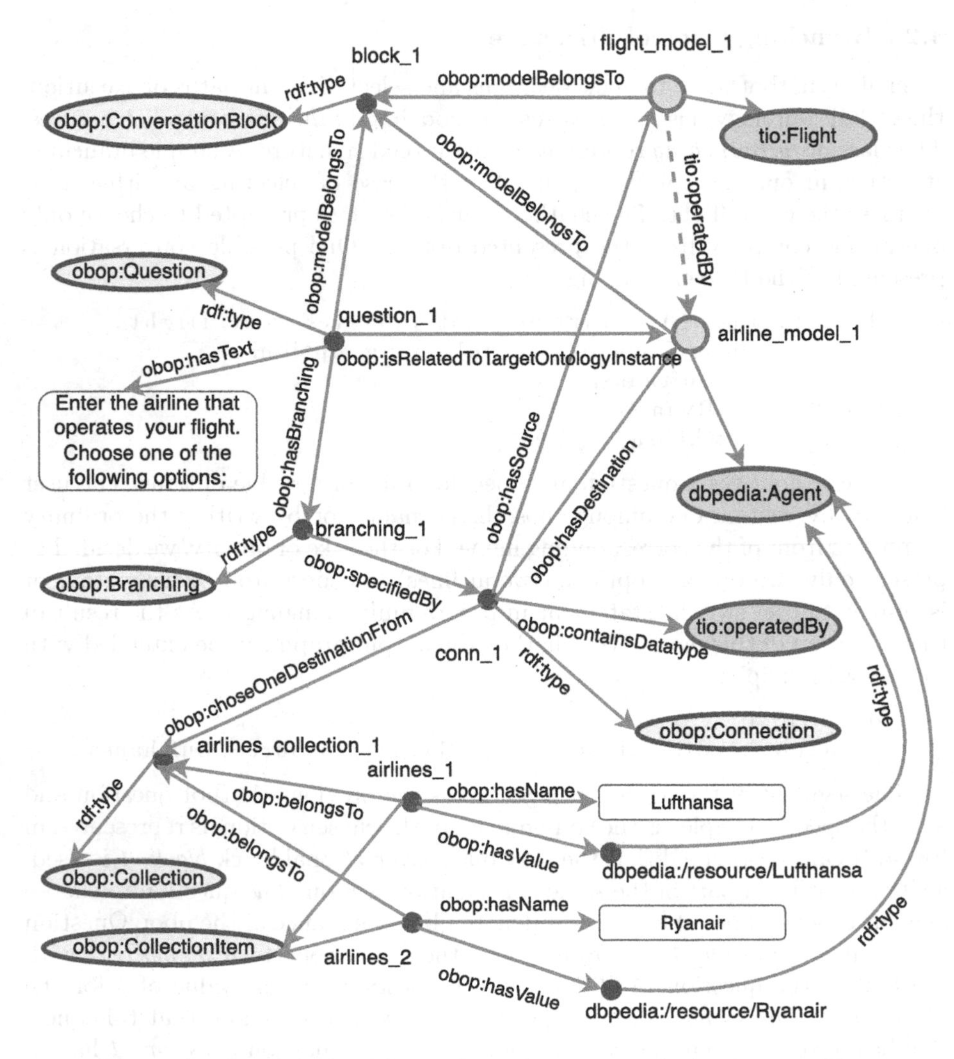

Fig. 4. A part of the model specifying the process of selecting an airline

It can be seen that the connection instance specifies both a predicate (object property) and an object (instance) of a new triple. The user, interacting with the chatbot, is required to choose a single airline from the list of possible airlines. In the lower section of Fig. 4 is defined an instance of the *obop:Collection* class, named *airlines_collection_1.* The connection instance conn_1 is related to this collection by the object property *obop:choseOneDestinationFrom*, which specifies exactly how to form the object of a new triple. In this case, our simple collection contains two instances of the class *obop:CollectionItem* corresponding to Lufthansa and Ryanair airlines. OBOP collections are not defined using rdf:Bag, rdf:Seq or by using restrictions like *owl:someValuesFrom.* Collections are basically used by the chatbot program in the runtime phase and they are not meant to be used in the process of ontology reasoning.

An alternative, that can be specified in our model, which brings more generality to this example, is to allow the user to directly input an IRI of the desired airline, which must belong to the *dbpedia:Agent* class. However, we presented the selection from the list of available options to show that the model designer has this option.

4.3 Iteration Control Structure

To effectively manage control flows, the chatbot model has to provide definition of iteration structures (loops). As the iteration structures in programming languages allow the repetitive execution of a specific code block, iteration structures in chatbots enable the repetition of a set of questions within a conversation block or the repetition of other activities. For this purpose, the OBOP ontology introduces the *obop:Loop* class.

An example of iterations can be demonstrated within our flight scenario. The user has to define various ticket types for the flight, which could initiate the following conversation:

```
chatbot: Do you want to define one more ticket type?
user:    Yes
chatbot: Please enter the label of the ticket?
user:    Economy tickets from Frankfurt to London Heathrow
chatbot: Choose the service level of the ticket?
user:    Economy
chatbot: Do you want to define one more ticket type?
```

The result of executing the previous chatbot conversation is the following part of the graph:

```
foo:ticket5 a tio:TicketPlaceholder ;
  rdfs:label "Economy tickets from Frankfurt to London
              Heathrow"@en ;
  tio:scope [ a tio:ScopeOfAccess ;
     tio:accessTo foo:LH1234 ;
     tio:eligibleServiceLevel tio:Economy ] .
```

The part of the model that generates the preceding section of the chatbot conversation is illustrated in Fig. 5. The question *question_2*, which ask for a new ticket type is related to an instance of the *obop:Loop* class called *loop_1*. Each iteration of the loop specifies adding a new ticket which is represented by a conversational block *block_3*. If the user answers positively to the chatbot's question, a new ticket instance is generated according to the *ticket_model_1*. Together with the ticket is generated a blank node corresponding to the *_:bnode* instance. The new blank node is related to the flight instance, which is already created in the previous examples according to the *flight_model_1*.

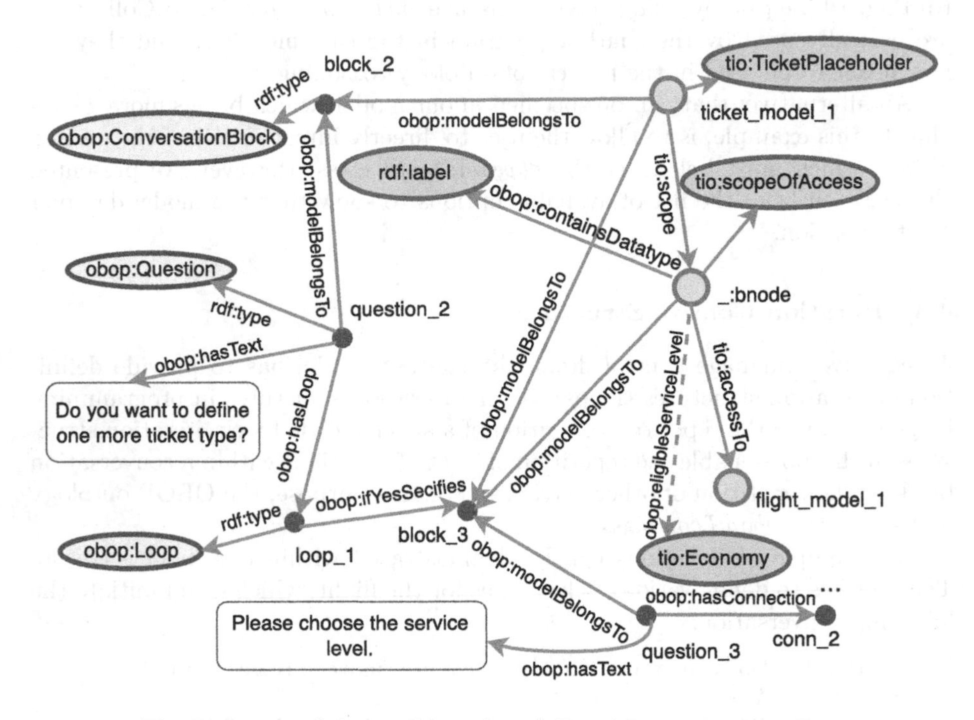

Fig. 5. A part of the model used to define tickets using iteration.

The object properties *tio:scope* and *tio:accessTo* are also automatically created. However, *rdfs:label* had to be specified using *obop:containsDatatype* object property. In order to choose between possible service classes (e.g., Business or Economy) a new question (*question_3*) and connection (*conn_2*) was necessary. *conn_2* is not completely presented and ontology classes of same instances are omitted in Fig. 5 for improved readability.

5 Implementation

To implement a prototype of our chatbot, we employed Python libraries, namely, *ChatterBot* as the conversational dialogue engine and *Owlready* [8] for ontology-oriented programming, to facilitate the creation and manipulation of OWL ontologies. The chatbot prototype can be tested as a REST application implemented using Flask framework[3] with a simple JavaScript frontend. The source code can be found on GitHub[4].

Chatbots generated using the ChatterBot library answer user questions based on the functionalities of logic adapters. For example, there are adapters like the *Time adapter* that can answer questions like "What time is it?", the *Math adapter* that can handle arithmetic operations, etc. When the user submits a question, all logic adapters assess whether they are capable of providing a response. The ability of a logical adapter to answer the question is measured by a confidence factor, expressed as a decimal value between 0 and 1. The answer from the adapter with the highest confidence is chosen as the reply. The architecture of our chatbot is shown in Fig. 6.

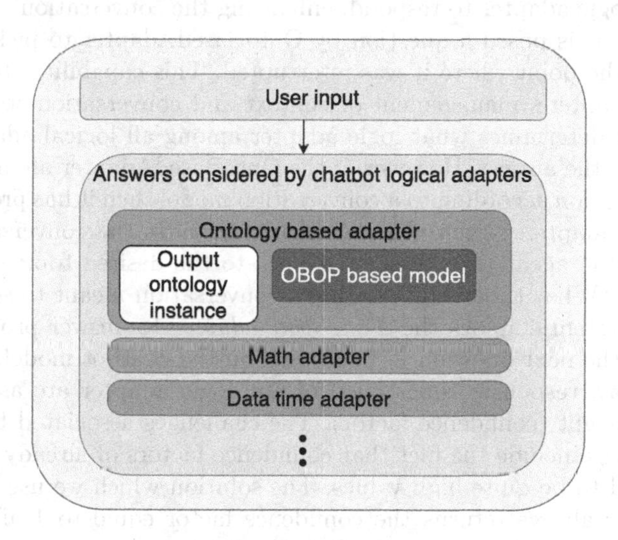

Fig. 6. Architecture of the chatbot system.

Our model-based chatbot functionalities are implemented as a logic adapter for the ChatterBot library, named the *OntoBasedAdapter*. Unlike other logic adapters in ChatterBot, the OntoBasedAdapter keeps track of the current state of conversation. Once the OntoBasedAdapter starts to gather information and populate the corresponding output knowledge graph, it needs to store information of the conversation's current state. It includes tracking visited nodes to enable system to make reverse steps and undo entries if necessary. Chatbot stores information on the current context, which is comprised of the entire array of nodes that have been visited and created during the conversation up to the current state.

[3] https://flask.palletsprojects.com.
[4] https://github.com/ontosoft/ontochatbot.

Keeping track of the state of an ongoing dialogue is quite challenging task. In the study [4], authors dealt with this problem by exploiting ontologies for both the knowledge base and the dialogue manager in domain-driven conversation, creating a banking chatbot. Our chatbot, on the other hand, is a domain independent conversational agent, with the domain specified in the corresponding chatbot model.

Answering Questions and Generating Replies. When the OntoBasedAdapter takes charge of the dialogue, it maintains the conversation status. The system asks questions in the order that is defined by the chatbot model. If the user asks a question that is not related to the current task specified in the model the conversation might take a different direction, potentially allowing another logic adapter to respond, enhancing the conversation's natural flow. Subsequently, it is posed a question by OntoBasedAdapter to pick up the conversation at the point where it was interrupted. This capability stems from the OntoBasedAdapter's management of context and conversation state. The confidence factor determines what logic adapter among all logical adapters will be chosen to give the answer. However, if the OntoBasedAdapter already started to gather information according to a conversation model then it has precedence over other logical adapters. Even if a user abruptly shifts the conversation's focus, the conversation agent tries to steer it back to the desired topic. For instance, if a user asks "What time is it?" during a conversation meant to schedule flight details, the system employs the *Time logic adapter* to answer promptly, before returning to the next question in line based on the chatbot model.

In this way, responses (questions) of any logic adapter are assigned a corresponding weight (confidence factor). The challenges associated to this way of choosing replies include the fact that confidence factors of already implemented adapters tend to be quite high values. The solution which we use is that OntoBasedAdapter always returns the confidence factor equal to 1 after it started to collect information. At present, OntoBasedAdapter responses are manually embedded into the model by the model designer.

6 Contribution

In this paper, we introduced models for conversational agents (chatbots). These chatbots can accomplish domain-specific tasks such as generating flight descriptions, booking flight tickets or creating restaurant menus that could be used for restaurant reservations in a user-friendly manner. In order to show how to create a chatbot model, we designed a chatbot intended for flight descriptions. The chatbots we propose follow the rule-based approach, in which replies are generated using predefined rules integrated into the models of respective conversations. These conversation models represent templates according to which replies (questions) are created.

The requirements for the approach outlined in the introduction of the paper are met with our implementation. The conversation model and the structure of the knowledge graph are both included in the same knowledge graph according to the first requirement. The limitations of this approach are the high complexity of the model graphs and challenging maintenance of the generated graphs. The second and third requirements are addressed by the explicit definition of specific classes, object properties, and data properties in the OBOP ontology. These definitions correspond to fundamental programming control structures, ensuring the capability to capture and manage various aspects of the conversation's flow and logic.

Describing chatbot models based on OWL ontologies might offer significant benefits, including the potential for OWL reasoning applied to these models. The OWL reasoning capability could identify flawed models that might result in inconsistent knowledge graphs. Another advantage lies in the potential for reusing existing models, enabling their sharing and querying through SPARQL. Moreover, the application of machine learning algorithms to existing models could streamline the process of automatically or semi-automatically generating new chatbots for description of similar problems.

References

1. Agarwal, R., Wadhwa, M.: Review of state-of-the-art design techniques for chatbots. SN Comput. Sci. **1**(5), 246 (2020)
2. Ait-Mlouk, A., Jiang, L.: KBot: a knowledge graph based chatbot for natural language understanding over linked data. IEEE Access **8**, 149220–149230 (2020)
3. Al-Zubaide, H., Issa, A.A.: OntBot: ontology based chatbot. In: International Symposium on Innovations in Information and Communications Technology, pp. 7–12. IEEE (2011)
4. Altinok, D.: An ontology-based dialogue management system for banking and finance dialogue systems. arXiv preprint arXiv:1804.04838 (2018)
5. Bocklisch, T., Faulkner, J., Pawlowski, N., Nichol, A.: Rasa: open source language understanding and dialogue management. arXiv preprint arXiv:1712.05181 (2017)
6. Hepp, M.: GoodRelations: an ontology for describing products and services offers on the web. In: Gangemi, A., Euzenat, J. (eds.) EKAW 2008. LNCS (LNAI), vol. 5268, pp. 329–346. Springer, Heidelberg (2008). https://doi.org/10.1007/978-3-540-87696-0_29
7. Hitz, M., Radonjic-Simic, M., Reichwald, J., Pfisterer, D.: Generic UIs for requesting complex products within distributed market spaces in the internet of everything. In: Buccafurri, F., Holzinger, A., Kieseberg, P., Tjoa, A.M., Weippl, E. (eds.) CD-ARES 2016. LNCS, vol. 9817, pp. 29–44. Springer, Cham (2016). https://doi.org/10.1007/978-3-319-45507-5_3
8. Lamy, J.B.: Owlready: ontology-oriented programming in python with automatic classification and high level constructs for biomedical ontologies. Artif. Intell. Med. **80**, 11–28 (2017)
9. Pérez-Soler, S., Guerra, E., de Lara, J.: Model-driven chatbot development. In: Dobbie, G., Frank, U., Kappel, G., Liddle, S.W., Mayr, H.C. (eds.) ER 2020. LNCS, vol. 12400, pp. 207–222. Springer, Cham (2020). https://doi.org/10.1007/978-3-030-62522-1_15
10. Radonjic-Simic, M., Pfisterer, D., Rutesic, P.: Arising internet of everything: Business modeling and architecture for smart cities in recent developments in engineering research, vol. 8, chap. 7 (2020)

11. Rutesic, P., Radonjic-Simic, M., Pfisterer, D.: An enhanced meta-model to generate web forms for ontology population. In: Villazón-Terrazas, B., Ortiz-Rodríguez, F., Tiwari, S., Goyal, A., Jabbar, M.A. (eds.) KGSWC 2021. CCIS, vol. 1459, pp. 109–124. Springer, Cham (2021). https://doi.org/10.1007/978-3-030-91305-2_9
12. Turnip, T.N., Silalahi, E.K., Sinulingga, Y.A.V., Siregar, V.: Application of ontology in semantic web searching of flight ticket as a study case. J. Phys. Conf. Ser. **1175**, 012092 (2019)
13. Wallace, R.S.: The anatomy of A.L.I.C.E. In: Epstein, R., Roberts, G., Beber, G. (eds.) Parsing the Turing Test, pp. 181–210. Springer, Dordrecht (2009). https://doi.org/10.1007/978-1-4020-6710-5_13

An Ontology for Reasoning About Fairness in Regression and Machine Learning

Jade S. Franklin(✉), Hannah Powers, John S. Erickson, Jamie McCusker, Deborah L. McGuinness, and Kristin P. Bennett

Rensselaer Polytechnic Institute, Troy, NY 12180, USA
frankj6@rpi.edu

Abstract. As concerns have grown about bias in ML models, the field of ML fairness has expanded considerably beyond classification. Researchers now propose fairness metrics for regression, but unlike classification there is no literature review of regression fairness metrics and no comprehensive resource to define, categorize, and compare them. To address this, we have surveyed the field, categorized metrics according to which notion of fairness they measure, and integrated them into an OWL2 ontology for fair regression extending our previously-developed ontology for reasoning about concepts in fair classification. We demonstrate its usage through an interactive web application that dynamically builds SPARQL queries to display fairness metrics meeting users' selected requirements. Through this research, we provide a resource intended to support fairness researchers and model developers alike, and demonstrate methods of making an ontology accessible to users who may be unfamiliar with background knowledge of its domain and/or ontologies themselves.

Keywords: ontology · fairness · metrics · regression · reasoning · SPARQL · OWL · machine learning

1 Introduction

As machine learning (ML) models have advanced, their application in a variety of domains has become widespread. In medicine, ML models are used to predict patient cancer risk and prognosis from complex datasets [32]; in the business world, algorithmic screening of job candidates has become commonplace [34]; and in the criminal justice system, some jurisdictions have adopted the use of ML models to predict likelihood of recidivism when determining which inmates to release on parole [6]. In each of these domains and more, ML-based predictions have tangible effects on people's lives. Growing concern about the potential for biases in ML models has driven researchers to study to what extent ML models are biased, the ways they are biased, and what can be done to make ML models more fair. As an example, a 2016 news article argued that COMPAS,

F. Ortiz-Rodriguez et al. (Eds.): KGSWC 2023, LNCS 14382, pp. 243–261, 2023.
https://doi.org/10.1007/978-3-031-47745-4_18

an ML-based predictor of recidivism risk and other "criminogenic needs" in use by Florida courts, was biased against African-American defendants [2]. Additional research on COMPAS has led to conflicting results, with some arguing that COMPAS's results do not show bias [26], while others argue that it shows bias under some definitions of fairness, but not all [19]. It is in this context that research in ML fairness has expanded, and several frameworks have been proposed to analyze metrics for measuring fairness and classify them according to which conceptual notions of fairness they measure [29,35,45].

Most frameworks for ML fairness focus on classification [20], yet regression models for predicting continuous values are used extensively, such as in the prediction of disease risk probabilities in medicine [43,47] and estimation of future incomes in finance [9,37]. While classification and regression may in some cases solve similar problems (e.g., risk may be predicted as a value or as one of "low," "medium," or "high" classes), regression provides the benefit of increased interpretability: for example, a doctor may be more trusting of a linear model that clearly shows trends and slight differences in response to various inputs than a model that only outputs an opaque label [33]. Thus, regression-specific approaches to ML fairness are increasingly identified as an important area of emerging research [13]. Although regression models can be analyzed using standard classification metrics by grouping the results of the model into classes, this leads to a loss in precision. The better approach is to build fairness metrics and bias mitigation methods which are specifically targeted towards regression. A variety of such metrics have been proposed (see Sect. 3.1).

In contrast to classification, there are not yet frameworks for organizing fairness metrics for regression, resulting in inconsistent terminology in the literature and making comparisons between different fairness metrics difficult to conduct. Frequently, authors propose new fairness metrics as part of a broader approach to bias mitigation in regression models, but they don't specifically name the metric, analyze which conceptual notion of fairness the metric corresponds to, or compare it against other authors' techniques for measuring fairness in regression models. In some cases, multiple authors come up with the same mathematical formula but give it different names [1,22,23], or give the same name to different metrics or notions [38]. An ontology that formalizes the notions and metrics in fair regression and links these to their associated bias mitigation methods could help support creation of more equitable and accurate regression models, assessment of the fairness of these models, and support research in ML fairness by identifying related research as well as gaps in research.

To address these deficiencies we provide the Fairness Metrics Ontology for Regression and Machine Learning (FMO-R). FMO-R is the latest version of the Fairness Metrics Ontology (FMO) [27], an OWL 2 ontology intended to help ML developers find and understand fairness metrics, notions, or bias mitigation methods that meet user requirements, and to help researchers with analysis and understanding of existing work in the ML fairness domain. FMO as originally published focused solely on classification, so to create the FMO-R update we conducted a literature review of existing works that propose techniques for fair regression; this review focused on identifying, defining, and organizing any fairness metrics defined in these papers, including some unnamed metrics. We determined the conceptual notions of fairness that each metric measures and adapted categorization schemes from classification fairness metrics in order to determine which high-level idea of fairness each notion is extending. Where applicable, we specified whether a given notion of fairness is a stronger or weaker version of another, or derived from a similar notion in classification, and also determined how various other paradigms used to categorize classification notions applied. The resulting ontology defines the various fairness notions, fairness metrics, and bias mitigation methods proposed for regression, organizing these into a class hierarchy and annotating these with definitions, mathematical formulas, alternative names, and provenance.

Our main contribution is extending FMO for regression, but we also provide updates to the FMO framework as a whole to assist with general use and development. The most important of these improvements is the addition of categorization schemes incorporating basic graph patterns as annotations; these annotations enable practical applications of the ontology such as a web-based faceted search interface that dynamically populates its content from the ontology and builds queries from user-selected categorizations to filter results. Previously we demonstrated the use of FMO through description logic (DL) queries to reason about fairness concepts; we continue to support this method, but we hope that the novel annotation-based search interface will enable provide accessibility to bring FMO-R's reasoning capabilities to a broader audience.

2 Background

In Sect. 2.1, we provide an overview of foundational concepts in ML fairness. In Sect. 2.2, we provide a summary of the pre-existing work we have done on making an ontology for fairness metrics in classification. In Sect. 2.3, we provide a brief summary of related work.

2.1 Fairness in Machine Learning

Fundamentally, a fair machine learning model should treat different *protected classes* of people fairly. What exactly is meant by "fair treatment" varies—it could mean equal outcome rates, equal error rates, or any number of other ideal scenarios—but in general means avoiding *biases* towards or against groups of people that disproportionately receive better or worse outcomes than their peers. The groups are identified by one or more *protected attributes*, also known as *sensitive attributes*, such as race or gender. Fairness metrics and bias mitigation techniques generally rely on the presence of a protected attribute in the data to measure and mitigate bias in the model, but differ in how they calculate unfairness and in what fairness notions are used to determine how fair treatment of different groups is defined.

Fairness notions are the ideals a model must meet in order to be considered fair [35]. Notions usually place a requirement on the performance or outcome of a model. For example, the fairness notion "Independence" requires that a model's output should be statistically independent from the protected attributes of its inputs. A example of a fairness notion for regression is "Equal Accuracy," which requires the expected loss of any two groups of the protected class to be equivalent [18]. Fairness notions can be categorized in several ways, such as whether they are group-level or individual-level: *group-level* notions place requirements on how models treat different groups of people in aggregate, while *individual-level* notions place requirements on how models treat individuals regardless of group (such as "similar individuals should be treated similarly") [8,25]. Determining which fairness notion(s) to use depends on situational factors, and is made more complicated by the fact that some fairness notions are incompatible with others–for example, it is generally impossible for a model to simultaneously achieve all three group-level notions of "Independence," "Separation," and "Sufficiency" in real-world examples [31], and external biases arising from data collection biases, longstanding historical inequalities, and/or medical differences may be dealt with differently by different notions [35].

Fairness metrics measure how closely a model satisfies a given fairness notion. Notions are a quality—either a model satisfies a fairness notion or it doesn't—so fairness metrics are a quantity to express the distance between a model's ideal and actual performance. For the simpler fairness notions in classification, fairness metrics are generalized—most of these notions can be calculated with a difference metric, a ratio-based metric, or a statistical p-test. As many regression fairness notions tend to be less flexible, we have identified fairness metrics specific to each notion. For example, Equal Accuracy is measured with "Error Gap," which computes the absolute difference in expected loss between the protected and other groups [18].

As perfect performance is generally not possible to achieve in real-world scenarios, so *thresholds* are used to determine when a model is broadly good enough. How a threshold is defined depends on which fairness notion or fairness metric is selected. For the Equal Accuracy notion, "Bounded Group Loss" requires the loss for all groups to be below a selected value [1]. We also have "Pairwise Equal Accuracy," another metric of the Equal Accuracy notion, which when thresholded requires the probability of correctly predicting the better outcome of any two individuals in a group to be above a set value for all groups [38].

Bias mitigation techniques reduce bias in a model typically with respect to some defined or implicit fairness metric. They are categorized according to when they are applied in the process of training and testing a model: *pre-processing* methods are applied to the data prior to training and testing, *in-processing* methods are used for the design and training of models, and *post-processing* methods are applied to the predictions of the model. Most mitigation techniques directly use fairness metrics, and will target one or more notions as well even when they do not explicitly use a metric. If we again consider Equal Accuracy and Pairwise Equal Accuracy, the related bias mitigation method places a lower bound on

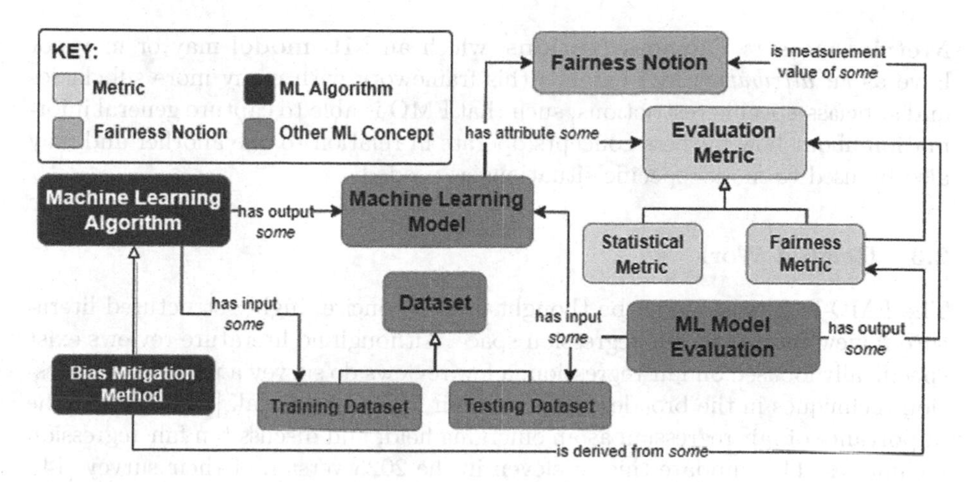

Fig. 1. A high-level concept map of the originally published FMO. Properties shown are defined in SIO [24], and are placed within OWL 2 existential restrictions (indicated with "*some*").

the Pairwise Equal Accuracy model using a constrained optimization [38]. The resulting model should have a good Pairwise Equal Accuracy value.

2.2 The Fairness Metrics Ontology Framework

We previously introduced the fairness metrics ontology for classification only[1] [27], and maintain the basic FMO class structure for the FMO-R extension. We show the pre-existing class structure in Fig. 1. FMO defines four kinds of classes: metrics, fairness notions, ML algorithms (including bias mitigation methods), and other ML modeling concepts. FMO uses the Semanticscience Integrated Ontology (SIO) as its upper-level ontology [24], and defaults to use of SIO superclasses and predicates wherever possible. FMO also imports statistical metrics from the statistics ontology (STATO) [30], and a list of ML algorithms from the machine learning ontology (MLO) [10]. Additionally, FMO is intended to support modeling of instances of its classes for user-specific situations, so we defined classes rather than instances whenever possible and made design choices to ensure users could flexibly extend FMO as needed.

Broadly, FMO classes are defined in relation to the **ML Model** class[2]. **ML models** are the output of **ML Algorithms**, which accept **training datasets** as *input* and, if they are also **Bias Mitigation Methods**, are *derived from* **Fairness Metrics. ML models** are themselves *input* along with **testing datasets** into **ML evaluations. ML Evaluations** *output* **Evaluation Metrics**, including **Statistical Metrics** (such as accuracy, recall, etc.) as well as **Fairness Metrics** (usually) *derived from* one or more **Statistical Metrics. Fairness**

[1] Since its publication, we have also begun adding some concepts for fair clustering.
[2] Classes are shown in **bold**, and properties are shown in *italics*.

Metrics *measure* **Fairness Notions**, which an **ML model** may or may not have as an *attribute*. FMO extends this framework with many more subclasses and subclass-specific restrictions, such that FMO is able to capture general information about how fairness concepts operate in relation to one another and may also be used to model specific situations as needed.

2.3 Related Work

The FMO-R extension can be thought of as a concise, highly structured literature review across the fair regression space. Although no literature reviews exist specifically focused on fair regression, a few reviews do survey a portion of regression techniques in the broader context of fair ML. Caton et al. [13] highlight the importance of fair regression as an emerging field, and discuss ten fair regression techniques. They update this to eleven in the 2023 version of their survey [14], but again stress the need for more work to focus on this area. Del Barrio et al. [3] analyze several fairness metrics for both classification and regression within a mathematical framework, although they only focus on the statistical parity and equalized odds notions. A few other reviews [35,36,46] describe a couple of fair regression notions or metrics. We incorporated terminology, mathematical definitions, and categorizations from these reviews into FMO-R as appropriate, though we still needed to conduct our own review to ensure the ontology was comprehensive (see Sect. 3.1).

Besides its organizational contribution of a standardized taxonomy, FMO-R is a machine-readable knowledge resource for semantically-driven recommendation and question-answering. While we are unaware of any ontology other than FMO-R that addresses fair ML, we can identify other related prior work: a machine learning ontology (MLO) [10] that provides a taxonomy of ML algorithms, and a metrics ontology [42] for knowledge-based representation and analysis of program measurement. Although we were unable to build off of the metrics ontology–it was intended for use with clinical programs, and largely incompatible with machine learning metrics–we were able to interface with a version of MLO, adjusted to work with our DL reasoner. Other ontology-based methods of analyzing ML models have been proposed, including the explanation ontology [17] and Doctor XAI [39] for explainability. Tools to help with fairness analysis use non-semantic approaches: these include the documentation-driven approach of AI Fairness 360 [5] and a proposal for the algorithmic selection of fairness metrics [11].

3 Methods

In creating the FMO-R update, we had two main goals:

(1) Increase clarity, organization, and standardization for existing techniques in regression in ML fairness, and

(2) Make the benefits of the fairness metrics ontology as a whole more accessible to ML developers who may not be familiar with either ML fairness or ontologies, aiding them in selecting fairness metrics, analyzing their results, and understanding what their underlying fairness notions actually imply.

To achieve goal **(1)**, we focused on ensuring FMO-R would be a detailed, well-organized, and comprehensive knowledge resource of all relevant fairness notions and metrics. Section 3.1 describes our process of gathering, organizing, and categorizing knowledge for the ontology. To achieve goal **(2)**, we upgraded the original structure of FMO to support faceted search via our browser-based interface, while also improving its depth of representation. We describe the technical details of upgrading the ontology's representation in OWL in Sect. 3.2, and demonstrate the function of the web interface in Sect. 4.2.

3.1 Development Process

Existing ML fairness surveys that include regression, but none are a comprehensive study of the field and do little to organize all the work done. Our goal was to do a complete survey of fair regression bias mitigation methods, capture the notions and metrics explicitly or implicitly contained in those works. We rewrote them in a consistent mathematical and probabilistic notation for better understanding and comparison. We renamed distinct fairness metrics of the same name to distinguish them, and for papers that proposed the same fairness notion or metric as others, we credited all creators and used the definition which most closely matched our chosen notation. Once gathered and rewritten, we organized the fairness metrics under their related notions and categorized bias mitigation techniques by the metrics they targeted. By providing uniform notation and organization, we hope to clarify understanding of the field as it stands and highlight the gaps in the existing work for ML fairness developers.

Research papers were initially found via Google Scholar by combining keywords such as "fairness", "bias", and "mitigation" with "regression." We used the bibliographies of these papers and those works which cited them to expand our collection of relevant work until we had exhausted our search. From this corpus we excluded papers which only described logistic regression since, although logistic regression models use regression internally, they output classifications instead of continuous values. From the selected papers we gathered fairness notions, including both natural language and probabilistic definitions; and fairness metrics, as natural language and mathematical definitions, along with references to all associated bias mitigation techniques. Some papers did not explicitly define fairness metrics but directly applied novel fairness metrics in their bias mitigation models and algorithm [38] . These metrics were also included in FMO-R, and we chose a name for them if the authors did not originally provide one. We also made note of the papers that had implementations in public repositories or packages to list as a resource in FMO-R.

A primary goal of our survey was to organize fairness metrics and notions. The fairness notions were organized under their own high-level statistical notions:

Table 1. Fairness metrics sorted by corresponding fairness notion and high-level notion

High-Level Notion	Fairness Notion	Metric
Independence	Statistical Parity [1,21–23]	KS Statistical Parity [22,23]
Independence	Statistical Parity [1,21–23]	Cross-Pair Group Fairness [7]
Independence	Pairwise Statistical Parity [38]	Pairwise Statistical Parity [38]
Independence	Mean Difference [12]	Group Mean Difference [12]
Independence	Pairwise Eq. Opportunity [38]	Pairwise Eq. Opportunity [38]
Loss	Equal Accuracy [18]	Accuracy Parity [18]
Loss	Equal Accuracy [18]	Error Gap [23]
Loss	Equal Accuracy [18]	Bounded Group Loss [1]
Loss	Equal Accuracy [18]	AUC [12]
Loss	Pairwise Eq. Accuracy [38]	Pairwise Eq. Accuracy [38]
Loss	Symmetric Eq. Accuracy [38]	Symmetric Eq. Accuracy [38]
Individual Fairness	Individual Fairness [7]	Fairness Degree [40]
Individual Fairness	Individual Fairness [7]	Fairness Penalty [7]

"Independence," "Loss," and "Individual Fairness." Independence states that the predicted outcomes and protected attributes are independent of each other. Individual fairness is a weaker form of independence, where the predicted outcome and protected attribute are independent of each other given the remaining features. The Loss notion requires that the loss of a model is independent of the protected attribute. Individual fairness requires that similar data should have similar outcomes. Independence and Loss capture Group Fairness, because they are concerned how the overall distributions of predictions vary between groups. When a notion is satisfied, so is the high-level notion it falls under. These high-level notions are ultimately what determine if and how a model is fair. Examples of the notions and metrics we collected, categorized by high-level notion, are listed in Table 1.

We linked all fairness metrics with a corresponding notion, whether from their original work, another work which defined notions that related to the metric in question, or through the development of a novel notion which best described the metric. In some cases, fairness metrics are common derivations of their corresponding notions, e.g. KS Statistical Parity [22,23] and Statistical Parity [1,21–23]. Kolmogorov-Smirnov test is commonly used so is a clear choice for comparing distributions of different groups. In contrast, metrics like Pairwise Statistical Parity and the other pairwise notions had well-defined notions but were never explicitly defined as a metric in the source paper [38]. The bias mitigation technique proposed instead creates a generic thresholded metric to be used for all the pairwise notions which we use to define the fairness notions as fairness metrics. The original papers for some fairness metrics define the metrics well mathematically but did not explicitly refer to fairness notions. An example of this is such as Bounded Group Loss [1] which we describe in Sect. 2.1 as being a thresholded metric that requires the losses for each group of a protected class to be bounded above by a set value. By the definition, we see the expectation that the losses by group should be equivalent and small in an ideal model, which is the requirement of Equal Accuracy [18].

By putting these metrics into a common notational framework, we could readily determine that they corresponded to previously-defined fairness notions. In rare circumstances, we developed new fairness notions when no explicit notions and the existing notions from other papers did not satisfy the measures they provided. The Fairness Degree [40] and Fairness Penalty [7] metrics were defined as related metrics with the common idea that identical individuals should receive similar predicted outcomes regardless of group. Then the ideal scenario would be equivalent outcomes for identical individuals of different groups, a fairness notion which we named Individual Fairness after the criteria which metrics satisfy by demonstrating similar individuals are treated consistently.

In many cases, multiple metrics exist to measure bias for the same notion. We gave Equal Accuracy [18] as an example in Sect. 2.1 to explain how a metric relates to a notion and in doing so provided three separate metrics which satisfy the notion; the only difference is how the notion is measured and the values it must be to be fair. Bounded Group Loss [1] and Pairwise Equal Accuracy [38] are thresholded metrics, whereas Risk Measure [23] looks to produce values near zero. The variety of metrics allows flexibility in how we enforce a notion. Some metrics are better options for implementing for bias mitigation, as they are more easily modeled or less computationally intensive. With the organization we did, we hoped to draw connections between the various fairness notions and metrics which exist and relate them to broader concepts.

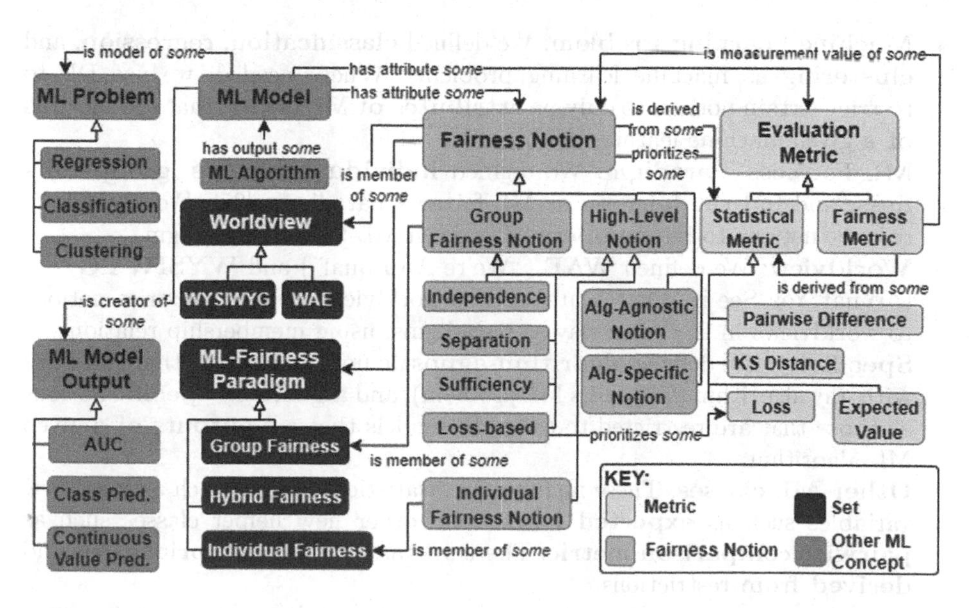

Fig. 2. A concept diagram of the new classes and predicates added for FMO-R.

3.2 Structure

FMO-R expands on the original FMO by adding both the new class hierarchy of notions, metrics, and mitigation methods for fair regression as well as upgrades to the existing annotations, OWL restrictions, and class structure. The new additions of FMO-R are highlighted in Fig. 2 and described below. We bold the first use of any annotation, property, or new class.

Annotations. The original FMO annotated each class with a **label**, a **definition**, a **source** (as a DOI when possible, or a URL if not), and any number of **synonyms**. It also included a **probabilistic definition** for each statistical metric and high-level notion. For the FMO-R update, we assigned a probabilistic definition for every notion, and created the new **mathematical formula** annotation for each metric. Whereas probabilistic definition is a Unicode string defining the probabilistic constraint that a notion imposes upon a model, a mathematical formula is a MathJax-compatible [15] encoding that describes the formula to compute a metric (and by extension, estimate how well the metric's corresponding notion is satisfied). Several additional classes have also been annotated with their **preferred mathematical notation**, to ensure each mathematical formula and probabilistic definition has its variables defined.

Classes and Restrictions. In addition to defining the new regression-specific classes listed in Table 1, along with the **subclass of** and **derived from** DL restrictions used to determine when they can be inferred, we also defined several new classes to aid with the selection and filtering of different categories:

- **Machine Learning Problem**: We defined **classification**, **regression**, and **clustering** as machine learning problems. Where needed, we use DL to restrict certain notions to only be **attributes of** ML models that are **models of** a given machine learning problem.
- **ML-Fairness Paradigm**: We defined **individual fairness**, **group fairness**, and **hybrid fairness** as ML-fairness paradigms [25]. We use DL to restrict notions to be **members of** a given ML-fairness paradigm.
- **Worldview**: We defined **WAE** ("We're All Equal") and **WYSIWYG** ("What You See is What You Get") as worldviews [28]. We restrict notions to worldviews in the same way as paradigms, using membership relations.
- **Specificity**: We defined **algorithm-agnostic** notions as those that can work with any algorithm (within its ML problem), and **algorithm-specific** notions as those that are restricted to only ML models that are **outputs of** a given ML algorithm.
- **Other ML classes**: These include new statistical metrics such as **loss**, new variables such as **expected value**, and other new helper classes such as **pairwise comparison metric**. These are mainly used with **prioritizes** and **derived from** restrictions.

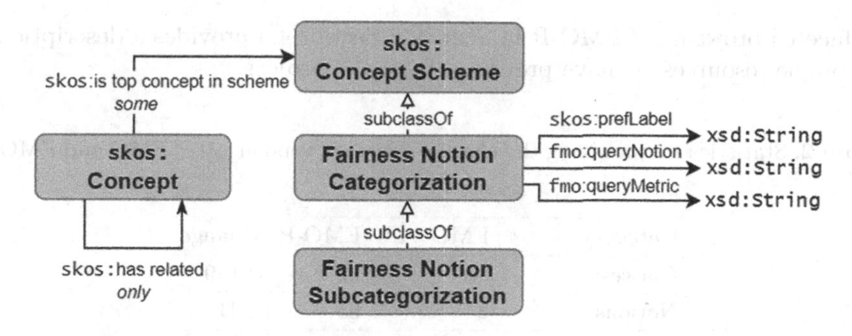

Fig. 3. The SKOS conceptual hierarchy, used to model the concept schemes that organize the ontology and enabling faceted browsing via selective filtering. Visit https://github.com/frankj-rpi/fairness-metrics-ontology/blob/main/skos-diagram.md to view the full-size diagram.

Concept Schemes. In order to enable the faceted search interface, we also needed a way to identify specific classes as facets that can be used for selection. In order to avoid unintended side effects during reasoning with DL, we used punning to instantiate facets from the simple knowledge organization scheme (SKOS) instead (see Fig. 3). We defined two subclasses of the SKOS **Concept Scheme**, **Fairness Notion Categorization** and **Fairness Notion Subcategorization**, and created instances of them with the same IRI as each of the categorizations above, (ML Problem, ML Fairness Paradigm, etc.). Then, we created instances of the SKOS **Concept** with the same IRI as each category, and asserted that they were a **top concept in the scheme** of their corresponding categorization (e.g., Classification, Clustering, and Regression are each instances of Concept and a top concept in the scheme of ML Problem). Then, we annotated each categorization with **skos:prefLabel** to indicate the name to use when displaying it, as well as one or both of the following newly defined sub-properties of **skos:note**: **Query Notion** and **Query Metric**. Each of these annotations is a basic graph pattern intended to be used to construct a SPARQL query to filter out notions or metrics that correspond to a specific category. Lastly, we also use the **skos:hasRelated** property to link categories to lower-level categories in subcategorizations, such as the **Independence** subcategory of **Group Fairness**.

4 Evaluation and Results

We present statistics summarizing the changes between the original version of FMO and the FMO-R update in Table 2.

To verify the function of the ontology, we developed a series of competency questions for FMO-R to answer via DL queries. We describe this evaluation in Sect. 4.1. Additionally, in order to make the content of the ontology accessible to those without familiarity with DL or ML fairness, we present a visual interface for faceted browsing of FMO-R in Sect. 4.2. Section 4.3 provides a description of the online resources we have produced from the project.

Table 2. Statistics summarizing the changes between version 1.0 of FMO and FMO-R.

Category	FMO v1.0	FMO-R	Change
Classes	80	229	+149
Notions	24	65	+41
Metrics	28	80	+52
Other classes	28	84	+56
Individuals	13	100	+87
DL Restrictions	62	193	+131
Triples	899	2672	+1773

Table 3. A table of competency questions and the answers given by FMO-R. **Bold** indicates a class name, and *italics* indicates a property name. See the full set of questions, and the DL queries used to produce these answers, in our GitHub repository.

Competency Question	Answer
Which **notion of fairness** is a *member of* the **WAE worldview** and is *derived from* **expected value**?	**Mean Difference**
Which **fairness metrics** *measure* the **high-level notion** of **Independence** and are *derived from* a **pairwise comparison** between each individual?	**Pairwise Statistical Parity** **Pairwise Equal Opportunity** **Cross-Pair Group Fairness**
Which **bias mitigation methods** are *derived from* a **metric** that *prioritizes* **Equal Accuracy**?	**Constrained Optimization** **Robust Optimization**

4.1 Competency Questions

The DL restrictions in FMO-R support reasoning about fairness concepts, so that the ontology may guide a user in determining which fairness metrics, fairness notions, or bias mitigation methods should be used for a given situation. We created a set of competency questions to verify that each DL restriction works as intended; we display a subset of these questions, along with results of asking them of FMO-R, in Table 3. Each of these competency questions was encoded in DL and queried of the ontology, running in Protege 5.5.0 [44] with the Pellet 2.2.0 [41] reasoner plugin. The full set of competency questions, along with the specific DL queries and responses, are viewable in our GitHub repository[3]. The results of these questions validate the reasoning ability of FMO-R and demonstrate its ability to assist developers in the field of ML fairness, and we provide a method for these questions to be answered in practice via the Fairness Metrics Explorer.

4.2 Fairness Metrics Explorer

The Fairness Metrics Explorer enables an ML developer or researcher to find which fairness notions and fairness metrics meet the requirements of their specific use case. The user may select different facets to filter results by, such as filtering by ML problem, fairness paradigm, or whether algorithm-agnostic methods are required or algorithm-specific methods are acceptable (see Fig. 4). Then the developer can choose among a relevant set of fairness notions and metrics, funneling to probabilistic definitions and mathematical formulas. This results in a high degree of transparency and provenance when it becomes necessary to justify the choices made when ensuring a model is fair. The user interface is generated from the ontology: each group of checkboxes in the left pane corresponds to a categorization in SKOS. Multiple options can be chosen in each categorization. As the researcher chooses an item, the right pane displays information about the result. The user can click on any class name (including categories) to select and view its information, and hover over any class name to temporarily view that class's info.

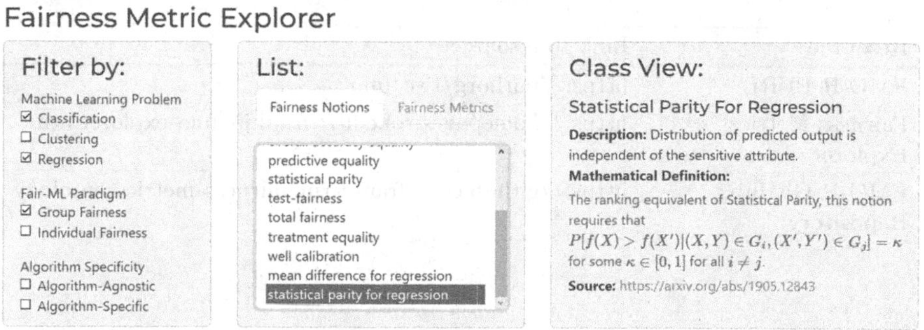

Fig. 4. The Fairness Metrics Explorer. The left panel filters by category; it is set to display classification or regression notions which are a form of Group Fairness. The middle panel is set to display a list of notions (as opposed to metrics). The right panel is showing details about the currently selected "Statistical Parity for Regression" notion.

[3] https://github.com/frankj-rpi/fairness-metrics-ontology/blob/main/competency-questions.md.

The browser is built as an R Shiny app [16], querying an instance of the ontology hosted in Blazegraph triplestore [4]. Whenever it needs to populate a pane with information, the application sends a SPARQL query to the ontology, and it relies on the SKOS hierarchy to determine which categories to display and how. Using the Query Notion and Query Metric annotations, it is able to build a list of notions or metrics that satisfy the given category. These annotations are snippets of SPARQL code assigned to each categorization, with variables such as `?notion_uri`, `?metric_uri`, `?category_uri`, and so on. When a given category is selected, its URI is substituted into the `?category_uri` slot in the corresponding annotation, and a list of these annotations is added to the interface's standard SPARQL query for listing classes. In this way, the query is dynamically generated based off of metadata in the ontology, so updates to the ontology are immediately reflected in the browser.

4.3 Resource Contributions

We contribute the following publicly available artifacts: the Fairness Metrics Ontology for Regression and Machine Learning, as well as the Fairness Metrics Explorer web application. The ontology and the browser have been made available as open-source artifacts under the Apache 2.0 license, and are hosted along with documentation on our GitHub repository. We also maintain a canonical reference to our ontology in the form of a persistent URL hosted on the PURL service. Links to the persistent URL, the faceted fairness browser, and the public GitHub repository are shown in Table 4.

Table 4. Links to resources we have released and refer to in the paper.

Resource	Link to resource
FMO-R PURL	https://purl.org/twc/fmo
Fairness Metrics Explorer	https://homepages.rpi.edu/~frankj6/fmo-explorer.html
FMO-R GitHub Repository	https://github.com/frankj-rpi/fairness-metrics-ontology

5 Discussion and Future Work

Fairness must be a consideration in the deployment of trustworthy effective AI models, and many research questions yet remain. The ontology-based standardization offered by FMO and its extension help address critical questions: What does it means to be fair? How do we quantify fairness? What mitigation methods are available to ensure fairness of deployed AI models? To fairness researchers, we provide FMO-R: a clear motivating example for how the ontology can help stabilize an emerging field and be used as a tool to assist in research. By capturing both implicit and explicit notions, metrics, and associated mitigation methods in research papers, FMO-R serves as a dynamic survey of fairness research which is critically needed for regression. We will translate this into a survey paper as future work. The original ontology's focus on classification limited its scope and potential. By adding an entire fairness domain that is still relatively understudied, we intend to demonstrate to fairness researchers the effectiveness of semantics in scientific resources and the benefits that adoption of ontologies can bring. By providing a detailed and comprehensive set of standardized representations for fairness concepts, we hope to assist fairness researchers in navigating the literature, disambiguating between similar metrics, and determining what holes exist in current research where new scientific effort is needed.

The upgraded and new resources reported in this paper are designed to increase the accessibility that is necessary for real adoption of FMO-R. The primary benefit here is in the online interface: for those unfamiliar with either fairness or ontologies, the interface provides a simple and highly approachable context for users to access FMO-R's content and directly learn about what fairness metrics do and how they should be used in different situations. ML modelers can find and understand potential metrics and mitigation methods to potentially improve the fairness of real-world deployed regression and other ML models. As the semantic web community makes frequent use of AI and ML models, we hope to encourage the adoption of fair ML techniques by those other than ML fairness researcher. At the same time, the semantic web audience presents the additional benefit of familiarity with ontologies, and may decide to make use of FMO-R directly if given the opportunity to experiment with it and learn a little about the depth of knowledge present in the ML fairness domain. We furthermore hope that the SKOS-based design patterns to support a reactive ontology interface are of value to those in the community interested in making the benefits of other ontologies more readily available to the general public.

In the future, we plan to investigate the integration of the ontology directly with AI Fairness 360 [5] and related packages. Combining structured knowledge about fairness with code to actually run these metrics is a clear area of future interest, and suggests a compelling workflow: first the system would reason about a user's actual situation to recommend possible fairness metrics, and then would compute the metrics and reason about their results to assist with analysis. An additional benefit of using an ontology in this scenario is that the system could provide explanations for its reasoning, natural language definitions and examples for all relevant concepts, and a natural method of representing specifics of the user's situation via a knowledge graph instantiating FMO-R classes.

6 Conclusion

Creation of fair ML models is extremely important to ensure that all realize the benefits of AI and to mitigate its potential harms. We provide FMO-R to assist users in the process of learning about ML fairness, selecting appropriate fairness metrics based on user requirements, and interpreting their results through the context of the fairness notions that these metrics measure. This paper builds off of the initial work about the fairness metrics ontology, describing the improvements that have been made to the resource to increase its depth of knowledge via the additional domain of fair regression, to increase its reasoning capabilities through the introduction of new conceptual schemes, and to increase its overall accessibility via an online interface for faceted search. We believe our ontology can be a valuable resource for the semantic web community, the ML fairness community, and the ML community at large, and plan to continue to support it through increased depth of knowledge and integration with ML fairness software.

Acknowledgements. This work is partially supported by IBM Research AI through the AI Horizons Network. We thank Mohamed Ghalwash and Ching-Hua Chen, Ioana Baldini and Dennis Wei (IBM Research) and Lydia Halter (RPI) for their research assistance.

References

1. Agarwal, A., Dudík, M., Wu, Z.S.: Fair regression: quantitative definitions and reduction-based algorithms. In: International Conference on Machine Learning (2019)
2. Angwin, J., Larson, J., Mattu, S., Kirchner, L.: Machine bias: there's software used across the country to predict future criminals. and it's biased against blacks (2016). https://www.propublica.org/article/machine-bias-risk-assessments-in-criminal-sentencing
3. del Barrio, E., Gordaliza, P., Loubes, J.M.: Review of mathematical frameworks for fairness in machine learning. arXiv preprint arXiv:2005.13755 (2020)
4. Bebee, B.: Blazegraph wiki. https://github.com/blazegraph/database/wiki
5. Bellamy, R.K.E., et al.: AI fairness 360: an extensible toolkit for detecting, understanding, and mitigating unwanted algorithmic bias (2018)
6. Berk, R.: An impact assessment of machine learning risk forecasts on parole board decisions and recidivism. J. Exp. Criminol. **13**, 193–216 (2017)
7. Berk, R.A., et al.: A convex framework for fair regression. ArXiv abs/1706.02409 (2017)
8. Binns, R.: On the apparent conflict between individual and group fairness. In: Proceedings of the 2020 Conference on Fairness, Accountability, and Transparency, pp. 514–524 (2020)
9. Bloise, F., Brunori, P., Piraino, P.: Estimating intergenerational income mobility on sub-optimal data: a machine learning approach. J. Econ. Inequal. **19**(4), 643–665 (2021)
10. Braga, J., Dias, J.L.R., Regateiro, F.: A machine learning ontology, October 2020. https://doi.org/10.31226/osf.io/rc954,osf.io/preprints/frenxiv/rc954
11. Breger, C.: Criteria for algorithmic fairness metric selection under different supervised classification scenarios. Master's thesis, Pompeu Fabra University (2020). http://hdl.handle.net/10230/46359

12. Calders, T., Karim, A., Kamiran, F., Ali, W., Zhang, X.: Controlling attribute effect in linear regression. In: 2013 IEEE 13th International Conference on Data Mining, pp. 71–80 (2013). https://doi.org/10.1109/ICDM.2013.114
13. Caton, S., Haas, C.: Fairness in machine learning: a survey. arXiv preprint arXiv:2010.04053 (2020)
14. Caton, S., Haas, C.: Fairness in machine learning: a survey. ACM Comput. Surv. (2023). https://doi.org/10.1145/3616865. Just Accepted
15. Cervone, D.: MathJax: a platform for mathematics on the web. Not. AMS **59**(2), 312–316 (2012)
16. Chang, W., et al.: shiny: Web Application Framework for R (2022). https://CRAN.R-project.org/package=shiny. r package version 1.7.4
17. Chari, S., Seneviratne, O., Gruen, D.M., Foreman, M.A., Das, A.K., McGuinness, D.L.: Explanation ontology: a model of explanations for user-centered AI. In: Pan, J.Z., et al. (eds.) ISWC 2020, Part II. LNCS, vol. 12507, pp. 228–243. Springer, Cham (2020). https://doi.org/10.1007/978-3-030-62466-8_15
18. Chi, J., Tian, Y., Gordon, G.J., Zhao, H.: Understanding and mitigating accuracy disparity in regression (2021)
19. Chouldechova, A.: Fair prediction with disparate impact: a study of bias in recidivism prediction instruments. Big Data **5**(2), 153–163 (2017)
20. Chouldechova, A., Roth, A.: The frontiers of fairness in machine learning (2018)
21. Chzhen, E., Denis, C., Hebiri, M., Oneto, L., Pontil, M.: Fair regression via plug-in estimator and recalibration with statistical guarantees. In: Neural Information Processing Systems (2020)
22. Chzhen, E., Denis, C., Hebiri, M., Oneto, L., Pontil, M.: Fair regression with Wasserstein barycenters. In: Larochelle, H., Ranzato, M., Hadsell, R., Balcan, M., Lin, H. (eds.) Advances in Neural Information Processing Systems, vol. 33, pp. 7321–7331. Curran Associates, Inc. (2020). https://proceedings.neurips.cc/paper_files/paper/2020/file/51cdbd2611e844ece5d80878eb770436-Paper.pdf
23. Chzhen, E., Schreuder, N.: A minimax framework for quantifying risk-fairness trade-off in regression (2022)
24. Dumontier, M., et al.: The semanticscience integrated ontology (SIO) for biomedical research and knowledge discovery. J. Biomed. Semant. **5**, 1–11 (2014)
25. Dwork, C., Hardt, M., Pitassi, T., Reingold, O., Zemel, R.: Fairness through awareness. In: Proceedings of the 3rd Innovations in Theoretical Computer Science Conference, pp. 214–226 (2012)
26. Flores, A.W., Bechtel, K., Lowenkamp, C.T.: False positives, false negatives, and false analyses: a rejoinder to machine bias: there's software used across the country to predict future criminals. and it's biased against blacks. Fed. Probation **80**, 38 (2016)
27. Franklin, J.S., Bhanot, K., Ghalwash, M., Bennett, K.P., McCusker, J., McGuinness, D.L.: An ontology for fairness metrics. In: Proceedings of the 2022 AAAI/ACM Conference on AI, Ethics, and Society, pp. 265–275 (2022)
28. Friedler, S.A., Scheidegger, C., Venkatasubramanian, S.: On the (im)possibility of fairness. CoRR abs/1609.07236 (2016). http://arxiv.org/abs/1609.07236
29. Friedler, S.A., Scheidegger, C., Venkatasubramanian, S.: The (im) possibility of fairness: Different value systems require different mechanisms for fair decision making. Commun. ACM **64**(4), 136–143 (2021)
30. Gonzalez-Beltran, A., Rocca-Serra, P., Burke, O., Sansone, S.A.: Stato: an ontology of statistical methods (2012). http://stato-ontology.org/

31. Kleinberg, J.M., Mullainathan, S., Raghavan, M.: Inherent trade-offs in the fair determination of risk scores. CoRR abs/1609.05807 (2016). http://arxiv.org/abs/1609.05807
32. Kourou, K., Exarchos, T.P., Exarchos, K.P., Karamouzis, M.V., Fotiadis, D.I.: Machine learning applications in cancer prognosis and prediction. Comput. Struct. Biotechnol. J. **13**, 8–17 (2015)
33. Kruppa, J., et al.: Probability estimation with machine learning methods for dichotomous and multicategory outcome: applications. Biom. J. **56**(4), 564–583 (2014)
34. Liem, C.C., et al.: Psychology meets machine learning: interdisciplinary perspectives on algorithmic job candidate screening. explainable and interpretable models in computer vision and machine learning, pp. 197–253 (2018)
35. Makhlouf, K., Zhioua, S., Palamidessi, C.: Machine learning fairness notions: bridging the gap with real-world applications. Inf. Process. Manage. **58**(5), 102642 (2021)
36. Mehrabi, N., Morstatter, F., Saxena, N., Lerman, K., Galstyan, A.: A survey on bias and fairness in machine learning. ACM Comput. Surv. **54**(6) (2021). https://doi.org/10.1145/3457607
37. Mhlanga, D.: Financial inclusion in emerging economies: the application of machine learning and artificial intelligence in credit risk assessment. Int. J. Financ. Stud. **9**(3), 39 (2021)
38. Narasimhan, H., Cotter, A., Gupta, M., Wang, S.: Pairwise fairness for ranking and regression. In: Proceedings of the AAAI Conference on Artificial Intelligence, vol. 34, pp. 5248–5255, April 2020. https://doi.org/10.1609/aaai.v34i04.5970
39. Panigutti, C., Perotti, A., Pedreschi, D.: Doctor XAI: an ontology-based approach to black-box sequential data classification explanations. In: Proceedings of the 2020 Conference on Fairness, Accountability, and Transparency, FAT* 2020, pp. 629–639. Association for Computing Machinery, New York (2020). https://doi.org/10.1145/3351095.3372855
40. Perera, A., et al.: Search-based fairness testing for regression-based machine learning systems. Empir. Softw. Eng. **27** (2022). https://doi.org/10.1007/s10664-022-10116-7
41. Sirin, E., Parsia, B., Grau, B.C., Kalyanpur, A., Katz, Y.: Pellet: a practical OWL-DL reasoner. J. Web Semant. **5**(2), 51–53 (2007)
42. Soergel, D., Helfer, O.: A metrics ontology. an intellectual infrastructure for defining, managing, and applying metrics. In: Knowledge Organization for a Sustainable World: Challenges and Perspectives for Cultural, Scientific, and Technological Sharing in a Connected Society: Proceedings of the Fourteenth International ISKO Conference, Ergon Verlag, Rio de Janeiro, Brazil, 27–29 September 2016 Ri, vol. 15, p. 333. NIH Public Access (2016)
43. Steyerberg, E.W., van der Ploeg, T., Van Calster, B.: Risk prediction with machine learning and regression methods. Biom. J. **56**(4), 601–606 (2014)
44. Tudorache, T., Noy, N.F., Tu, S., Musen, M.A.: Supporting collaborative ontology development in Protégé. In: Sheth, A., et al. (eds.) ISWC 2008. LNCS, vol. 5318, pp. 17–32. Springer, Heidelberg (2008). https://doi.org/10.1007/978-3-540-88564-1_2
45. Verma, S., Rubin, J.: Fairness definitions explained. In: 2018 IEEE/ACM International Workshop on Software Fairness (FairWare), Gothenburg, Sweden, pp. 1–7. IEEE (2018). https://doi.org/10.23919/FAIRWARE.2018.8452913

46. Wan, M., Zha, D., Liu, N., Zou, N.: Modeling techniques for machine learning fairness: a survey. CoRR abs/2111.03015 (2021). https://arxiv.org/abs/2111.03015
47. Xie, F., Chakraborty, B., Ong, M.E.H., Goldstein, B.A., Liu, N., et al.: Autoscore: a machine learning-based automatic clinical score generator and its application to mortality prediction using electronic health records. JMIR Med. Inform. **8**(10), e21798 (2020)

SCIVO: Skills to Career with Interests and Values Ontology

Neha Keshan(✉) and James A. Hendler

Department of Computer Science, Rensselaer Polytechnic Institute, Troy, NY 12180, USA
keshan@rpi.edu, hendler@cs.rpi.edu
https://tw.rpi.edu/

Abstract. While getting a doctorate degree, new skills are acquired, opening up multiple traditional and non-traditional avenues of future employment. To help doctoral students explore the available career paths based on their skills, interests and values, we built a findable, accessible, interoperable, and reusable Skills to Career with Interests and Values ontology (SCIVO). It is a compact ontology of seven classes to harmonize the heterogeneous resources available providing information related to career paths. We demonstrate the interoperability and usability of SCIVO through building a knowledge graph using the web scraped publicly available data from the Science Individual Development Plan tool for current doctoral students - myIDP and the National Science Foundation Survey of Earned Doctorates (NSF SED) 2019 data. The generated knowledge graph (named SCIVOKG) consists of one hundred and sixteen classes and one-thousand seven-hundred and forty instances. An evaluation is conducted using application-based competency questions generated by analyzing data collected through surveys and individual interviews with current doctoral students. SCIVO provides an ontological foundation for building a harmonized resource as an aid to doctoral students in exploring the career options based on their skills, interests and values.

Resource Website:
https://tetherless-world.github.io/sciv-ontology/.

Keywords: Ontology · Knowledge Graph · Graduate Mobility · Skill based Career exploration · Interoperability

1 Introduction

One of the major challenges faced by doctoral students is figuring out the career they want to pursue after obtaining a degree. For example, a STEM doctoral student might have information about choosing between academe or industry but

Supported by the RPI-IBM AI Research Collaboration, a member of the IBM AI Horizons network.

F. Ortiz-Rodriguez et al. (Eds.): KGSWC 2023, LNCS 14382, pp. 262–276, 2023.
https://doi.org/10.1007/978-3-031-47745-4_19

has limited information regarding the multiple positions available and how to select a role to gain maximum career satisfaction. This leads to the newly-minted graduates making uninformed decisions that may lead to unwanted outcomes. This conclusion was also supported by study results run under Rensselaer Polytechnic Institute's IRB #2081 which surveyed graduate students about career paths. The survey suggested that students had little to no idea regarding the requirements of the various potential roles on a day-to-day basis, adding to the already existing confusion. A deeper analysis suggests that not having the skills required and the activities of interest that a particular career path requires is one of the major reasons behind this confusion. Having such uncertainties in today's web world can be unimaginable, especially when one would assume at least parts of the required information to be accessible through job posts over various social platforms, such as University, Company and LinkedIn Job posts, or through standard search. For example, a preliminary google search on the question "What options do I have as a Computer Science doctoral student working in Artificial Intelligence, my focus being the use of natural language processing in the health domain?" [9] demonstrated that information is sparse and even hours of search might not lead to the specific information required for understanding the needs of various career paths.

Possible corrective actions fall into four main areas: 1) understanding the skills required in a particular career path; 2) understanding the activities that are necessary in a specific position on a daily basis and the percentage of those activities that are interesting for an individual; 3) understanding the values related to the profession, which are consistent with the personal values of the individual; and 4) accessing a variety of resources to find and understand the full career landscape. This information can help eliminate uncertainty and provide users with a foundational resource for exploring the various career opportunities open to graduate students after graduation. Producing such a foundational resource requires the production of a knowledge system built by the harmonization of silo-ed information pulled from various heterogeneous resources available across the web. This knowledge system requires the creation of an ontology for making proper connections between information across heterogeneous resources, and one that could be expanded as and when required to incorporate other helpful resources that may become available.

A concise seven class ontology, the Skills to Career with Interests and Values Ontology (SCIVO) was built to address the above problem. The classes were added based on a current literature review and the general requirements seen across survey analysis and posts across the web. SCIVO is modelled around the skills class and the proficiency level of skills required in a particular career path. It also provides a mechanism to add the interest and the level of these interests required in the career path, while providing heterogeneous resources helpful in exploring these paths. Additionally, the ontology provides a mechanism to add the questions one could ask to understand if the career path connects with users' perceived values. Protégé 5.5.0 was used to develop the required interoperable and reusable ontology. Its feasibility was tested by creating a knowledge graph using SCIVO to ingest data from two different sources. The first is from the Science Individual plan development tool-myIDP [5], and the second is from the National Science Foundation Survey of Earned Doctorates Survey 2019 [4].

1.1 Contribution

We make the following contributions in this paper as an aid to graduate students, especially Science, Technology, Engineering and Math (STEM) graduate students:

1. A consolidated Skills to Career with Interests and Values Ontology to allow users to extract both general and specific information.
2. A knowledge graph integrating data points from Science myIDP, NSF 2019 Doctoral survey, and the web.
3. A resource website[1] for more information on our ontology, its creation, and usage. The resource website also hosts the created ontology and knowledge graph for our users to download and use for various purposes based on the provided guidelines. In addition, the website provides links to the tools used and the detailed SPARQL queries to acquire the data presented in Sect. 7.

2 Background and Related Works

In recent years, we have seen a wave of semantic use in the job search/match domain where the focus has been predominantly on matching the job seekers to job postings and vice-versa [1–3,6,7,12–16,18–21].

One hybrid approach uses Description Logic in a human resource recruiting system based on skills and competencies of individuals [3]. The recruiting is based on skills possessed, posted jobs, formal and non-formal learning activities, degree program learning activities and the experience of the individual. We can see how the matching for recruitment is done both based on the skills required and the skills possessed which is also the main theme of the Job Description Ontology [1] which introduces a dichotomy by segregating the job description from the job position.

Some domain work focuses on demonstrating the gap between the theoretical and the practical requirements. One such ontology is the Job-Know ontology [12] that was created to represent the connection between the content taught in the vocational education and training to what knowledge, skills and abilities are required in the actual job. The other one is the Educational and Career-Oriented Recommendation Ontology (EduCOR) [7] which is built to be an online resource for personalized learning systems. The ontology also addresses the gap between the theoretical and the practical requirement during online learning. This ontology is divided into multiple patterns or components to help interconnect domains and resources. These patterns include the educational resource, skill, knowledge topic, test, learning path, recommendation and user profile patterns.

[1] https://tetherless-world.github.io/sciv-ontology/.

We see that most of the current research does include the skills but rarely includes the interests or values an individual possesses. As demonstrated in [9, 10], an individual's interests and values play a vital role in the success of the individual in the chosen career path. Hence, in this paper we provide a findable, accessible, interoperable, and reusable (FAIR) Skills to Career with Interests and Values (SCIV) ontology to help individuals explore different careers paths and explore a particular career path through various resources.

3 Skills to Career with Interests and Values Ontology (SCIVO) Modeling

As discussed above, there is considerable work related to the use of semantics in the career and job domains. We used the information on how these current ontologies are modeled and then used the top-down modeling approach [8] to design SCIVO. This allowed us to make sure that the ontology could then be extended and reused as required. Table 1 provides the prefix used and the resource links for the generated SCIVO ontology and knowledge graph.

Table 1. List of ontology and knowledge graph prefixes used in the paper.

Prefix	Ontology/Knowledge Graph	URI
scivo	Skills to career with Interests and Values Ontology	http://semanticscience.org/resource
scivokg	Skills to career with Interests and Values Knowledge Graph	http://www.semanticweb.org/neha/ontologies/2021/5/indo#

3.1 Ontology Composition

The current modeling approach allowed us to create a simple and expandable seven-class SCIVO ontology connected with object properties (Fig. 1). Class "scivo:Skills" is connected with class "scivo:CareerPath" through "scivo:Proficiency", while "scivo:CareerPath" is directly connected with "scivo:Interests", "scivo:Values" and "scivo:Resources". This provides a mechanism for users to explore career paths based on their skill proficiency levels. Additionally, the model allows users to investigate the explored career path through multiple resources that provide information about the respective career path.

4 Methodology

The web data was manually annotated on a spreadsheet. The data was then prepared for ingestion using some pre-processing steps. The conversion of data

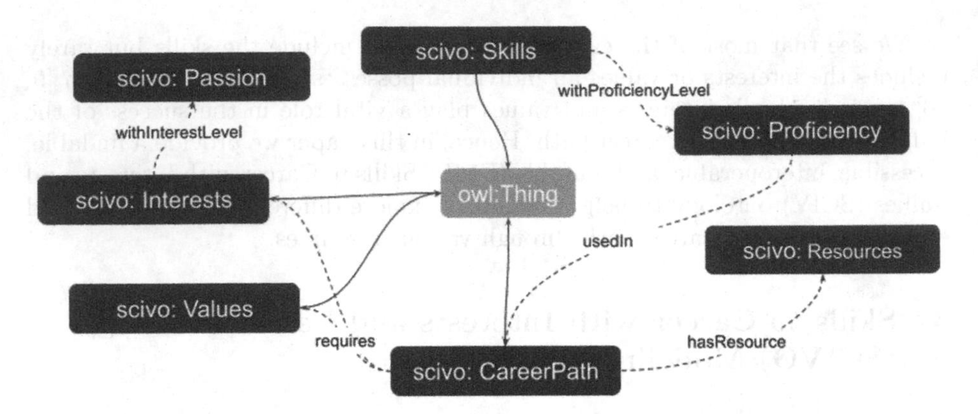

Fig. 1. A conceptual diagram depicting the classes of our Skills to Career with Interests and Values Ontology.

to triples was done using the Protégé Cellfie plugin. This method was followed for creating the main SCIVO ontology and then demonstrate its various uses through the two use cases discussed in Sects. 5 and 6. The Cellfie master mapping (M^2) rules to import spreadsheet data into OWL ontologies [17] were built using the information provided at the Protégé project github repository[2]. This M^2 language provides concise expressive mechanisms for creating both class and individual axioms when generating OWL ontologies. All of our Cellfie rules are available on our website.

5 Science myIDP Use Case

The use of SCIVO is demonstrated by building a knowledge graph[3] with Science myIDP data [5]. The individual development plan, a method to help students choose their future employment, was proposed by the Federation of American Societies for Experimental Biology (FASEB) for postdoctoral fellows in sciences. AAAS/Science, with the help of FASEB and other experts, expanded on this framework to create myIDP for doctoral students in science. Anyone can create a free account and take different tests to get the percentage of skill and interest matching the twenty scientific careers (Fig. 2) that graduate students typically pursue. The match is done with expert average ratings from one to five for the various skills and interests each of these career paths requires. Figure 3 represents an overview of how SCIVO has been extended to incorporate the required classes for myIDP data.

[2] https://github.com/protegeproject/mapping-master/wiki/MappingMasterDSL.

[3] A knowledge graph is referred to the graph created by expanding and instantiating SCIVO classes and referred to as SCIVOKG.

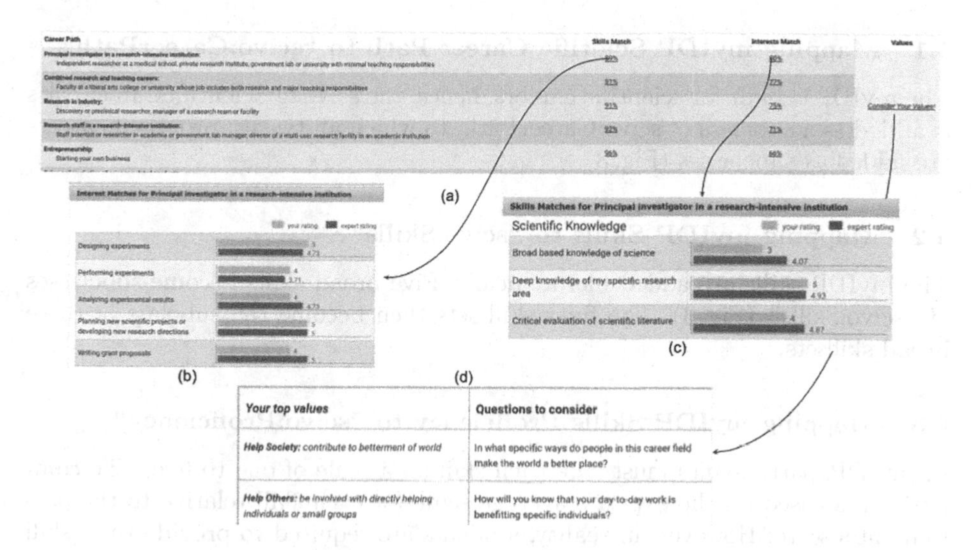

Fig. 2. A snippet of myIDP frame (a) The image depicts a screenshot of how the pop-up looks after the test is taken. Each cell under the skills, interests, and values column opens a new pop-up window. Image (b), (c), and (d) are parts of the screenshots demonstrating the interests match, skills match and values to consider for the career path 'Principal Investigator at a Research-Intensive Institute'.

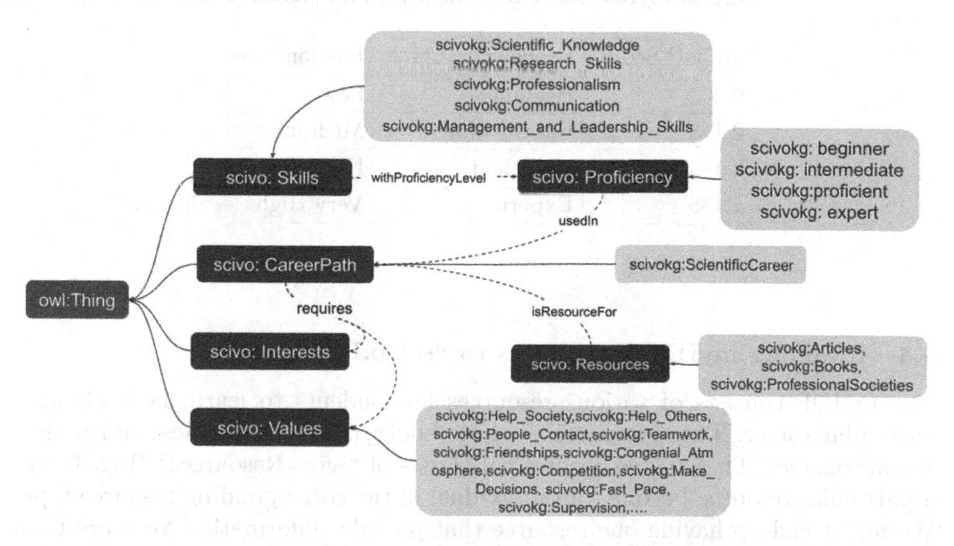

Fig. 3. A conceptual diagram representing a part of SCIVOKG created by extending the SCIVO ontology for the myIDP use case. Light blue boxes represent classes added for the knowledge graph as subclasses of SCIVO. (Color figure online)

5.1 Mapping myIDP Scietific Career Path to "scivo:CareerPath"

The myIDP is built for scientific careers, hence, the scivokg:ScientificCareer class is added as a subclass of scivo:CareerPath, to which all twenty proposed careers are added as subclasses (Fig. 3).

5.2 Mapping myIDP Skills to "scivo:Skills"

The myIDP skills are added hierarchically. Five broad skills become subclasses of "scivo:Skills" (Fig. 3). The finer skill-sets then become the subclass of these broad skillsets.

5.3 Mapping myIDP Skills Proficiency to "scivo:Proficiency"

In myIDP, participants must rate each skill on a scale of one to five. The comparison is based on the expert's average score for each skill relative to the participant's score. However, in reality, students are required to provide their skill proficiency levels. Therefore, for compatibility with many existing systems we mapped the myIDP five pointer to a four pointer skill levels (Fig. 3), as shown in Table 2.

Table 2. myIDP score to proficiency and passion levels

myIDP Scores	Proficiency Levels	Passion Levels
0–2	Beginner	Low
2.1–3	Intermediate	Medium
3.1–4	Proficient	High
4.1–5	Expert	Very High

5.4 Mapping myIDP Resources to "scivo:Resources"

The myIDP consists of various resources for students to learn more about a particular career. These resources include books/chapters, articles, and professional societies. These three become subclasses of "scivo:Resources" (Fig. 3) and a particular resource becomes an individual of the corresponding resource type. We might end up having one resource that provides information for more than one career path. This could have been an issue if the system was not built using semantics. With the power of semantics we can now simply add these connections in a cleaner way during the creation of our knowledge graph.

5.5 Mapping myIDP Interests to "scivo:Interests"

All the thirty-five plus interests discussed in myIDP become direct subclasses of "scivo:Interests" (Fig. 3).

5.6 Mapping myIDP Interests Level to "scivo:Passion"

Similar to the skill level, each participant is required to rate their interest level on a scale of one to five. This is then used to compare with the average expert values as a factor to calculate the percentage of career paths matching. We use the "scivo:Passion" class to connect the interest level with the passion level using the Table 2.

5.7 Mapping myIDP Values to "scivo:Values"

The myIDP supports our argument that one should select a career based on one's skills, interests, and values. They provide more than thirty-five common values that PhDs look at while finding jobs. Since these values are very personal, they provide one example question that could be asked during an interview to find out if the role matches their personal values. Each value becomes an individual of the "scivo:Values" (Fig. 3) having a data property of "correspondingQuestion".

5.8 Knowledge Graph Based on SCIVO

The use and interoperability of SCIVO was demonstrated by creating a knowledge graph. The current SCIVOKG consists of one-thousand seven-hundred forty individuals from both myIDP and NSF SED 2019 data, as discussed in Sect. 5 and Sect. 6. This generated knowledge graph was queried as part of the competency-question based evaluation method [8] discussed in Sect. 7

6 National Science Foundation Survey of Earned Doctorates (NSF SED) 2019

The National Science Foundation conducts an annual survey of earned doctorates. Part of this survey focuses on the postgraduate plans of the doctorates. The career paths are divided into 5 sectors, namely: academe, government, industry or business, non-profit organizations, and other or unknown. We simply added the five sectors as subclasses of "scivokg:Sector" added as subclass of "scivo:CareerPath". This helps us to then add properties to connect these sectors with the scientific career path from the myIDP. Hence, this use case helps us demonstrate the interoperability of our built ontology by helping connect heterogeneous datasets, leading to the possible connections of multiple ontologies and knowledge graphs for our users as discussed in Sect. 9.

7 Evaluation

We evaluate our ontology against a set of competency questions processed for the generated knowledge graph. The curation process included analysis of survey data and several face-to-face interviews with Rensselaer PhD students. For this article, we will present a set of competency questions for a STEM degree student enrolled in an American institution. A student will want to understand the different careers available after a STEM doctorate. The student progresses to select careers based on skills and interests with the appropriate skills and passion level. The student is also curious about the skills and the level of competence required for a particular profession. The student will want to know some of the initial resources that will help them understand the bigger picture of the role. With a better understanding of preferred career paths, the student next wants to know what questions to ask during an informational interview to understand how the role will fit their values.

Please refer to our resource website[4] for detailed SPARQL queries and corresponding responses obtained through SPARQL query on Blazegraph Workbench. The following are a set of competency questions with references to the candidate responses (in table format for readability) produced by our ontology:

1. Which are the different sectors that a $\langle domain \rangle$ doctorate could explore career in? For example, Which sectors can STEM doctorates could explore career in? (candidate response are the ones provided under Sect. 6)
2. What are the various $\langle domain \rangle$ career paths available to a $\langle domain \rangle$ doctorate? For example, What are the various scientific career paths available to a STEM doctorate? (snippet of the candidate response shown in Table 3)
3. I consider myself as an $\langle proficiency\_level \rangle$ in $\langle skill \rangle$. Which scientific career paths can I explore that requires this $\langle skill \rangle$ with my $\langle proficiency\_level \rangle$? For example, Which scientific career paths require an expert proficiency level in creativity/innovative thinking?(candidate response in Table 4)
4. How is $\langle scientific\_career \rangle$ defined as? For example, What does "Principal investigator in a research-intensive institute" mean? (candidate response shown in Table 5)
5. What level of proficiency is required for the required skills in $\langle scientific\_career\_path \rangle$? For example, what are the various skills with proficiency levels required for "Principal investigator in a research-intensive institute"? (snippet of the candidate response shown in Table 6)
6. What interests does $\langle scientific\_career\_path \rangle$ requires that may help one to succeed? For example, What are the interests that someone can possess to do well in "Principal investigator in a research-intensive institute"? (snippet of the candidate response shown in Table 7)
7. Which $\langle Resource\_type \rangle$ could be explored to get a better understanding regarding the career path $\langle scientific\_career \rangle$? For example, What are some of the articles that could help me understand "Principal investigator in a research-intensive institute" better? (candidate response shown in Table 8)
8. What questions to ask at an interview to understand if the role imbibes my values? For example, What are some of the values that these roles support? What questions can I ask during an informative interview to understand if the "Principal investigator in a research-intensive institute" role could support my values? (snippet of the candidate response shown in Table 9)

[4] Resource Website: https://tetherless-world.github.io/sciv-ontology/.

9. What questions can I ask to understand if the role would help me in my specific ⟨*value/s*⟩? For example, What questions could I ask to understand if the "Principal investigator in a research-intensive institute" role would support me in having a 'professional development', maintain a 'work-life balance' while understanding the various 'benefits available'? (candidate response shown in Table 10)

The SPARQL query uses both simple and complex pattern matching with some of them using more expensive commands such as Filter and Order By as seen in our resource website[5]

Table 3. Tabular format of the snippet of competency question 2 example SPARQL query output obtained on Blazegraph Workbench.

ScientificCareerPaths
Business of science
Clinical practice
Clinical research management
Combined research and teaching careers
Drug/device approval and production
Entrepreneurship
Intellectual property
Principal investigator in a research-intensive institution
Public health related careers
Research administration
Research in industry

Table 4. Tabular format of the competency question 3 example SPARQL query output obtained on Blazegraph Workbench.

ScientificCareerPaths
Combined research and teaching careers
Entrepreneurship
Principal investigator in a research-intensive institute
Research in industry

8 Discussion

To address the gap of 'what next' after a doctoral degree and of the resource accessibility for exploring the career paths that open up for a doctoral student

[5] Resource Website: https://tetherless-world.github.io/sciv-ontology/.

Table 5. Tabular format for better readability for Competency question 4 example SPARQL query output obtained on Blazegraph Workbench.

label	Definition
Principal investigator in a research-intensive institution	Independent researcher at a medical school, private research institute, government lab or university with minimal teaching responsibilities

Table 6. Tabular format of the snippet of competency question 5 example SPARQL query output obtained on Blazegraph Workbench.

Skills	ProficiencyLevel
Time management	Expert
Writing grant proposals	Expert
Writing scientific publications	Expert
Presenting to nonscientists	Intermediate
Teaching in a classroom setting	Intermediate
Writing for nonscientists Contributing to institution	Intermediate
(e.g. participate on communities)	Proficient
Demonstrating workplace etiquette	Proficient

Table 7. Tabular format of the snippet of competency question 6 example SPARQL query output obtained on Blazegraph Workbench.

Interests	PassionLevel
Analyzing experimental results	VeryHigh
Attending conferences or scientific meetings	VeryHigh
Creating presentations	VeryHigh
Designing experiments	VeryHigh
Discussing science with others	VeryHigh
Giving presentations about science	VeryHigh
Keeping up with current events in science	VeryHigh
Leading or supervising other	VeryHigh
Mentoring or teaching one-on-one Planning new scientifics projects or	VeryHigh
developing new research directions	VeryHigh
Reading papers in your fields	VeryHigh
Representing data in figures/illustrations	VeryHigh
Thinking abour science Using quantitative methods in understanding	VeryHigh
science (e.g., statistics, mathematical modeling)	VeryHigh
Writing grant proposals	VeryHigh
Writing scientific manuscripts	VeryHigh

Table 8. Tabular format of the competency question 7 example SPARQL query output obtained on Blazegraph Workbench.

Resources	Link
Academic Scientist's Toolkit (collection of articles)	http://sciencecareers.sciencemag.org/career_magazine/previous_issues/articles/2004_04_29/noDOI.9200335458625575487
Are you PI material? Assess yourself	http://sciencecareers.sciencemag.org/career_magazine/previous_issues/articles/2003_01_17/noDOI.4924325304480736142
Competition and Careers in Biosciences	http://sciencecareers.sciencemag.org/career_magazine/previous_issues/articles/2001_12_21/noDOI.3752964224463541678
Getting a Job with the Federal Govern-ment (presentations and handouts)	https://www.training.nih.gov/events/view/_2/88/Getting_a_Job_with_the_Federal_Government
Listing of federal government research centers in the U.S	http://www.ornl.gov/info/library/govt/labs.html
Making the leap to Independence	http://sciencecareers.sciencemag.org/career_magazine/previous_issues/articles/2007_03_02/caredit.a0700029
Working in a Government Lab	http://chronicle.com/article/Working-in-a-Government-Lab/45212/

based on skills, interests, and values, we designed the FAIR OWL ontology, Skills to Career with Interests and Values Ontology (SCIVO). This ontology is modeled to be compact, reusable, and extendable to harmonize data from heterogeneous resources to help students get a better understanding of the different available career paths. This is demonstrated by creating a knowledge graph containing data from the Science myIDP tool and NSF SED 2019 publicly available data. Connecting these two resources using SCIVO helps students understand the different sectors one can pursue their selected career paths in and the skills and interests that these career paths may require with their respective proficiency and passion levels. One could also explore the career path through various resources and get one step closer to making an informed decision regarding 'where do I go?' after graduation. Based on our survey analysis, this helps our students, hopefully, make their career choices with more work at their interest level and less work that is not of their interest. Researchers could use a similar method to expand and/or instantiate the SCIVO classes to create a knowledge graph for their work.

Our biggest challenges included 1) making a generalized compact FAIR ontology to cater to various graduate student requirements for the questions 'what next?' and 'where do I go'? (discussed in Sect. 3) 2) creating mapping between the skills and interests required with the expert value range while keeping its

Table 9. Tabular format of the snippet of competency question 8 example SPARQL query output obtained on Blazegraph Workbench.

Values	QuestionsToConsider
Aesthetics	How would this career allow you to express your artistic side?
Benefits	What benefits generally accrue to employees in this career field?
Available	What are your minimum requirements?
Competition	Will your achievements be measured and recognized? Will the results be used for promotion and salary decision?
Congenial Atmosphere	Do you know enough about the culture?
Creativity	In this field how will you be able to be innovative?
Earning	What salaries are common for people entering this field?
Potential	Does this match your earnings expectations?

Table 10. Tabular format of the snippet of competency question 9 example SPARQL query output obtained on Blazegraph Workbench.

Values	QuestionsToConsider
Benefits	What benefits generally accrue to employees in this career field?
Available	What are your minimum requirements?
Professional	For the people in this field what is the next position?
Development	What is the path for continued promotions?
Work_Life	
Balance	What do people in this field say about their worklife balance?

authenticity intact (discussed in Sect. 5), and 3) working around individuals which were part of both skills and interests, for example, 'Writing Grant Proposals' as seen in Table 6 and Table 7. In principle, it is the same individual that we are talking about, but in reality, one could be good at writing grant proposals but might not have interest in spending time on them, or vice versa, creating a dichotomy. Hence, it had to be connected with both skills and interests. This is resolved based on what one is looking for. If the user is looking at writing grant proposals as a skill, then they get the information connected to it as a skill and as an interest if they are looking at it as an interest.

9 Conclusion

In this paper, we describe the modeling of the SCIVO ontology, which can be used to create a complete resource for doctoral students in making informed decisions regarding 'what next' and 'where do I go' after getting a doctoral degree. We discuss its usability and extensibility by creating an application using the myIDP and NSF SED 2019 data. The ontology and knowledge graph creation are done using Cellfie, a Protégé 5.5.0 plugin. The ontology is evaluated using the traditional application-based competency question method. The competency questions are based on the analysis of survey responses and personal interviews with current graduate students. Both the SCIVO and SCIVOKG, along with the Cellfie rules used to create them, are available on our resource website. It also includes all of the actual SPARQL queries run on the blazegraph workbench, along with candidate responses, as an example of the use of the ontology and knowledge graph.

Our future work includes extending SCIVO to connect with other available ontologies to provide a complete resource for the graduate journey. Since the sector information is coming from the NSF SED 2019 data, it also provides us room for connecting SCIVO with INDO: the Institute Demographic Ontology [10, 11], to help pursue graduate students not only make an informed decision based on the graduate program but also understand the different opportunities they might have after graduating from a particular graduate program.

Acknowledgement. This work is a part of "Building a Social Machine for Graduate Mobility". We would like to thank Dean Stanley Dunn, Dean of Graduate Education, who provided expert insights into this issue. We would like to thank the members of the Tetherless World Constellation Lab at Rensselaer Polytechnic Institute who provided insights and expertise that greatly assisted this research. This work was funded in part by the RPI-IBM AI Research Collaboration, a member of the IBM AI Horizons network.

References

1. Ahmed, N., Khan, S., Latif, K.: Job description ontology. In: 2016 International Conference on Frontiers of Information Technology (FIT), pp. 217–222. IEEE (2016)
2. Enăchescu, M.I.: Screening the candidates in it field based on semantic web technologies: automatic extraction of technical competencies from unstructured resumes. Informatica Economica **23**(4) (2019)
3. Fazel-Zarandi, M., Fox, M.S.: Semantic matchmaking for job recruitment: an ontology-based hybrid approach. In: Proceedings of the 8th International Semantic Web Conference, vol. 525, p. 2009 (2009)
4. Foley, D.: Survey of doctorate recipients, 2019. NSF 21-230.national center for science and engineering statistics (NCSES), Alexandria, VA: National Science Foundation (2021). https://ncses.nsf.gov/pubs/nsf21320/
5. Fuhrmann, C., Hobin, J., Lindstaedt, B., et al.: myIDP, an online interactive career planning tool. AAAS (American association for the advancement of science). Website launched 7 Sept 2012
6. Guo, S., Alamudun, F., Hammond, T.: Résumatcher: a personalized résumé-job matching system. Expert Syst. Appl. **60**, 169–182 (2016)
7. Ilkou, E., et al.: EduCOR: an educational and career-oriented recommendation ontology. In: Hotho, A., et al. (eds.) ISWC 2021. LNCS, vol. 12922, pp. 546–562. Springer, Cham (2021). https://doi.org/10.1007/978-3-030-88361-4_32

8. Kendall, E.F., McGuinness, D.L.: Ontology engineering. Synthesis Lectures on The Semantic Web: Theory and Technology, vol. 9, no. 1, p. i (2019)
9. Keshan, N.: Building a social machine for graduate mobility. In: 13th ACM Web Science Conference 2021, pp. 156–157 (2021)
10. Keshan, N., Fontaine, K., Hendler, J.A.: InDO: the institute demographic ontology. In: Villazón-Terrazas, B., Ortiz-Rodríguez, F., Tiwari, S., Goyal, A., Jabbar, M.A. (eds.) KGSWC 2021. CCIS, vol. 1459, pp. 1–15. Springer, Cham (2021). https://doi.org/10.1007/978-3-030-91305-2_1
11. Keshan, N., Fontaine, K., Hendler, J.A.: Semiautomated process for generating knowledge graphs for marginalized community doctoral-recipients. Int. J. Web Inf. Syst. **18**(5/6), 413–431 (2022)
12. Khobreh, M., et al.: An ontology-based approach for the semantic representation of job knowledge. IEEE Trans. Emerg. Top. Comput. **4**(3), 462–473 (2015)
13. Kumar, N.A., Ramu, K.: Ontology based website for job posting and searching. In: IOP Conference Series: Materials Science and Engineering, vol. 1042, p. 012006. IOP Publishing (2021)
14. Lv, H., Zhu, B.: Skill ontology-based semantic model and its matching algorithm. In: 2006 7th International Conference on Computer-Aided Industrial Design and Conceptual Design, pp. 1–4. IEEE (2006)
15. Mochol, M., Oldakowski, R., Heese, R.: Ontology based recruitment process. Informatik 2004, Informatik verbindet, Band 2, Beiträge der 34. Jahrestagung der Gesellschaft für Informatik eV (GI) (2004)
16. Ntioudis, D., Masa, P., Karakostas, A., Meditskos, G., Vrochidis, S., Kompatsiaris, I.: Ontology-based personalized job recommendation framework for migrants and refugees. Big Data Cogn. Comput. **6**(4), 120 (2022)
17. O'Connor, M.J., Halaschek-Wiener, C., Musen, M.A.: Mapping master: a flexible approach for mapping spreadsheets to OWL. In: Patel-Schneider, P.F., et al. (eds.) ISWC 2010. LNCS, vol. 6497, pp. 194–208. Springer, Heidelberg (2010). https://doi.org/10.1007/978-3-642-17749-1_13
18. Phan, T.T., Pham, V.Q., Nguyen, H.D., Huynh, A.T., Tran, D.A., Pham, V.T.: Ontology-based resume searching system for job applicants in information technology. In: Fujita, H., Selamat, A., Lin, J.C.-W., Ali, M. (eds.) IEA/AIE 2021. LNCS (LNAI), vol. 12798, pp. 261–273. Springer, Cham (2021). https://doi.org/10.1007/978-3-030-79457-6_23
19. Rentzsch, R., Staneva, M.: Skills-matching and skills intelligence through curated and data-driven ontologies. In: Proceedings of the DELFI Workshops (2020)
20. Senthil Kumaran, V., Sankar, A.: Towards an automated system for intelligent screening of candidates for recruitment using ontology mapping (expert). Int. J. Metadata Semant. Ontol. **8**(1), 56–64 (2013)
21. Usip, P.U., Udo, E.N., Umoeka, I.J.: Applied personal profile ontology for personnel appraisals. Int. J. Web Inf. Syst. **18**(5/6), 487–500 (2022)

Topic Modeling for Skill Extraction from Job Postings

Ekin Akkol[1(✉)], Muge Olucoglu[2], and Onur Dogan[1]

[1] Department of Management Information Systems, Izmir Bakircay University, 35665 Izmir, Turkey
ekin.akkol@bakircay.edu.tr

[2] Department of Computer Engineering, Izmir Bakircay University, 35665 Izmir, Turkey

Abstract. With an increase in the number of online job posts, it is becoming increasingly challenging for both job searchers and employers to navigate this large quantity of information. Therefore, it is crucial to use natural language processing techniques to analyze and draw inferences from these job postings. This study focuses on the most recent job postings in Turkiye for Computer Engineering and Management Information Systems. The objective is to extract skills for the job postings for job seekers wanting to apply for a new position. LSA is a statistical method for figuring out the underlying characteristics and meanings of sentences and words in natural language. Frequency analysis was also performed in addition to the LSA analyses with the goal of conducting a thorough examination of job postings. This study was conducted to determine and evaluate the skills that the sector actually needs. Thus, job seekers will have the chance to develop themselves in a more planned way. The findings indicate that the job postings for the two departments reflect various characteristics in terms of social and technical abilities. A higher requirement for social skills is thought to exist in the field of Management Information Systems rather than Computer Engineering. It has been discovered that employment involving data have been highly popular in recent years, and both departments often list opportunities involving data analysis.

Keywords: latent semantic analysis · job postings · natural language processing · social web

1 Introduction

Discovering hidden patterns from large text data is of great importance in today's digitalized world. With the digitalization of the world, much larger data is produced and this data needs to be processed and analyzed in order to become usable valuable information. Information extraction techniques enable implicit information that is impossible to distinguish in a short time with the human eye in the data to be easily revealed and the data to be transformed into information.

F. Ortiz-Rodriguez et al. (Eds.): KGSWC 2023, LNCS 14382, pp. 277–289, 2023.
https://doi.org/10.1007/978-3-031-47745-4_20

Professional networking platforms such as LinkedIn offer valuable insights into the industry's workforce needs as the places where employers and job seekers meet. It is important to accurately extract and analyze the secret information contained in these job posts in order to make more efficient recruiting and career decisions. It is critical for job seekers to identify and disclose the skills needed in job postings. Job seekers will get information on how to train themselves if skill is extracted from job postings.

This study focuses on the recent job postings of companies in Turkiye for Computer Engineering and Management Information Systems on LinkedIn. It conducts an in-depth analysis of the contents of these job posts using a process known as Latent Semantic Analysis (LSA). LSA is a statistical method for inducing and representing features of the meanings of natural language words and passages [11]. The LSA method is used to extract large amounts of text data in order to discover and understand hidden correlations [14]. Due to the volume of big data generated by digitalization, businesses and organizations are looking for new ways to make better and faster decisions. At this point, it is of great importance to reveal the confidential information in the text data. Traditional data analysis methods may be incapable of detecting latent meanings or relationships in text content. This is where LSA comes into play, helping in the analysis of data patterns by finding semantic relationships between texts.

By emphasizing the power of LSA analysis on big text data, this study aims to provide a valuable source of information to businesses and educational institutions. This study provides a road map for both job applicants and educational institutions that train employees for the sector. It is meant to be a valuable resource for students and new graduates interested in Computer Engineering and Management Information Systems as they consider their career possibilities.

The paper structure is as follows. Section 2 provides a brief overview of the literature review of other studies on LSA and the analysis of text in job postings. Section 3 gives the methodology of the technical infrastructure for all processes from obtaining LinkedIn job postings to analysis. Section 4 consists of presenting and interpreting the results of the steps of the proposed methodology. Section 5 concludes the study by discussing the results and future directions.

2 Literature Review

Studies using Latent Semantic Analysis and studies on job postings with various methods are examined in the literature review section.

2.1 Studies on Latent Semantic Analysis

LSA is often used in social sciences for topics such as semantic analysis, text summarization, fake news and e-mail detection. For example, the 2016 US presidential debates between Hillary Clinton and Donald Trump were analyzed using the LSA method [25]. The results accurately reflect the individual political attitudes of the two politicians. In another study, LSA and LDA methods were used to critically extract semantic analysis and sentiment analysis of the text. Thus, they tried to contribute to the design of the translation of computers more appropriate to human language [19]. As the amount of information in the news texts increases, the viewership of the news increases. On the other hand, news with a large amount of information have longer text content and it may happen that readers lose their focus in long texts. Therefore, a group of researchers in Indonesia used LSA to auto-summarize Indonesian news texts [17]. News is an important way to obtain information. The increasing spread of fake news causes serious social problems by misleading readers. They detected fake news by using common concepts in fake news with LSA [27]. In this study [18], first of all, e-mails are categorized according to their fields (finance, education, health, etc.). LSA is used to detect spam of e-mails.

In social sciences such as psychology and education, LSA is used to measure unobserved latent traits [4]. The Liebowitz Social Anxiety Scale (LSAS) is a widely used measure in the field of social anxiety [13]. The LSAS is a semi-structured interview tool designed to assess anxiety and avoidance in social and performance circumstances. A self-report version of the LSAS (LSAS-SR) has recently been developed and adapted in several languages, demonstrating strong internal consistency and test-retest reliability [24]. The LSAS-SR's psychometric qualities have been tested and determined to be trustworthy and valid.

2.2 Studies on Job Postings

LSA has been widely used in the analysis of job postings, such as developing a job recommendation system, matching job postings with candidates according to their abilities, and categorizing job types.

Early job recommendation systems used Boolean search techniques, which frequently failed to satisfy the complex information requirements of job searchers [12]. Latent Semantic Analysis (LSA) was used in a study by Khaouja et al. [9] to identify employment needs from postings between 2010 and 2020. Then, on online hiring sites, machine learning algorithms were used to match job postings with candidates. This required collecting semantic concepts from resumes and using feature combinations to comprehend candidate and job scopes. Long Short-Term Memory (LSTM) networks were then used to match jobs to candidates based on past data [7].

In addition, Luo et al. [10] presented a deep learning-based method for matching talent profiles with job criteria, computing matching scores for applicants with various information kinds for the same position. This gave recruiters the ability to rank prospects in order of matching score.

These developments were crucial in putting people in the right jobs who had the right interviewing skills. A clever grading system was put into place in the context of employment interviews to combat unfair or irregular interview procedures. With the use of LSA, Shen et al. [22] created a collaborative learning model that included job descriptions, applicant resumes, and interview ratings.

Similar to this study, Verma et al. [26] used LSA to examine the skill sets needed for AI and machine learning roles on indeed.com. Another study by Qin et al. [16] examined previous job applications, using word-level semantic analysis and recurrent neural networks to determine the compatibility of candidate skills and job criteria. In another study, job postings collected in the field of information and communication technologies in Jordan in 2020 were examined. LSA approach has been applied to these job postings. And open positions in Jordan have been identified. This study revealed that academic institutions should update their traditional programs according to the gap in the sector [1].

In order to categorize job kinds and extract student profiles appropriate for job postings, the author used logistic regression and word frequency analysis [20]. In order to emphasize the differences between employer and university requirements, hierarchical clustering methods were used to compare job postings and academic results [15]. Additionally, Chen et al. [3] used cosine similarity and the K-means++ clustering technique to evaluate the information provided in job posts about requirements.

There are many studies in the literature where LSA has been used successfully in a variety of domains, including text mining, content analysis, document classification, and semantic relation identification. This method is recognized as a powerful tool to understand the meaning of large and complex text data and to process it more meaningfully. This study was carried out to discover and quantify the talents that are really needed in the sector. Thus, job seekers can develop themselves in a more planned way. Although there are many studies on job postings in the literature, there are deficiencies in the studies written for this purpose. The main contribution of this study lies in its aim to address existing gaps in the literature and generate valuable insights. Unlike previous research in this field, this study stands out by conducting a specific methodology tested with two main departments. Moreover, the results obtained from these analyses are subjected to rigorous frequency analysis and visual representation, offering a novel approach to interpreting the data. By adopting this methodology, the study contributes to the advancement of knowledge in the field and offers a fresh perspective that can inform future research and decision-making processes.

3 Methodology

In this study, LSA method was used to extract topic models from job postings. Information retrieval, information filtering, natural language processing, and machine learning are just a few of the areas where this technique has been widely used [6]. The problem of words with distinct spellings but similar meanings has been successfully addressed using LSA, allowing for the discovery of similarities among such terms [8]. LSA is based on the assumption of an underlying vector space spanned by an orthogonal set of latent variables closely related to the semantics/meanings of the particular language [23]. LSA provides a latent semantic space in which texts and individual words are represented as vectors as a meaning theory. LSA employs linear algebra as a computational technique to extract the dimensions that represent that space. This representation enables automated procedures that mirror the way people conduct similar cognitive tasks to calculate the similarity between terms and documents, categorize terms and documents, and summarize massive collections of documents [5]. Singular Value Decomposition (SVD) is a technique distinctive to LSA. SVD is a linear algebra technique with numerous functions in document and text processing [2]. The SVD factorizes a matrix A into

$$A = U \times S \times V^T \tag{1}$$

where U is the new orthonormal basis for A, S is a diagonal matrix indicating how prevalent each column is in the basis, and V^T is the original documents' coordinates using the new basis [21].

LSA uses the SVD to decrease the dimensionality of the term-document matrix and capture the latent semantic structure. LSA creates a lower-dimensional version of the original matrix by keeping the top k singular values and their accompanying singular vectors. This structure enables for the discovery of semantic similarities between words and texts. The formula for transforming a document into its LSA representation consists of multiplying the document vector by the matrix V^T, where V^T comprises the appropriate singular vectors. This transformation projects the document vector onto the latent semantic space, resulting in a more abstract representation of the underlying semantic structure.

After the topics were extracted from the job postings with the LSA method, the keywords containing the technical competencies that could be related to the topics were determined, and the number of occurrences of these keywords in the text was revealed by frequency analysis using the N-gram model. According to these numbers, the most sought-after competencies in the two selected departments were determined. All the steps performed in the study are summarized in Fig. 1.

In Step 1, the requirements for the Computer Engineering and Management Information Systems departments were extracted and data were collected in order to conduct analysis. In Step 2, data preprocessing was done, such as cleaning, organizing and normalizing the data. In Step 3, word frequencies are calculated using the tf/idf method. In this way, common words are extracted for the requirements of the analyzed sections. In the 4th step, statistical analysis of the calculated frequencies is performed using the LSA method. In the 5th step, the analysis results were interpreted.

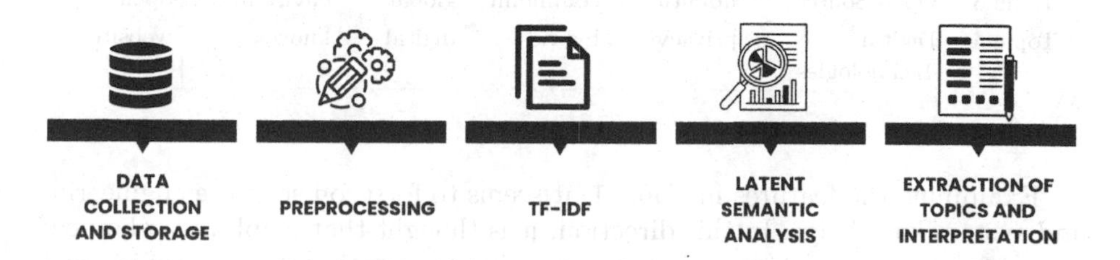

Fig. 1. Methodology for Job Posting Analysis

4 Case Study

4.1 Data Preprocessing

Data scraping methods on Linkedin were used to acquire job posts for Computer Engineering and Management Information Systems graduates within the scope of the study. Job postings were filtered by typing "Computer Engineering" and "Management Information Systems" as keywords in the Linkedin search bar, and the text of the postings was obtained. In the data set used in the study, 1000 of the job postings related to Computer Engineering in Turkiye were used. On the other hand, the last 667 job postings related to the Management Information Systems department were used in the study. The recorded texts were first subjected to data preprocessing stages. Unnecessary characters, numbers and special signs were removed from the texts. All texts were converted to lower case and the texts were divided into words by tokenization. Using the stop-word list, common and meaningless words were removed to prevent them from negatively affecting the results. Then, TF-IDF vectorization was performed to reveal the importance of the words in the texts. The Latent Semantic Analysis method represents texts more effectively by using TF-IDF vectorization to reveal more meaningful and representative themes and words. The data set prepared for analysis was analyzed with the LSA method to reveal the themes of job postings.

4.2 Results

Job postings related to Computer Engineering were analyzed with the LSA model, and four different topics were extracted. The model was run in such a way that each topic was expressed in five words. The results from the model are shown in Table 1.

Table 1. Computer Engineering LSA Topics

Topic ID	Representation	Feature 1	Feature 2	Feature 3	Feature 4	Feature 5
Topic 1	Development	experience	software	system	knowledge	engineer
Topic 2	Mobile Applications	android	experience	mobile	secondary	promote
Topic 3	Open Source	ubuntu	community	global	environment	open
Topic 4	Digital Technologies	privacy	hybrid	digital	know	website

Examining the features in Topic 1, it seems to focus on software engineering and system knowledge. In this direction, it is thought that employers who need a workforce on the software and systems side are posting jobs on topics such as software development, system knowledge, database management, software architecture and design principles, object-oriented programming and testing and debugging skills.

When the features in Topic 2 are examined, it is seen that the topic is for Android platform and mobile application development. The words "android" and "mobile" represent the field of mobile application development, while the word "experience" can indicate a focus on user experience. Considering this topic and its features, it is seen that employers need a workforce in areas such as Android application development, mobile application interface design, mobile application testing and debugging, user experience (UX) design, mobile devices and cross-platform compatibility, and application marketing and promotion strategies.

Topic 3 is about open-source software and the Ubuntu operating system. The term "ubuntu" is connected with this operating system, while the terms "community" and "open" highlight the open-source community and approach. The words "global" and "environment" both signify support and contribution from a wide user base. When this topic is explored, it is clear that there is a need for employment in areas such as Linux system management, open-source software development, open-source software licenses, and open-source software applications.

Topic 4 appears to be about privacy and websites in the digital world. While the word "privacy" draws attention to the issue of privacy, the word "website" refers to websites operating on digital platforms. The words "hybrid" and "digi-

Table 2. Computer Engineering Frequency Analyzes

Skills	Frequency	Percentage	Technical/Social
Testing	360	36 %	Technical
Database	330	33 %	Technical
Agile Software Development	327	32,7 %	Technical
Object Oriented Programming	270	27 %	Technical
Cloud Computing	250	25 %	Technical
Web Programming	235	23,5 %	Technical
Python	213	21,3 %	Technical
Linux	193	19,3 %	Technical
Java	171	17,1 %	Technical
Mobile	166	16,6 %	Technical
Analytic Skills	149	14,9 %	Social
Problem Solving Skills	125	12,5 %	Social
Project Management	112	11,2 %	Social
Machine Learning	110	11 %	Technical
Cyber Security	84	8,4 %	Technical
Digital Transformation	84	8,4 %	Technical
Telecommunication	43	4,3 %	Technical

tal" refer to digital technologies. Areas such as data privacy and security, digital security and penetration testing, hybrid and cloud-based infrastructures, digital communication and marketing strategies, website design and development, SEO (Search Engine Optimization) and analytics are the areas that can come to the fore in job postings on this topic.

After examining all the topics and their features produced by the LSA model, frequency analyzes of the words in the job postings related to these topics were carried out in order to reveal more in-depth and concrete results. Analyzes were made in three different ways as unigram, bigram and trigram and the words that could be related to the topics are shown in Table 2.

When the frequency analysis data is evaluated, it has been determined that the testing skill is the most sought after skill with a rate of 36% in the Computer Engineering positions examined. Testing is followed by database and agile software development skills. When the emerging abilities were investigated, just three of them were social abilities, with the remainder being entirely about technical capabilities. Python programming is the most popular programming language.

Job postings for graduates of Management Information Systems were also analyzed with the LSA model and four different topics with five features were extracted. Extracted topics and their features are shown in Table 3.

Table 3. Management Information Systems LSA Topics

Topic ID	Representation	Feature 1	Feature 2	Feature 3	Feature 4	Feature 5
Topic 1	Business Areas	management	process	business	project	customer
Topic 2	Financial Data	control	financial	bank	audit	risk
Topic 3	Data Analysis	data	career	marketing	customer	sale
Topic 4	Data Protection	security	network	information	personal	data

In job postings, Topic 1 highlights essential business areas such as business process management, project planning and execution, business operations, and customer relations. Management Information Systems graduates are widely sought after in job postings requiring abilities such as managing business processes, project management, and customer interactions.

Topic 2 explains how MIS graduates can be placed in financially oriented job postings such as financial management, risk assessment, audit processes, and banking. Graduates will make strategic decisions based on financial data analysis, manage audit activities, and work on topics such as risk management.

In Topic 3, the use of features like "customer, data, career, marketing, sale" demonstrates that MIS graduates can work in areas like customer relationship management, data analysis, marketing strategies, and sales activities. Graduates will analyze consumer data to understand target markets, conduct marketing initiatives, and work to increase product or service sales.

Topic 4 shows that MIS graduates are in high demand in crucial areas such as information security, data protection, personal information management, and network security. Graduates should be able to ensure the security of sensitive information, analyze risks to security, and implement various security methods to secure network infrastructure.

Following the analysis of the job post texts collected for the Management Information Systems department with the LSA model, the results were enhanced by doing frequency analyses, as in the Computer Engineering department. The goal of these analyses is to show the superficial results in further detail. Table 4 shows frequency analyses of job posts from the Management Information Systems department.

Table 4. Management Information Systems Frequency Analyzes

Skills	Frequency	Percentage	Technical/Social
Data Analysis	276	41,4 %	Technical
MS Office Programs	232	34,8 %	Technical
Problem Solving	156	23,4 %	Social
Design	131	19,6 %	Technical
Communication Skills	119	17,8 %	Social
Analytical Thinking	81	12,1 %	Social
SQL	77	11,5 %	Technical
Project Management	75	11,2 %	Social
Human Resources	61	9,1 %	Social
Software Development	53	7,9 %	Technical
Business Economics	38	5,7 %	Technical
Information Security	37	5,5 %	Technical
SAP Experience	31	4,6 %	Technical
Cyber Security	25	3,7 %	Technical
Digital Transformation	24	3,6 %	Social
Social Media	23	3,4 %	Social
Financial Analysis	23	3,4 %	Technical

When the job posts dataset frequency analysis of Management Information Systems is analyzed, it is discovered that social skills are preferred over Computer Engineering. Although it appears that social skills are more commonly featured in job postings than Computer Engineering, it has been discovered that the rate of seeking technical talents in the Management Information Systems department is slightly higher. With a serious rate of 41.4%, job postings are focussed on data analysis. It has been concluded that a thorough understanding of Microsoft Office programs is also essential for a Management Information Systems graduate.

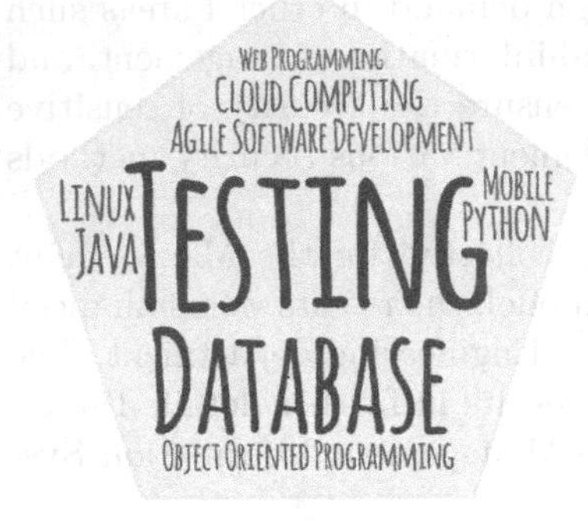

(a) Computer Engineering

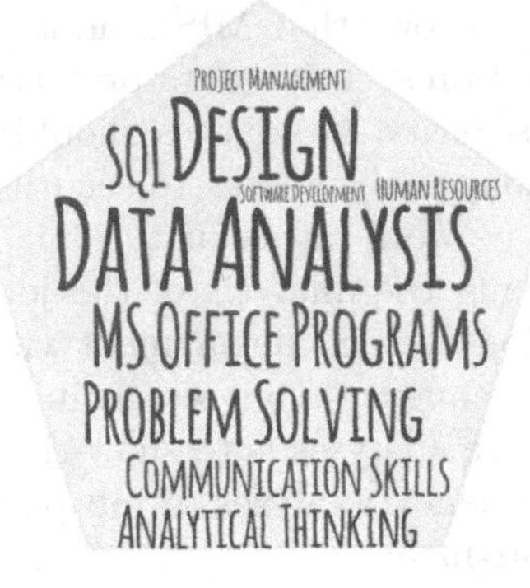

(b) Management Information Systems

Fig. 2. Word Clouds For Job Postings

Figure 2 presents word clouds generated by a frequency analysis of job postings in the departments of Computer Engineering (2a) and Management Information Systems (2b). The size of the words is proportional to how frequently they appear in job postings.

5 Conclusion

With the development of information and communication technologies, various sorts of data are generated from a variety of sources, including web pages, blogs, social media platforms, and sensors. These data obtained from many sources are utilized in a variety of applications, including advertising, promotion, marketing, and resource planning for institutions and organizations. Aside from that, researchers collect and use this data in various studies, evaluate it, and share their findings with the public.

To make the acquired data stacks meaningful, the data must pass through different phases before becoming valuable information. In their raw form, data often does not make sense unless it is processed and evaluated. To make the data meaningful, it must be pre-processed, filtered, and analyzed using various methods to disclose the valuable information contained within it. This study demonstrates how big text data analysis and the extraction of confidential information play an important role in knowledge-driven decisions and understanding of future workforce needs.

Within the scope of the study, job postings for graduates of Computer Engineering and Management Information Systems in Turkiye were obtained from the LinkedIn professional social network and social sharing platform. The LSA method was used in the text analysis to discover hidden information through topic modeling. LSA and other natural language processing techniques are effective at discovering hidden patterns and links in text content. In addition to the topic modeling, frequency analyses of job postings from departments, as well as topics issued by LSA, were performed and presented within the scope of the study. The results obtained are also grouped as technical and social skills.

When the results are examined, it can be seen that job postings for Computer Engineering graduates focus on technical skills rather than social skills, whereas the situation is a little more equal in the Management Information Systems department, but the technical skills sought are still in the majority. According to the findings of both departments, data-related jobs are at the forefront of newly published job postings in Turkiye. Problem solving and analytical thinking are also seen as essential skills for graduates of both departments. When the programming language capabilities were analyzed, it was discovered that Python was the most sought-after programming language, followed by Java, and SQL, which is utilized in data management, was also commonly listed in job postings. Given the current state of the digitalizing world, the fact that skills such as digital transformation, cloud computing, social media, and web programming are frequently mentioned in job postings emphasizes the importance of graduates keeping up with current technologies and trends. Besides all these, the accuracy of LinkedIn data can change because it is created and maintained by individuals. Analyses could be negatively impacted by inaccurate or out-of-date data.

This study provides a resource for individuals, educational institutions, and companies interested in pursuing careers in Computer Engineering and Management Information Systems. These departments research and report on which talents are most in demand by companies in Turkiye. Effective use of LSA analysis and confidential information extraction is crucial for adapting to job market demands and effectively analyzing/understanding the market. Although LSA is good at discovering semantic relationships between documents and words, it expresses topics planarly. In other words, it addresses each issue completely independently of the others. This can create incompleteness in cases where some documents contain multiple topics. While different methods can be used in future studies, hybrid methods can also be preferred to minimize the disadvantages of all methods.

This study can be used as a reference to make decision-making processes more informed and data-driven for individuals who want to anticipate future trends and keep up with evolving skill needs. In future studies, global scale data can be used to evaluate global patterns instead of being limited to job postings in Turkey. Studies can be conducted in different fields and sectors, and a system that makes curriculum recommendations to university departments can be developed based on the results obtained.

References

1. Alsmadi, D., Omar, K.: Analyzing the needs of ICT job market in Jordan using a text mining approach. In: 2022 International Conference on Business Analytics for Technology and Security (ICBATS), pp. 1–5. IEEE (2022)
2. Amalia, A., Gunawan, D., Fithri, Y., Aulia, I.: Automated Bahasa Indonesia essay evaluation with latent semantic analysis. J. Phys. Conf. Ser. **1235**, 012100 (2019)

3. Chen, J., et al.: Data analysis and knowledge discovery in web recruitment-based on big data related jobs. In: 2019 International Conference on Machine Learning, Big Data and Business Intelligence (MLBDBI), pp. 142–146. IEEE (2019)
4. Chen, Y., Li, X., Zhang, S.: Structured latent factor analysis for large-scale data: Identifiability, estimability, and their implications. J. Am. Stat. Assoc. **115**(532), 1756–1770 (2020)
5. Evangelopoulos, N.E.: Latent semantic analysis. Wiley Interdiscip. Rev. Cogn. Sci. **4**(6), 683–692 (2013)
6. Han, K., Chien, W.T., Chiu, C., Cheng, Y.: Application of support vector machine (SVM) in the sentiment analysis of twitter dataset. Appl. Sci. **10**, 1125 (2020). https://doi.org/10.3390/app10031125
7. Jiang, J., Ye, S., Wang, W., Xu, J., Luo, X.: Learning effective representations for person-job fit by feature fusion. In: Proceedings of the 29th ACM International Conference on Information & Knowledge Management, pp. 2549–2556 (2020)
8. Jirasatjanukul, K., Nilsook, P., Wannapiroon, P.: Intelligent human resource management using latent semantic analysis with the internet of things. Int. J. Comput. Theory Eng. **11**, 23–26 (2019). https://doi.org/10.7763/ijcte.2019.v11.1235
9. Khaouja, I., Kassou, I., Ghogho, M.: A survey on skill identification from online job ads. IEEE Access **9**, 118134–118153 (2021)
10. Luo, Y., Zhang, H., Wen, Y., Zhang, X.: ResumeGAN: an optimized deep representation learning framework for talent-job fit via adversarial learning. In: Proceedings of the 28th ACM International Conference on Information and Knowledge Management, pp. 1101–1110 (2019)
11. Maletic, J., Marcus, A.: Using latent semantic analysis to identify similarities in source code to support program understanding. In: Proceedings 12th IEEE Internationals Conference on Tools with Artificial Intelligence. ICTAI 2000, pp. 46–53 (2000). https://doi.org/10.1109/TAI.2000.889845
12. Malinowski, J., Keim, T., Wendt, O., Weitzel, T.: Matching people and jobs: a bilateral recommendation approach. In: Proceedings of the 39th Annual Hawaii International Conference on System Sciences (HICSS 2006), vol. 6, p. 137c. IEEE (2006)
13. Oakman, J., Van Ameringen, M., Mancini, C., Farvolden, P.: A confirmatory factor analysis of a self-report version of the Liebowitz social anxiety scale. J. Clin. Psychol. **59**(1), 149–161 (2003)
14. Olney, A.M.: Large-scale latent semantic analysis. Behav. Res. Methods **43**(2), 414–423 (2011)
15. Piróg, D., Hibszer, A.: Utilising the potential of job postings for auditing learning outcomes and improving graduates' chances on the labor market. Higher Educ. Q. (2023)
16. Qin, C., et al.: Enhancing person-job fit for talent recruitment: an ability-aware neural network approach. In: The 41st international ACM SIGIR Conference on Research & Development in Information Retrieval, pp. 25–34 (2018)
17. Rofiq, R.A., et al.: Indonesian news extractive text summarization using latent semantic analysis. In: 2021 International Conference on Computer Science and Engineering (IC2SE), vol. 1, pp. 1–5. IEEE (2021)
18. Saidani, N., Adi, K., Allili, M.S.: A semantic-based classification approach for an enhanced spam detection. Comput. Secur. **94**, 101716 (2020)
19. Salloum, S.A., Khan, R., Shaalan, K.: A survey of semantic analysis approaches. In: Hassanien, A.-E., Azar, A.T., Gaber, T., Oliva, D., Tolba, F.M. (eds.) AICV 2020. AISC, vol. 1153, pp. 61–70. Springer, Cham (2020). https://doi.org/10.1007/978-3-030-44289-7_6

20. Salm, V.: Student Success in Co-operative Education: An Analysis of Job Postings and Performance Evaluations. Master's thesis, University of Waterloo (2023)
21. Sellberg, L., Jönsson, A.: Using random indexing to improve singular value decomposition for latent semantic analysis. In: 6th International Conference on Language Resources and Evaluation, Marrakech, Morocco, May 26-June 1, 2008. European Language Resources Association (2008)
22. Shen, D., Zhu, H., Zhu, C., Xu, T., Ma, C., Xiong, H.: A joint learning approach to intelligent job interview assessment. In: IJCAI, vol. 18, pp. 3542–3548 (2018)
23. Shivakumar, P.G., Georgiou, P.G.: Confusion2vec: towards enriching vector space word representations with representational ambiguities. PeerJ Comput. Sci. **5**, e195 (2019). https://doi.org/10.7717/peerj-cs.195
24. Srisayekti, W., Fitriana, E., Moeliono, M.F.: The Indonesian version of the Liebowitz social anxiety scale-self report (LSAS-SR-Indonesia): psychometric evaluation and analysis related to gender and age. Open Psychol. J. **16**(1), e221227 (2023)
25. Valdez, D., Pickett, A.C., Goodson, P.: Topic modeling: latent semantic analysis for the social sciences. Soc. Sci. Q. **99**(5), 1665–1679 (2018)
26. Verma, A., Lamsal, K., Verma, P.: An investigation of skill requirements in artificial intelligence and machine learning job advertisements. Ind. High. Educ. **36**(1), 63–73 (2022)
27. Zeng, Z., Ye, L., Liu, R., Cui, Z., Wu, M., Sha, Y.: Fake news detection by using common latent semantics matching method. In: 2021 IEEE 33rd International Conference on Tools with Artificial Intelligence (ICTAI), pp. 1059–1066. IEEE (2021)

Author Index

F. Ortiz-Rodriguez et al. (Eds.): KGSWC 2023, LNCS 14382, pp. 291–292, 2023.
https://doi.org/10.1007/978-3-031-47745-4